Effects of Climate Change and Anthropogenic Disturbances on Water Ecological Environment and the Coping Strategies

Effects of Climate Change and Anthropogenic Disturbances on Water Ecological Environment and the Coping Strategies

Editors

Xun Wang
Zhiyuan Wang
Xin Zhao

MDPI

Basel • Beijing • Wuhan • Barcelona • Belgrade • Novi Sad • Cluj • Manchester

Editors

Xun Wang
College of Environment
Hohai University
Nanjing
China

Zhiyuan Wang
Center for Eco-Environment Research
Nanjing Hydraulic Research Institute
Nanjing
China

Xin Zhao
Department of Environmental Engineering
Northeastern University
Shenyang
China

Editorial Office
MDPI
St. Alban-Anlage 66
4052 Basel, Switzerland

This is a reprint of articles from the Special Issue published online in the open access journal *International Journal of Environmental Research and Public Health* (ISSN 1660-4601) (available at: https://www.mdpi.com/journal/ijerph/special_issues/water_ecological_environment).

For citation purposes, cite each article independently as indicated on the article page online and as indicated below:

Lastname, A.A.; Lastname, B.B. Article Title. *Journal Name* **Year**, *Volume Number*, Page Range.

ISBN 978-3-0365-9126-1 (Hbk)
ISBN 978-3-0365-9127-8 (PDF)
doi.org/10.3390/books978-3-0365-9127-8

Contents

About the Editors

Xun Wang

Dr. Xun Wang has been an Associate Professor in the College of Environment, Hohai University, since 2020. She received a BS degree from Hohai University in 2012, an MS from Ecole des Mines de Douai of France in 2013 and a PhD from Hohai University in 2019. Her research focuses on environmental microbial ecology, water resource protection and ecological restoration. Dr. Wang has published more than 60 scientific research papers.

Zhiyuan Wang

Prof. Zhiyuan Wang has been a principle investigator for the Center for Eco-Environmental Research, Nanjing Hydraulic Research Institute, since 2015. Prof. Wang received a BS degree from Hohai University in 2010, an MS from Ecole des Mines de Douai of France in 2011 and a PhD from Hohai University in 2015. Dr. Wang is a pioneer who has explored the risk of emerging contaminants in water environments, the conservation of river and lake ecosystems and the adaptive management of lake eutrophication. Prof. Wang has published more than 40 papers in SCI journals and two books. He has also received the first prize at provincial or ministerial awards four times.

Xin Zhao

Dr. Xin Zhao has been an Associate Professor in the Department of Environmental Engineering, School of Resources and Civil Engineering, Northeastern University, since 2014. He received a BS degree from Liaoning University in 2005, and an MS and PhD from the Harbin Institute of Technology in 2011. His main research directions are environmental pollution control and waste recycling. Dr. Zhao has published more than 80 scientific research papers, including 50 papers in SCI journals and two books.

International Journal of
Environmental Research and Public Health

MDPI

Article

Benthic Biofilm Bacterial Communities and Their Linkage with Water-Soluble Organic Matter in Effluent Receivers

Longfei Wang [1], Yutao Wang [1], Yi Li [1,*], Wenlong Zhang [1], Huanjun Zhang [1], Lihua Niu [1] and Nuzahat Habibul [2]

[1] Key Laboratory of Integrated Regulation and Resource Development on Shallow Lakes, Ministry of Education, College of Environment, Hohai University, Nanjing 210098, China; lfwang@hhu.edu.cn (L.W.); yubao15@163.com (Y.W.); 1223zhangwenlong@163.com (W.Z.); zhanghuanjun@hhu.edu.cn (H.Z.); niulh@hhu.edu.cn (L.N.)

[2] College of Chemistry and Chemical Engineering, Xinjiang Normal University, Urumqi 830054, China; nuzahat@163.com

* Correspondence: envly@hhu.edu.cn; Tel.: +86-13776417768

Abstract: Benthic biofilms are pioneering microbial aggregates responding to effluent discharge from wastewater treatment plants (WWTPs). However, knowledge of the characteristics and linkage of bacterial communities and water-soluble organic matter (WSOM) of benthic biofilms in effluent-receiving rivers remains unknown. Here, we investigated the quality of WSOM and the evolution of bacterial communities in benthic biofilm to evaluate the ecological impacts of effluent discharge on a representative receiving water. Tryptophan-like proteins showed an increased proportion in biofilms collected from the discharge area and downstream from the WWTP, especially in summer. Biofilm WSOM showed weak humic character and strong autochthonous components, and species turnover was proven to be the main factor governing biofilm bacteria community diversity patterns. The bacterial community alpha diversity, interspecies interaction, biological index, and humification index were signally altered in the biofilms from the discharge area, while the values were more similar in biofilms collected upstream and downstream from the WWTP, indicating that both biofilm bacterial communities and WSOM characters have resilience capacities. Although effluent discharge simplified the network pattern of the biofilm bacterial community, its metabolic functional abundance was basically stable. The functional abundance of carbohydrate metabolism and amino acid metabolism in the discharge area increased, and the key modules in the non-random co-occurrence network also verified the important ecological role of carbon metabolism in the effluent-receiving river. The study sheds light on how benthic biofilms respond to effluent discharge from both ecological and material points of view, providing new insights on the feasibility of utilizing benthic biofilms as robust indicators reflecting river ecological health.

Keywords: biofilm resilience; water-soluble organic matter; microbial community; co-occurrence network

Citation: Wang, L.; Wang, Y.; Li, Y.; Zhang, W.; Zhang, H.; Niu, L.; Habibul, N. Benthic Biofilm Bacterial Communities and Their Linkage with Water-Soluble Organic Matter in Effluent Receivers. *Int. J. Environ. Res. Public Health* **2022**, *19*, 1994. https://doi.org/10.3390/ijerph19041994

Academic Editor: Paul B. Tchounwou

Received: 15 January 2022
Accepted: 7 February 2022
Published: 10 February 2022

Publisher's Note: MDPI stays neutral with regard to jurisdictional claims in published maps and institutional affiliations.

1. Introduction

The ecological health of urban rivers and internal lakes have been cumulatively affected by anthropogenic activities. A major impact of urbanization is inputs from wastewater treatment plants (WWTPs) [1]. With the increasing scale of treated sewage, effluent discharge has become one of the most important sources of river replenishment. WWTPs release a multitude of nutrients, dissolved organic matter (DOM), and micropollutants, e.g., pharmaceuticals and personal care products [2,3]. Synthetic chemicals and nutrients mix and enter the receiving waterbodies, triggering eutrophication and altering biogeochemical cycling in fluvial ecosystems [4]. Effluent-receiving waterbodies have been regarded as environmental sensitive regions and have attracted intensive research concerns [5].

Effluent discharge has multidimensional impacts on the ecosystem and functioning of receiving waters. Primarily, effluent input alters the trophic level and constituents of DOM molecules [6], whereafter pelagic algae and communities change in response to trophic level variation and affect primary production. The effects of effluent discharge on other biota, including benthic biofilm, macrophytes, and invertebrates, have also been reported [7]. These effects are likely to be governed by hydrological parameters such as wetting/drying alteration [8], seasonal variation [9]. and dilution of effluent discharge [10]. IN the study [11], the impacts of effluent and hydrological stresses on river functioning were investigated, and it was observed that even highly diluted WWTP effluents can affect the structure of the biofilm community and river ecosystem functions.

Benthic biofilms are assemblages of living and dead algae, microbes, and organic debris, constituting the basis of the benthic food web [12,13]. Benthic biofilms have been considered pioneer microbial aggregates in response to effluent discharge [12]. The architecture and functioning of biofilms are constantly changing following variations in dissolved oxygen, organics, hydrodynamics, etc. [14]. The total biomass, microbial composition, photosynthesis of algae, and assimilation of organic matter are accordingly modified as a consequence [15]. Existing studies have shown that benthic biofilms can restrain pollutants, e.g., heavy metals, pharmaceuticals and brominated flame retardants released from historical events and transfer the contaminants to higher trophic levels [16,17]. Hence, biofilms are often used as indicators to evaluate ecological changes in aquatic environments by assessing their respiratory rate [18] or soluble reactive phosphorus uptake capacity [11]. Biofilms adapted to anthropogenic disturbances typically show higher resistance to effluent discharge [4,11]. Water-soluble organic matter (WSOM) is the most active component in biofilm organic matter, comprising carbohydrates, amino acids, and organic acids. Effluent discharge has been verified to promote extracellular enzyme activities, e.g., leucine aminopeptidase and amino glucosidase in benthic biofilms, facilitating the conversion of complex organic matter and its subsequent uptake and utilization by microbes [19,20].

A number of studies have demonstrated the complicated and bidirectional relationship between organic carbon and bacteria communities in rivers. River organic carbon serves as a carbon source and nutrient for heterotrophic bacteria and some algae, and can be metabolized by micro-organisms in the aquatic environment [21]. Different from the organic contents in other matrices, such as sediments, the organic compounds in biofilm may affect the microbial community structure and modify the food web character and energy transfer efficiency [12]. Chromophore DOM (CDOM) is an optically active part in the bulk DOM pool that can intensely absorb light in the ultraviolet and blue spectral regions, and exhibits remarkable changes in the quantity and quality of diverse biogeochemical processes [6]. Avila et al. unveiled the dynamic succession between CDOM and the microbial community in a small river dominated by effluent discharge and confirmed a remarkable response of actinomycetes and protein components [22]. Zhang et al. explored the relationships between phytoplankton communities and CDOM in a tropical lake and found that CDOM could affect bacterial community structure by participating in the metabolism of specific bacterial communities. Similar results were observed by analyzing the connection between phytoplankton community and CDOM in a eutrophic lake [23]. Nonetheless, most studies have focused on the characteristics and association between planktonic communities and DOM in streams, while the bacterial community structure and interspecific interactions of organic compounds in adherent aggregates, e.g., benthic biofilms, have been largely ignored. An analysis linking the microbial community and WSOM components in benthic biofilms may help to bring forward biological indicators for assessing freshwater quality and ecosystem fitness.

In the past decade, the rapid development in bioinformatics has afforded technical support to decipher responses of microbial community assembly and metabolic functions to environmental stresses [23,24]. Burdon et al. confirmed the resilience of microbial communities to effluent discharge and found that the buffering capacity of microbial communities is vulnerable to impact by environmental fluctuations [2]. Recently, co-occurrence network analysis has been employed to explain the interspecific interactions of microbial communities in suspended particulates, soils, and sediments. These studies manifest that microbial communities usually have non-random co-occurrence patterns and a modular structure, implying the vital role of biological interactions in adjusting the fluvial ecosystem functioning [25–27]. Unfortunately, no studies have been performed to evaluate microbial interactions in river biofilms in response to effluent discharge; thus, the co-occurrence patterns between bacterial communities and biofilm WSOM have not been elucidated yet.

In this case study, we investigated how benthic biofilm bacterial communities and biofilm WSOM alter in response to effluent discharge in an effluent-receiving river. We hypothesized that the stress of effluent discharge has a considerable impact on the co-occurrence patterns of benthic biofilm bacterial communities, as well as their linkage with WSOM, especially CDOM. We also hypothesized a higher proportion of metabolic functions on account of nutrient and micropollutant inputs in the discharging area. These results will contribute to unraveling the overall impacts of allochthonous inputs such as effluent discharge on river biofilm properties and ecological functions, providing insights in the search for appropriate pollution indicators in effluent impacted areas and revealing the potential of benthic biofilm as an indicator of the ecological response of effluent-receiving rivers.

2. Materials and Methods

2.1. Study Site, Experimental Design, and Water Characteristics

The North City WWTP of Jiangning District (31°58′ N, 118°50′ E) uses an oxidation ditch process that treats the sewage of >40 thousand P.E. from domestic sources. On average, 70 thousand cubic meters of wastewater are treated per day and the effluent is discharged into the Qinhuai River, located in Nanjing, Jiangsu Province, China, as illustrated in Figure 1. The water level and temperature of the Qinhuai River show significant seasonal differences (Table S1). Nearly 75% of the annual precipitation is concentrated during the May-to-September rainy season (summer) [28]. The area of interest has a straight channel with few tributary confluences, and the WWTP effluent discharge is the main external input to the reach, endowing the area with superiority for evaluating the effects of wastewater effluent on ecosystem functioning.

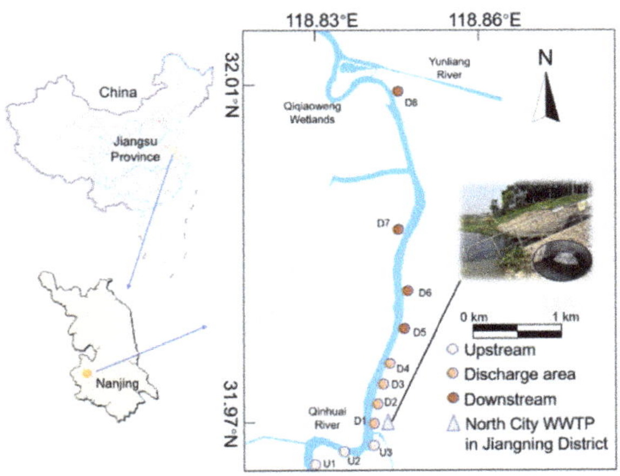

Figure 1. Location of sampling sites in Qinhuai River receiving effluent from North City wastewater treatment plant (WWTP) in Jiangning District, Nanjing, China.

Eleven sampling sites were selected along the reach. The sites were categorized into three areas according to their relative positions to the effluent outfall shown in Figure 1. U1 to U3, 1, 0.5, and 0.2 km upstream from WWTP effluent discharge, are referred to as the Upstream group; D1 to D4, 0, 0.2, 0.5, and 0.7 km downstream the discharge, are referred to as the Discharge area group; D5 to D8, 1, 1.5, 2, 3 km, downstream from the discharge, are referred to as the Downstream group. The values of pH, temperature (T, °C). and dissolved oxygen (DO) were measured using a portable meter (HQ30d, HACH Company, Loveland, CO, USA) at each sampling site in December 2019 and September 2020, respectively. Water and benthic biofilms were collected on 26 December 2019 (winter, averaged water temperature: 10.3 °C) and 8 September 2020 (summer, averaged water temperature: 26.8 °C) (Table S1). Total nitrogen (TN) and total phosphorus (TP) were measured to reflect the trophic conditions at varying sampling sites [29]. Three parallel water samples were collected and analyzed at each site.

2.2. Biofilm Harvesting and Water-Soluble Organic Matter (WSOM) Extraction

Benthic biofilms were collected following the protocols reported by Wang et al. [30]. The main steps of collection include: rock selection, biofilm scraping, sample preservation, and transportation. The detailed collection method is available in SI. Prior to extraction, 20 g of biofilm samples in wet weight were lyophilized using a freeze dryer (Christ ALPHA 1-2 LD plus, Marin Christ Co., Osterode am Harz, Germany). To improve the extraction efficiency, we ground the lyophilized biofilm and passed it through a 100-mesh sieve to remove impurities. Biofilm WSOM was extracted using a leaching-centrifugation method according to the protocols of previous studies [31,32]. Pre-treated biofilm was packed into sterilized conical bottles at a material-to-water (g:g) ratio of 1:3. Then the samples were shaken for 16 h at ambient temperature. Afterwards the leachate was transferred to 50 mL sterilized centrifugal tubes and then centrifuged at $4000\times g$ r/min for 30 min. The supernatant was filtered through a 0.45 μm sterile acetate membranes, and the generated filtrate was defined as biofilm WSOM and stored at −4 °C prior to analytical approaches.

2.3. Spectral Analyses of Biofilm WSOM

Dissolved organic carbon (DOC) concentrations in bulk WSOM solutions were measured using a total organic carbon analyzer (Multi N/C2100, Analytik Jena, Jena, Germany). The DOC concentration in biofilm WSOM were normalized to mg/g shown in Table S2. UV-vis absorption spectra were measured by a spectrophotometer (UV-1800, Shimadzu, Japan). The excitation–emission matrices (EEMs) of CDOM in WSOM were measured using a fluorescence spectrophotometer (F7000, Hitachi, Japan). Detailed information on spectral analysis is provided in SI. Parallel factor analysis (PARAFAC) was performed using the DOM Fluor toolbox in MATLAB (R2017a) software [33]. The relative contents of fluorescent components were obtained via F_{max} values analyzed by PARAFAC. Several UV-visible spectra-derived parameters were calculated to demonstrate the aromaticity ($SUVA_{254}$), hydrophobicity (E_{254}/E_{204}), and molecular weight (S_R) of biofilm WSOM. We also calculated parameters including biological index (BIX), humification index (HIX), and fluorescence index (FI) to describe the fluorescent characteristics of biofilm WSOM as described in the Table S3.

2.4. Microbial Community Analysis

2.4.1. DNA Extraction and PCR Amplification

For each biofilm sample, DNA extraction was executed through the E.Z.N.A.® Soil DNAKit (Omega Bio-tek, Norcross, GA, USA), following the manufacturer's instructions. Agarose gel electrophoresis was used to analyze the quality of the extracted DNA. 16S rRNA gene amplification and Illumina MiSeq sequencing were performed at Biozeron Science and Technology Ltd. (Shanghai, China). The bacterial primers 341F (5′-CCTAYGGGRBGCASCAG-3′) and 806R (5′-GGACTACNNGGGTATCTAAT-3′) were used to amplify the V3−V4 regions of the bacterial 16S rRNA gene. The PCR conditions were as follows: DNA denaturation for 5 min at 95 °C with 27 cycles, at 95 °C for 30 s, 55 °C for 30 s, and 72 °C for 45 s, with a final extension of 10 min at 72 °C.

2.4.2. Sequence Analyses and Functional Prediction

Sequence analyses were carried out via Quantitative Insights Into Microbial Ecology version 2 (QIIME2 version 18.6) software [34]. A single-end sequence data denoising method called the Divisive Amplicon Denoising Algorithm program (DADA2, v1.10) was implemented for processing valid data from BIOZERON Co., Ltd. (Shanghai, China) [35]. We then used the 'classify-sklearn' option to assign classification identities to these representative sequences via the 'qiime feature-classifier' command, referred to here as Amplicon Sequence Variants (ASVs). These ASVs have more than 99% similarity to the SILVA128 reference comparison database used in classification identities. Information on the sequence reads corresponding to each sample has been uploaded to the NCBI SRA database for public access (bioengineering number: PRJNA717165). Prediction of potential microbial function was performed by an improved metagenome inference method of PICRUSt [36]. Functional gene predictions were performed based on the Kyoto Encyclopedia of Genes and Genomes (KEGG) database. The Nearest Sequenced Taxon Index (NSTI) was used to evaluate the prediction accuracy of PICRUSt, with lower values indicating higher prediction accuracy.

2.5. Statistical Analyses

The richness, Pielou, and Shannon indices were calculated in R (version 3.6.2) using the vegan and picante packages. Non-metric multidimensional scaling (NMDS) based on Bray–Curtis distance was performed to decipher the clustering of benthic biofilm bacterial communities among different groups, together with non-parametric multivariate analysis of variance (Adonis) analyzing the significant differences of microbial communities. The dissimilarity indices including the Sørensen dissimilarity index (β_{SOR}), the Simpson dissimilarity index (β_{SIM}), and the nestedness resultant dissimilarity index (β_{NES}) of benthic biofilm bacterial communities were calculated in R, employing the function 'beta-multi.R' [37]. The null model was used to quantify the contribution of ecological processes to the microbial assemblage by vegan, ape, and picante packages [38,39].

One-way ANOVA and Tukey's post hoc test were performed to uncover the differences between groups using SPSS v26 software. STAMP (v.2.1.3) software was used to perform a two-sided Welch's *t*-test on the functional abundance map predicted by KEGG to discover the metabolic pathways with significant differences between groups. The 'ggcor' package was applied to test the correlation between microbial community and spectral indicators via mantel analysis. To integrate spectral information and biological data in the biofilm, redundancy analysis (RDA) was performed using the vegan package in R.

Molecular ecological networks (MENs) corresponding to different seasons were constructed to elucidate the correlation between CDOM fluorophores and bacterial communities using an online MENA pipeline based on a Random Matrix Theory (RMT) bioinformatics approach. To reduce the network complexity, we only selected ASVs that are present in all samples of the same group for network construction. A random network of 100 ASVs corresponding to each empirical network was built to test the statistical significance of the empirical networks [40,41]. The details of network construction are referred to in [42]. Gephi was applied for analyzing network visualization and modularity.

3. Results

3.1. Spectral Characteristics of Biofilm WSOM

Table S2 summarizes the organic carbon concentrations in biofilm WSOM at varying sampling sites. The normalized DOC concentration of biofilm WSOM ranged from 141.4 to 360.4 mg C/g. The averaged DOC concentration of biofilm WOSM in the Upstream group, Discharge area group, and Downstream group were 175.5, 223.9, and 296.3 mg C/g, respectively. In terms of the UV-visible spectra-derived parameters, the averaged values of $SUVA_{254}$ and S_R were 2.96 and 1.91 in winter and were 0.52 and 1.27 in summer, respectively, with significant seasonal differences ($p < 0.05$ and $p < 0.001$) (Table S4).

PARAFAC modeling can identify and verify four fluorescent components, providing a total of 95.72% variability within the data (Figure 2a). The model was compared with available models in the Openflour database, finding a 95% similarity. We found that C1 presented characteristics such as terrestrial humic-like fraction, with high-molecular-weight and photo-labile character [43,44]. C2 also exhibits humus-like properties associated with microbial activity and can be reprocessed in situ by the microbial community [45]. C3 was classified as an intermediate, between humic-like and amino acid-like moiety [46]. C4 serves as a tryptophan-like protein material associated with microbial activity or wastewater discharge [47]. The variation of the percentage of each fluorescent component is shown in Figure 2b. In winter, humic-like fraction (C1 and C2) dominantly accounted for 70% of the total fluorescent components, while in summer, the Upstream group was dominated by humic-like and amino acid-like intermediates (63%), and the Discharge area and Downstream groups were dominated by tryptophan-like proteins (45% and 39%). The percentage of tryptophan-like proteins (C4) in the three areas was 12%, 45%, and 39% in winter and 11%, 14%, and 27% in summer, respectively.

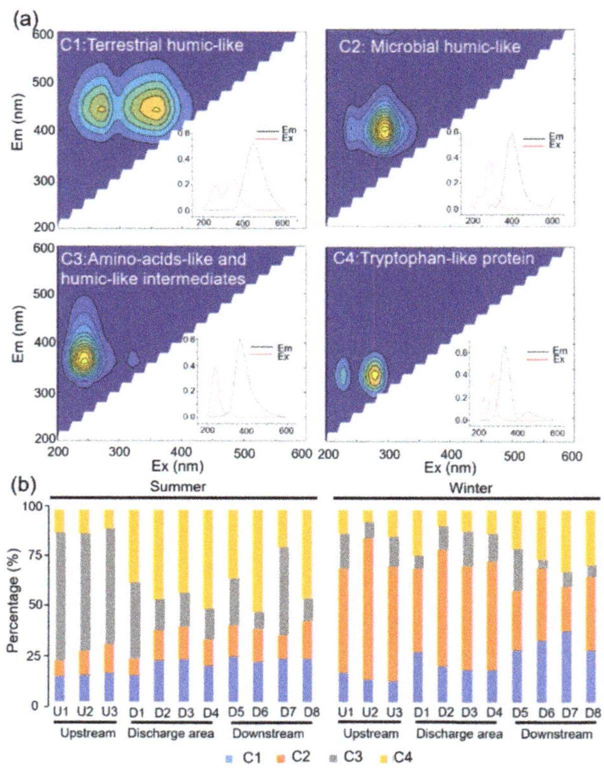

Figure 2. (**a**) Four main components of the spectrum identified by parallel factor analysis (PARAFAC). Insets visualize the excitation and emission loadings of the four components and (**b**) the relative percentages of each component based on F_{max} value.

The fluorescent characters of biofilm WSOM exhibit different variations in response to effluent discharge shown in Figure 3a–c. BIX and HIX are significantly different among three areas ($p < 0.001$ and $p < 0.05$), but FI is not. In addition, we found that there was a distinct boundary of BIX-HIX values among three areas (Figure 3d), with the Upstream group clearly isolated from the other two areas, while overlapping was observed between the Discharge area and Downstream groups.

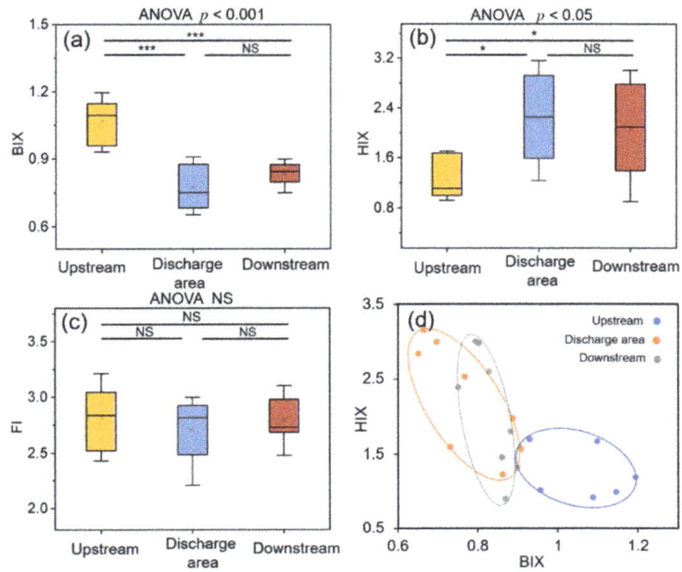

Figure 3. (**a–c**) Fluorescence parameters of biofilm water-soluble organic matter (WSOM). Asterisks represent the difference between two groups (* $p < 0.05$; *** $p < 0.001$; NS represents no significant difference) and p-values represent the overall group difference. (**d**) Values of biological index (BIX) and humification index (HIX) of biofilm WSOM in different areas: different colors represent different areas, and the horizontal and vertical coordinates indicate the values of BIX and HIX, respectively.

3.2. Dynamics, Diversity and Assembly Mechanisms of Bacterial Communities

Figure S1 illustrates the taxonomic composition and relative abundance of bacterial communities at the phylum level in biofilm collected in varying seasons. The results show that Proteobacteria (42.7–65.9%), Actinobacteria (4.8–16.1%), Bacteroidetes (5.2–19.0%) and Chloroflexi (3.6–12.6%) represent the dominant phyla. The microbial α-diversity aspects in Figure 4a,b indicate that the Richness and Shannon indexes are remarkably different among the Upstream, Discharge area, and Downstream groups ($p < 0.05$), with the highest values in the Discharge area group. The averaged Peilou index was highest in the Discharge area group, with no remarkable difference observed among the three areas (Figure 4c). NMDS analysis presented the differences in bacterial communities grouped by sampling time (Figure 4d). Additionally, the Adonis analysis exhibited statistically significant differences in bacterial communities among sampling areas ($F = 0.02$) and between different seasons ($F = 0.031$).

The process of biodiversity change was clarified via two patterns of biome beta diversity: nestedness and turnover. The discrepancy indices of bacterial communities grouped by sampling area and season are shown in Table S5. The mean β_{SOR} value among the three areas was 0.82, with a strong contribution of spatial turnover ($\beta_{SIM} = 0.79$) and a small contribution of nestedness ($\beta_{NES} = 0.02$). Similar results were also observed for samples grouped by season. We subjected the samples from the Discharge area group and the Downstream group to a resampling procedure in which 100 random samples were taken from six inventories and the mean value β value was calculated, so that the different number of samples from different areas (eight vs. six) was comparable (Table S6). β'_{SOR} in the Upstream group (0.76) is lower than that in the Discharge area group (0.81) and the Downstream group (0.82), which is mainly systematic in the difference of β'_{SIM}, while β'_{NES} is almost constant (Table S6). Additionally, we quantified the relative contribution of each microbial ecological process in the assembly among seasons (Figure S2a) and areas (Figure S2b). The ecological processes include homogeneous selection, variable selection, dispersal limitation, homogenizing dispersal, and ecological drift. In winter, variable selection (45.5%) and homogeneous selection (52.7%) accounted for a comparable proportion, while in summer there was lower variable selection (27.3%) and higher homogeneous selection (69.1%). In the grouping by area, homogeneous selection was the dominant factor driving the assembly of bacterial communities in the Upstream group (73.3%) and the Discharge area group (78.6%), whilst in the Downstream group, variable selection (46.4%) and homogeneous selection (50.0%) performed comparably.

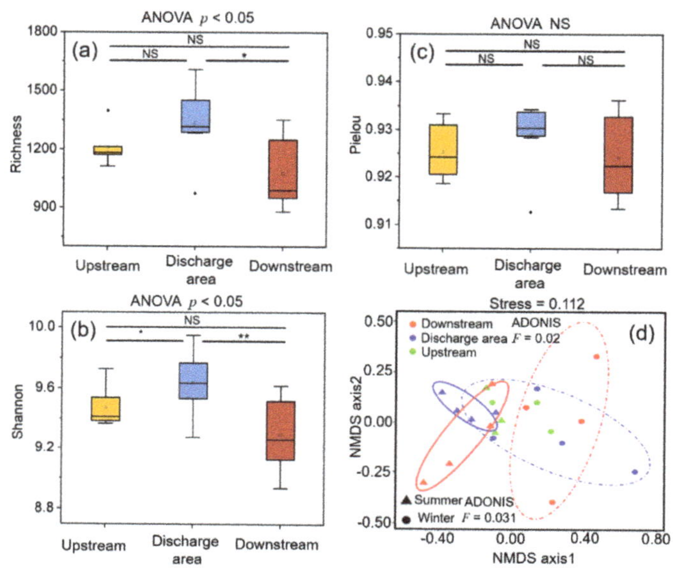

Figure 4. (**a**–**c**) Diversity of bacterial communities and mechanisms of assembly. Comparison of diversity indices in different areas. Asterisks represent the difference between two groups (* $p < 0.05$; ** $p < 0.001$; NS represents no significant difference) and p-values represent the overall group difference. (**d**) Non-metric multidimensional scaling (NMDS) analysis by areas and seasons; different colors represent different areas and different shapes represent different seasons.

3.3. Functional Prediction of Bacterial Communities

The average NSTI value of all samples was 0.18, indicating that these samples provided an appropriate data set for accurate predictions. By comparing the abundance of KEGG categories predicted by PICRUSt in level-2 metabolic pathways, significant functional differences in distinct sampling areas could be observed (Figure S3). Amino acid metabolism, carbohydrate metabolism, and membrane transport were the three predicted functions with the highest abundance in benthic biofilms, and the values were higher in the Discharge area group than in the groups collected upstream and downstream. We performed a two-by-two comparison of the predicted functions in metabolic pathways at level-3 and discovered that the carbohydrate metabolism (TCA cycle, C5-branched dibasic acid metabolism, and inositol phosphate metabolism), biosynthesis of other secondary metabolites (clavulanic acid biosynthesis), and metabolism of cofactors and vitamins (nicotinate and nicotinamide metabolism) in the Upstream group were significantly different from those in the Discharge area group and the Downstream group ($p < 0.05$, Figure 5a–c). In contrast, only the amino acid metabolism (phosphonate and phosphinate metabolism) was significantly different between the Discharge area group and the Downstream group.

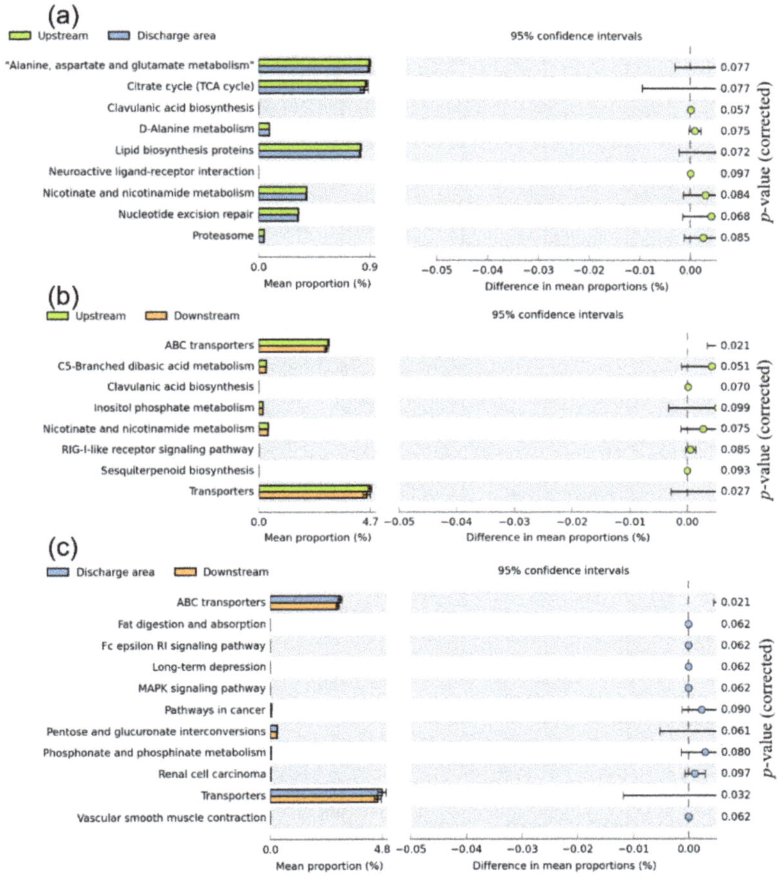

Figure 5. Significant differences in bacterial function predicted by PICRUSt within Kyoto Encyclopedia of Genes and Genomes (KEGG) grouped into level-3 functional categories between (**a**) the Upstream group and Discharge area group, (**b**) the Discharge area group and Downstream group, (**c**) the Discharge area group and Downstream group, using the response ratio method at a 95% confidence interval.

3.4. Key WSOM Parameters Affecting Bacterial Community Composition

Mantel and RDA analysis were employed to determine the association of biofilm WSOM parameters with bacterial community composition. As shown in Figure 6a, WSOM parameters show more diverse and remarkable correlations with taxonomic compositions in winter. FI and C3 have been confirmed as key factors affecting the composition of bacterial communities in winter ($p < 0.05$), while in summer the key factors are $SUVA_{254}$, S_R, and C4 ($p < 0.05$, Figure 6b). RDA analysis manifested similar results, as shown in Figure 6c,d.

The results of the both RDA models proved to be significant ($p < 0.05$) based on the screening of VIF < 5. In winter, humic-like and amino acid-like intermediates (C3) posed the greatest influence on bacterial community composition, while in summer, tryptophan-like proteins (C4) exhibited the strongest impact. We also found that FI was a common key factor influencing the composition of bacterial communities in both seasons.

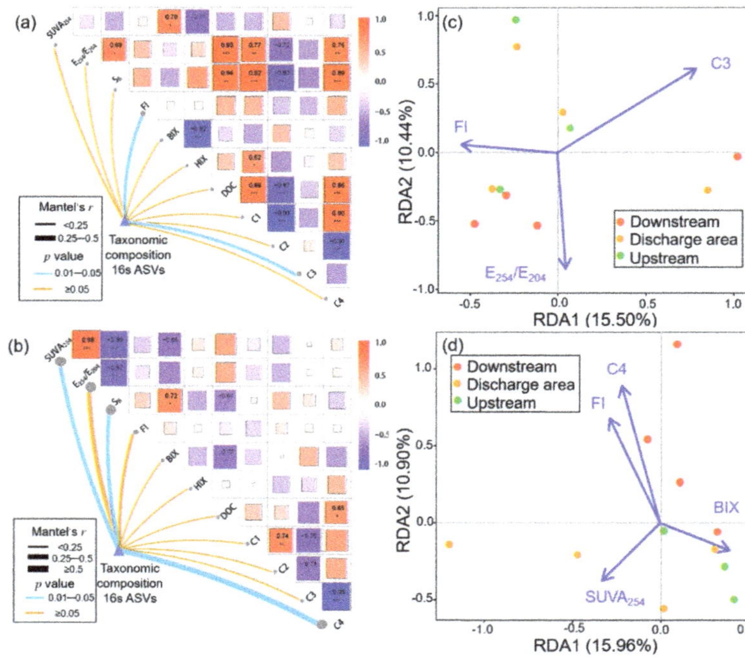

Figure 6. WSOM parameters driving the composition of bacterial communities in the effluent-receiving waterbodies. Mantel test between taxonomic community and biofilm WOSM parameters in (**a**) winter and (**c**) summer. RDA analysis based on bacterial community composition in (**b**) winter and (**d**) summer. (* $p < 0.05$; ** $p < 0.01$; *** $p < 0.001$).

3.5. Co-Occurrence Network Analysis

An RMT-based model was employed to analyze the phylogenetic characters and to determine the symbiotic relationship between bacterial communities and fluorophores. A total of 2564 edges and 443 nodes were obtained in the network derived from samples collected in winter, with a 0.515 modularity encompassing six modules (Figure 7a,b). For the network derived from samples collected in summer, 330 nodes and 1692 edges with a modularity of 0.434 was observed encompassing seven modules (Figure 7c,d). Positive correlations prevailed in both networks derived from winter (76%) and summer (74%). The parameters of the topological network and the random network are shown in Table S7, and the reliability and non-randomness of the empirical network structure is verified by comparing it with the random network analysis [48]. In both networks, Proteobacteria, Bacteroidetes, Acidobacteria, and Chloroflexi occupy the dominant nodes in the network (Figure 7a,c), and are also the predominant bacterial phylum in benthic biofilm bacterial composition, as shown in Figure S1.

Modularity reflects the connectivity within and between clusters, and nodes have closer interactions with each other within the module than with nodes in other modules. There are six modules in the winter network, among which C1, C2, and C4 belong to Module IV, and C3 is affiliated with Module III. Among the seven modules in the summer network, C1, C2, and C3 are grouped into Module IV, and C4 belongs to Module III. By integrating the observations in Figures 6 and 7, we speculate that Module II containing C3 in winter, and Module III comprising C4 in summer serve as the key modules reflecting the impacts of effluent discharge on benthic biofilms.

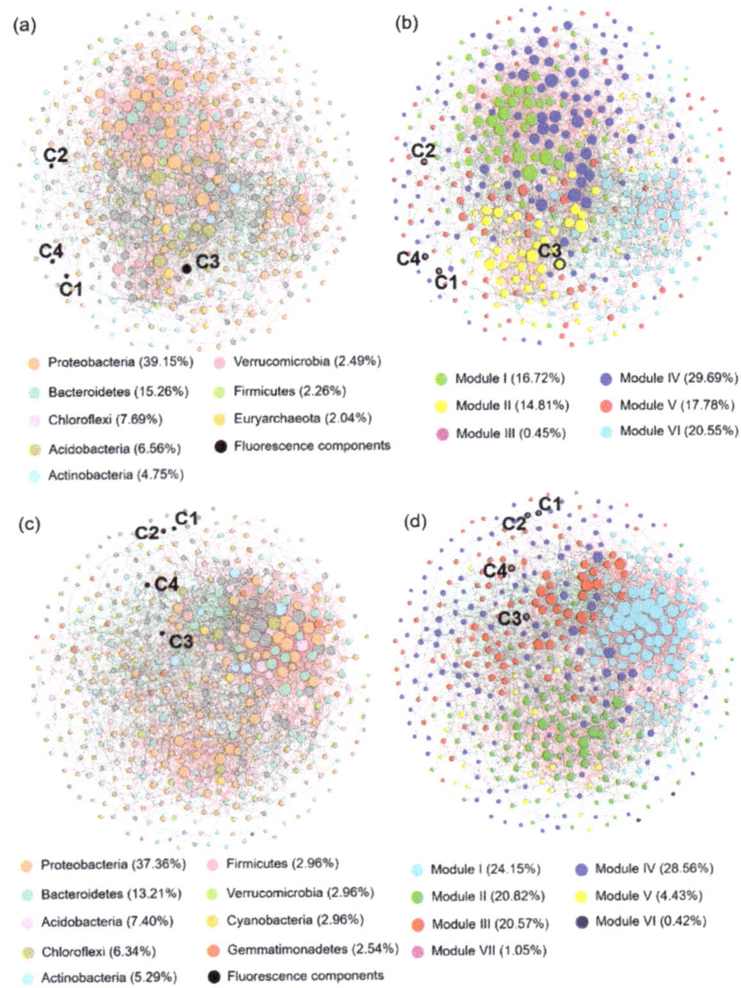

Figure 7. Co-occurrence network based on Random Matrix Theory (RMT) model in (**a**,**b**) winter and (**c**,**d**) summer. The size of each node in the network is proportional to the degree of connectivity, and the nodes are colored according to phylum (**a**,**c**) and different types of modularity classes (**b**,**d**), respectively. Red and green edges indicate positive and negative correlations, respectively.

4. Discussion

4.1. Effluent Discharge Alters the Nature of Biofilm WSOM

To date, the response of biofilm WSOM to effluent discharge in receiving waterbodies remains unknown. The content of WSOC in biofilm matrix in the discharge area displayed an increase in both seasons (Table S2), implying that effluent discharge could facilitate carbon storage in benthic biofilms. During conventional wastewater treatment processes, high-molecular-weight and aromatic substances are difficult to degrade and a certain amount will be stored in the effluent and discharged to receiving waterbodies [49]. Combined with the variations of $SUVA_{254}$, E_{254}/E_{204}, and S_R, it can be inferred that the effluent discharge changed the nature of benthic biofilm WSOM in the Discharge area and the Downstream groups, whereas there are no clear patterns of changes seasonally nor regionally (Table S4). It is difficult to explain such results within the context of an absence of knowledge about the nature of the effluent; therefore, we speculate that the phenomenon could be ascribed to the fluctuating operating conditions of the WWTPs [6,12,49]. Taken together, the findings confirm that the aromaticity, molecular weight, and hydrophobicity of biofilm WSOM in the Downstream group have difficulty recovering to the original state of the Upstream group after receiving effluent.

It is interesting to observe that the fluorescent properties of the biofilm WSOM exhibited a clear resilience responding to effluent discharge. HIX values were signally higher ($p < 0.05$) and BIX values were significantly lower ($p < 0.001$) in the Discharge area group than in the Downstream group (Figure 3). The effluent discharge increased the humification state and decreased the proportions of autochthonous component in benthic biofilms. Changes in protein-like and aliphatic fractions degraded by micro-organisms have been reported to facilitate the humification processes of organic compounds [50]. Meanwhile, the enzymatic processes during biodegradation may promote the enhancement of condensed aromatic structures or the production of structures with increased conjugation, bringing on an increase in HIX [51]. However, unlike the response of DOM in receiving waterbodies, the BIX values of biofilm WSOM exhibited a significant decrease, implying a decline in the input of autochthonous DOM sources [52]. The distinct change between the BIX values in these two studies can be attributed to the adsorption features of biofilm, implying that the biofilm may store allochthonous DOM from effluents. Although the discharge increased the HIX values while decreasing the BIX values, the biofilm in the effluent-receiving river was found to exhibit weak humic character and strong autochthonous components dominated by microbial metabolism [52]. Interestingly, the changes in BIX and HIX indices in the Downstream group show opposite trends (Figure 3d), suggesting that the benthic biofilm WSOM in the downstream area is inclined to recover to the original state in the absence of effluent discharge.

The fluorescent properties of biofilm WSOM differed significantly in response to seasonal changes. The benthic biofilm CDOM in winter is dominated by humic-like materials, whereas intermediate humic-like and amino acid-like dominate in the Upstream group and tryptophan-like proteins occupy the highest fluorescent fractions in the Discharge area and the Downstream groups in summer. The relatively lower proportion of humic-like fractions might be attributed to the high temperature and humid conditions in summer, during which humic-like substances are easily to be released from biofilm WSOM, and winter provides a perfect ecological niche for micro-organisms to release DOM characterized by native protein-like compounds [53]. The protein components of the effluent entrainment increased the amount of tryptophan-like proteins in the benthic biofilm, and this tendency was more pronounced in the warm season. Previous studies have shown that microbes can change the composition of DOM, and the proportion of humic substances is increased through the conversion of protein-like substances [54]. The higher microbial activity in summer may facilitate the conversion process and favor the release and detachment of humic-like fractions. Meanwhile, tryptophan-like proteins are usually associated with anthropogenic activities [6] and some studies have demonstrated a higher removal efficiency of tryptophan-like proteins by WWTP in warmer condition than in colder ones [55]. It must be pointed out that tryptophan-like compounds still dominate the biofilm WSOM in summer, so we consider that the impacts of effluent discharge on the ecological health of rivers cannot be ignored, especially in warmer seasons.

4.2. Response of Biofilm Bacterial Communities to Effluent Discharge

The bacterial diversity indices were found to increase in the discharge area and experienced a decrease downstream (Figure 4), with significant changes in the richness and Shannon indices ($p < 0.05$). Studies regarding the effects of effluent discharge on biofilm bacterial communities gave contradictory results, possibly due to either increasing or decreasing microbial diversity and enzyme activity [5,56,57]. In this study, the Discharge area and the Downstream groups had distinct responses to effluent discharge, reflected by the more vulnerable diversity and abundance of bacterial communities, while the diversity of bacterial communities in the Downstream group approached to the status observed in the Upstream group, implying that microbial ecological reconstruction of bacterial community occurred responding to effluent discharge [4]. Additional evidence on the effects of effluent discharge could be found in the co-occurrence network analysis (Figure S4). The discharge increased the diversity of the biofilm bacterial community, interfered with microbial interactions, and reduced the modularity in the Discharge area group. In the Downstream group, there was a recovery in both bacterial interspecies interaction and network modularity.

Figure 4d shows a specific clustering mode among seasons and sampling areas. Pascual-Benito et al. [57] reported a higher microbial diversity in effluent-receiving rivers under high-flow and low-temperature conditions in winter. However, we describe the results of greater differences in bacterial communities across areas under the influence of effluent discharge than seasonal differences (Figure 4d). Considering the relatively lower value of F (0.02) among sampling areas than among seasons (0.031), the ecological impacts of effluent discharge on biofilms in different areas are worthy of being discussed. The assembly pattern analysis suggests an estimated beta diversity β_{SOR} value of bacterial communities among different areas or seasons, with a strong contribution of spatial turnover and a small contribution of nestedness (Table S5). The results suggest that diversity patterns of biofilm in effluent-receiving rivers are primarily caused by species turnover rather than species loss [58]. Turnover is achieved through migration, attachment, and growth, and dispersal refers to the movement of microorganisms in space, especially those absorbed into the biofilms [59]. Typically, the attachment of planktonic cells to the media surface triggers biofilm formation and fundamentally regulates the microbial assembly process [60]. The higher beta diversity can be explained by more diverse environmental conditions. For example, water flow, turbulence, and bottom landscape topography regulate microbial dispersal and colonization patterns, while also producing microhabitats with distinct stresses and mass transfer efficiencies [13]. The increase in β_{SOR} suggests that effluent-borne microorganisms, DOM, and nutrients may be partially adsorbed into the benthic biofilm, resulting in the formation of more complex microhabitats in the discharge area and the area downstream WWTP.

4.3. Variations in Metabolic Functions of Bacterial Communities

Effluent discharge affects the original ecosystem functions of rivers, altering the availability of genes related to the carbon cycle and possibly carrying foreign co-energy genes, bringing unknown pressures to riverine ecosystems [23]. The metabolic function of bacteria communities in effluent-receiving rivers is more stable; however, it is important to note that the predicted functional abundance of carbohydrate metabolism and amino acid metabolism increased in the Discharge area group (Figure S3). This may imply that the benthic biofilm bacterial community has a higher rate of carbon turnover and enhanced utilization of carbon sources in areas directly receiving effluent [61]. The membrane transport function was also proven to exhibit a high abundance in this area, suggesting that bacterial cells had active transporter proteins that can transport organic matter and nutrients to facilitate bacterial metabolic processes [24].

In this study, significant differences were observed between alanine/aspartate/glutamate metabolism and nicotinate/nicotinamide metabolism in Upstream group and Discharge area group (Figure 5). Both metabolic pathways are closely related to the degradation of carbohydrates [62]. The involvement of carbohydrate metabolism, cofactors and vitamin metabolism, and other secondary metabolites was higher in the Discharge area and Downstream groups compared to the Upstream group, indicating that the biodegradation activity was higher in areas less affected by effluent discharge [61]. However, the effluent discharge ultimately enhanced the carbohydrate and amino acid metabolism in the Discharge area group, suggesting that the adverse effects of WWTP discharge on the metabolic functions of the benthic biofilm bacterial community may be limited to specific metabolic functions. Another promising finding was that little difference in the metabolic functions of the bacterial community could be found between the Discharge area and the Downstream groups. The results could be ascribed to the partial adaptation of biofilms in response to effluent discharge.

4.4. Roles of Bacteria in Shaping Biofilm WSOM

Here, we found significant seasonal patterns, including commonalities and anisotropies, associated with effluent discharge on biofilm WSOM properties and bacterial community composition. In winter, humic-like and amino acid-like intermediates were significantly correlated with bacterial community composition (Figure 6a,b). C3 is mainly comprised of low-molecular-weight and highly aromatic substances [63]. Existing studies have found that bacterial communities are inclined to utilize low-molecular-weight substances [6], and aromaticity is significantly associated with community succession dynamics [64]. The aforementioned two points may help explain why the C3 fraction acts as a key factor shaping the composition of the biofilm bacterial community harvested in winter. In summer, tryptophan proteins, aromaticity, and molecular weight were significantly associated with bacterial community (Figure 6c,d). The aromaticity and molecular weight of biofilm WSOM were remarkably lower in summer compared to winter ($p < 0.01$ and $p < 0.001$) (Table S4). We speculate that the seasonally variable biofilm WSOM properties may determine their distinct capacities in shaping biofilm bacterial community composition. The close association of tryptophan proteins with microbial activity has been reported in previous studies [6,47]. At this stage of understanding, we believe that certain factors, e.g., molecular weight and aromaticity in biofilm WSOM, might govern the bacteria community structure, regardless of the sampling seasons.

It is worth noting that FI values showed a significant correlation with changes in bacterial community composition in both seasons, although FI values did not show signal variations among seasons and areas. The results provide additional evidence that the microbial-derived organic matter of biofilm WSOM remains largely stable [65]. The value of FI has been strongly correlated to the relative contribution of microbial-derived versus plant-derived organic matter. The FI values of all samples ranged from 2.40 to 3.04, indicating that biofilm WSOM predominantly originated from microbial activity [66]. It can be assumed that microbial-derived organic matter produced by microbial metabolism occupies an important position in biofilms, especially in the process of shaping microbial communities.

4.5. Co-Occurrence Patterns Relate to Seasonal and Spatial Variation

The interactions between microbial communities in turn affect their adaptation to external environmental changes, and the co-occurrence networks constructed based on different areas can reveal the ecological interactions between biofilm bacterial communities [25]. In different areas of the network (Figure S4), the average path distance followed the sequence: Upstream group < Discharge area group < Downstream group (Table S8), reflecting more efficient information processing and material transfer among bacteria influenced by discharge [27]. The value of modularity was highest in the Upstream group and lowest in the Discharge area group, demonstrating that microbial interactions are stronger in the Upstream group and effluent discharge disrupts original interspecies interactions of bacteria, as in [67]. Compared to the Upstream group, the networks from the Discharge area and Downstream groups had fewer nodes and edges, and the average degree decreased along the effluent-receiving river, demonstrating that effluent discharge simplified the network pattern of benthic biofilms [48]. Previous studies have suggested that a more connected network could improve the efficient utilization of carbon and that a highly connected network may also provide more functional redundancy [68,69]. Effluent contains plentiful carbon, nitrogen and some toxic substances, which might possibly limit the complexity of microbial co-occurrence networks in the Discharge area and the Downstream groups. Consequently, effluent discharge may potentially diminish the stability and disturbance resistance of benthic biofilm communities.

Positive correlations in Figure 7 suggest that the interactions are chiefly symbiotic or mutually beneficial. The microbe-organic networks in different seasons were dominated by positive correlations. Generally speaking, a more positive microbe-organic correlation is beneficial for the degradation of refractory substances [70]. Proteobacteria, one of the dominant phyla in the network, can degrade humic substances and tend to form filamentous structures, facilitating the growth of such bacteria in biofilms [71]. Bacteroidetes, another dominant phylum, generally hold special importance for benthic biofilms and some members of these phylum can degrade biopolymers, contributing to the high-molecular-weight fraction of organic matters [72].

Moreover, modules can be considered as functional units, in which the same functional unit can perform the same ecological function with a high degree of in-connection between microbial communities within the same module [73]. Here, bacterial communities and biofilm WSOM components formed a strong network of modules, with a total of six modules in winter and seven modules in summer (Figure 7). By taking advantage of the key modules, we can acquire more information on the interactions between bacterial communities and fluorescent compounds. For example, in Module III of the network derived from samples collected in winter, genus *Kineosporiaceae* was able to convert cellulose and glucose to acetate, butyric acid, and carbon dioxide under anoxic conditions [74]. Genus *Lacihabitans* was proven capable of degrading a multitude of organic compounds including cellulose, chitin, and starch [75]. Similarly, in the network derived from samples collected in summer, genus *Nocardioidaceae* in Module IV was capable of depredating toxic pollutants, alkanes, crude oil, and derivatives [76]. The family *Beijerinckiaceae* is able to fix nitrogen and metabolize carbon. As a consequence, the pollutant degradation capacity of benthic biofilm bacterial communities in summer may also be amended in response to effluent discharge [77]. Genus *Aeromans* is an important pathogenic agent for fish and is harmful to aquatic ecosystems [78]. These findings support the notion that carbon metabolism remains a key ecological function of benthic biofilms in effluent-receiving rivers, and we speculate that there is an enhancement of the degradation of toxic pollutants in summer. Meanwhile, the production of pathogenic bacteria needs to be guarded against. Note that due to the lack of controlled experiments, these results need to be treated with caution. Simulation experiments need to be conducted to clarify the ecological impact of effluent discharge on effluent-receiving rivers in future work.

5. Conclusions

In highly urbanized areas, river benthic biofilms are pioneering microbial aggregates responding to effluent discharge from WWTPs. Our study reveals the optical properties of benthic biofilms WSOM in a representative effluent-receiving river. The diversity, function, and assembly of bacterial communities and their co-occurrence patterns were also investigated. After receiving effluent, WSOM in benthic biofilms showed weak humic character and strong autotrophic components. In the Discharge area, the fluorescence characteristics of CDOM and bacterial community diversity exhibited a signal alteration. Both the interspecies interaction of bacteria and the fluorescent nature of biofilm WSOM gradually recover to the conditions exhibited when less affected by effluent discharge. Species turnover was the main factor governing the formation of biofilm diversity patterns. Functional predictions showed that amino acid metabolism and carbohydrate metabolism increased significantly after receiving effluent discharge. Additionally, amino acid-like and humic-like intermediates and tryptophan proteins were found to be key factors affecting bacterial community composition in winter and summer, respectively. The key ecological functions present in the benthic biofilm in the effluent-receiving river were further elucidated by combining the key modules in co-occurrence networks. Future studies will be performed with a focus on WWTPs with different effluent standards to demonstrate the universality of benthic biofilms as an indicator of the ecological response of effluent-receiving rivers.

Supplementary Materials: The following supporting information can be downloaded at: https://www.mdpi.com/article/10.3390/ijerph19041994/s1: Figure S1: Taxonomic composition and relative abundance of bacterial communities at the phylum level. Proteobacteria is identified to class level; Figure S2: Mechanisms of microbial community assembly based on stochastic and deterministic models for different (a) seasons and (b) areas; Figure S3: Abundance of various predictive function samples on KEGG pathway level 2 based on PICRUSt. "S-" indicates sampling in summer and "W-" indicates sampling in winter; Figure S4: Co-occurrence network of bacterial communities in (a) Upstream group, (b) Discharge area group, and (c) Downstream group. The node colors represent major modules. The modularity for each network is shown below the figures; Table S1: Physicochemical parameters of water samples at each sampling site; Table S2: Normalized water-soluble organic carbon concentration in biofilm WSOM at varying sampling sites; Table S3: Description of UV-vis absorption spectrum parameters and EEMs parameters; Table S4: UV-visible spectral-derived parameters of biofilm WSOM; Table S5: The discrepancy of the spatial turnover and nestedness pattern indices in different areas and seasons; Table S6: The discrepancy of the spatial turnover and nestedness pattern indices in different areas (resampling); Table S7: Statistics of topological characteristics of empirical and random networks in different seasons; Table S8: Topological properties of benthic biofilm bacterial community networks in different areas [79–81].

Author Contributions: Conceptualization, L.W. and Y.L.; methodology, Y.W.; software, W.Z.; validation, H.Z.; formal analysis, L.N. and N.H.; investigation, L.W.; resources, Y.W.; data curation, L.N.; writing—original draft preparation, Y.L.; writing—review and editing, L.W. All authors have read and agreed to the published version of the manuscript.

Funding: This research was funded by the National Key R&D Program of China [grant number 2019YFC0408300], the National Natural Science Foundation of China [grant number 52170159], the Fundamental Research Funds for the Central Universities [grant number B210202054], the Open Fund of the State Key Laboratory of Hydraulics and Mountain River Engineering of Sichuan University, the Priority Academic Program Development of Jiangsu Higher Education Institutions, and the Top-Notch Academic Programs Project of Jiangsu Higher Education Institutions (TAPP).

Institutional Review Board Statement: Not applicable.

Informed Consent Statement: Not applicable.

Data Availability Statement: The data that support the findings of this study are available from the corresponding author, upon reasonable request.

Conflicts of Interest: The authors declare no conflict of interest.

References

1. Kamjunke, N.; Hertkorn, N.; Harir, M.; Schmitt-Kopplin, P.; Griebler, C.; Brauns, M.; von Tumpling, W.; Weitere, M.; Herzsprung, P. Molecular change of dissolved organic matter and patterns of bacterial activity in a stream along a land-use gradient. *Water Res.* **2019**, *164*, 114919. [CrossRef] [PubMed]
2. Burdon, F.J.; Bai, Y.; Reyes, M.; Tamminen, M.; Staudacher, P.; Mangold, S.; Singer, H.; Rasanen, K.; Joss, A.; Tiegs, S.D.; et al. Stream microbial communities and ecosystem functioning show complex responses to multiple stressors in wastewater. *Glob. Chang. Biol.* **2020**, *26*, 6363–6382. [CrossRef] [PubMed]
3. Waiser, M.J.; Tumber, V.; Holm, J. Effluent-dominated streams. Part 1: Presence and effects of excess nitrogen and phosphorus in Wascana Creek, Saskatchewan, Canada. *Environ. Toxicol. Chem.* **2011**, *30*, 496–507. [CrossRef] [PubMed]
4. Freixa, A.; Perujo, N.; Langenheder, S.; Romani, A.M. River biofilms adapted to anthropogenic disturbances are more resistant to WWTP inputs. *FEMS Microbiol. Ecol.* **2020**, *96*, 152. [CrossRef] [PubMed]
5. Drury, B.; Rosi-Marshall, E.; Kelly, J.J. Wastewater treatment effluent reduces the abundance and diversity of benthic bacterial communities in urban and suburban rivers. *Appl. Environ. Microbiol.* **2013**, *79*, 1897–1905. [CrossRef]
6. Yu, M.; Liu, S.; Li, G.; Zhang, H.; Xi, B.; Tian, Z.; Zhang, Y.; He, X. Municipal wastewater effluent influences dissolved organic matter quality and microbial community composition in an urbanized stream. *Sci. Total. Environ.* **2020**, *705*, 135952. [CrossRef]
7. Pereda, O.; Solagaistua, L.; Atristain, M.; de Guzman, I.; Larranaga, A.; von Schiller, D.; Elosegi, A. Impact of wastewater effluent pollution on stream functioning: A whole-ecosystem manipulation experiment. *Environ. Pollut.* **2020**, *258*, 113719. [CrossRef]
8. Romero, F.; Acuna, V.; Font, C.; Freixa, A.; Sabater, S. Effects of multiple stressors on river biofilms depend on the time scale. *Sci. Rep.* **2019**, *9*, 15810. [CrossRef]
9. Osorio, V.; Proia, L.; Ricart, M.; Perez, S.; Ginebreda, A.; Cortina, J.L.; Sabater, S.; Barcelo, D. Hydrological variation modulates pharmaceutical levels and biofilm responses in a Mediterranean river. *Sci. Total. Environ.* **2014**, *472*, 1052–1061. [CrossRef]

10. Sabater-Liesa, L.; Montemurro, N.; Font, C.; Ginebreda, A.; Gonzalez-Trujillo, J.D.; Mingorance, N.; Perez, S.; Barcelo, D. The response patterns of stream biofilms to urban sewage change with exposure time and dilution. *Sci. Total. Environ.* **2019**, *674*, 401–411. [CrossRef]
11. Pereda, O.; von Schiller, D.; Garcia-Baquero, G.; Mor, J.R.; Acuna, V.; Sabater, S.; Elosegi, A. Combined effects of urban pollution and hydrological stress on ecosystem functions of Mediterranean streams. *Sci. Total. Environ.* **2021**, *753*, 141971. [CrossRef] [PubMed]
12. Battin, T.J.; Besemer, K.; Bengtsson, M.M.; Romani, A.M.; Packmann, A.I. The ecology and biogeochemistry of stream biofilms. *Nat. Rev. Microbiol.* **2016**, *14*, 251–263. [CrossRef] [PubMed]
13. Besemer, K. Biodiversity, community structure and function of biofilms in stream ecosystems. *Res. Microbiol.* **2015**, *166*, 774–781. [CrossRef] [PubMed]
14. Risse-Buhl, U.; Anlanger, C.; Kalla, K.; Neu, T.R.; Noss, C.; Lorke, A.; Weitere, M. The role of hydrodynamics in shaping the composition and architecture of epilithic biofilms in fluvial ecosystems. *Water Res.* **2017**, *127*, 211–222. [CrossRef]
15. Sabater, S.; Guasch, H.; Ricart, M.; Romani, A.; Vidal, G.; Klunder, C.; Schmitt-Jansen, M. Monitoring the effect of chemicals on biological communities. The biofilm as an interface. *Anal. Bioanal. Chem.* **2007**, *387*, 1425–1434. [CrossRef]
16. Pu, Y.; Ngan, W.Y.; Yao, Y.; Habimana, O. Could benthic biofilm analyses be used as a reliable proxy for freshwater environmental health? *Environ. Pollut.* **2019**, *252 (Pt A)*, 440–449. [CrossRef]
17. Hobbs, W.O.; Collyard, S.A.; Larson, C.; Carey, A.J.; O'Neill, S.M. Toxic Burdens of Freshwater Biofilms and Use as a Source Tracking Tool in Rivers and Streams. *Environ. Sci. Technol.* **2019**, *53*, 11102–11111. [CrossRef]
18. Tlili, A.; Corcoll, N.; Arrhenius, Å.; Backhaus, T.; Hollender, J.; Creusot, N.; Wagner, B.; Behra, R. Tolerance Patterns in Stream Biofilms Link Complex Chemical Pollution to Ecological Impacts. *Environ. Sci. Technol.* **2020**, *54*, 10745–10753. [CrossRef]
19. Qiu, L.; Cui, H.; Wu, J.; Wang, B.; Zhao, Y.; Li, J.; Jia, L.; Wei, Z. Snowmelt-driven changes in dissolved organic matter and bacterioplankton communities in the Heilongjiang watershed of China. *Sci. Total. Environ.* **2016**, *556*, 242–251. [CrossRef]
20. Kamjunke, N.; Herzsprung, P.; Neu, T.R. Quality of dissolved organic matter affects planktonic but not biofilm bacterial production in streams. *Sci. Total. Environ.* **2015**, *506–507*, 353–360. [CrossRef]
21. Logue, J.B.; Stedmon, C.A.; Kellerman, A.M.; Nielsen, N.J.; Andersson, A.F.; Laudon, H.; Lindstrom, E.S.; Kritzberg, E.S. Experimental insights into the importance of aquatic bacterial community composition to the degradation of dissolved organic matter. *ISME J.* **2016**, *10*, 533–545. [CrossRef] [PubMed]
22. Avila, M.P.; Brandao, L.P.M.; Brighenti, L.S.; Tonetta, D.; Reis, M.P.; Staehr, P.A.; Asmala, E.; Amado, A.M.; Barbosa, F.A.R.; Bezerra-Neto, J.F.; et al. Linking shifts in bacterial community with changes in dissolved organic matter pool in a tropical lake. *Sci. Total. Environ.* **2019**, *672*, 990–1003. [CrossRef] [PubMed]
23. Zhang, L.; Fang, W.; Li, X.; Lu, W.; Li, J. Strong linkages between dissolved organic matter and the aquatic bacterial community in an urban river. *Water Res.* **2020**, *184*, 116089. [CrossRef] [PubMed]
24. Niu, L.; Li, Y.; Li, Y.; Hu, Q.; Wang, C.D.; Hu, J.; Zhang, W.; Wang, L.; Zhang, C.; Zhang, H. New insights into the vertical distribution and microbial degradation of microplastics in urban river sediments. *Water Res.* **2021**, *188*, 116449. [CrossRef] [PubMed]
25. Zhao, Q.; Chu, S.S.; He, D.; Wu, D.M.; Mo, Q.F.; Zeng, S.C. Sewage sludge application alters the composition and co-occurrence pattern of the soil bacterial community in southern China forestlands. *Appl. Soil Ecol.* **2021**, *157*, 103744. [CrossRef]
26. Zhou, H.; Gao, Y.; Jia, X.H.; Wang, M.M.; Ding, J.J.; Cheng, L.; Bao, F.; Wu, B. Network analysis reveals the strengthening of microbial interaction in biological soil crust development in the Mu Us Sandy Land, northwestern China. *Soil Biol. Biochem.* **2020**, *144*, 107782. [CrossRef]
27. Dong, Y.; Gao, J.; Wu, Q.; Ai, Y.; Huang, Y.; Wei, W.; Sun, S.; Weng, Q. Co-occurrence pattern and function prediction of bacterial community in Karst cave. *BMC Microbiol.* **2020**, *20*, 137. [CrossRef]
28. Fang, D.; Hao, L.; Cao, Z.; Huang, X.L.; Qin, M.S.; Hu, J.C.; Liu, Y.Q.; Sun, G. Combined effects of urbanization and climate change on watershed evapotranspiration at multiple spatial scales. *J. Hydrol.* **2020**, *587*, 124869. [CrossRef]
29. APHA. *Standard Methods for the Examination of Water and Wastewater*, 21st ed.; American Public Health Association: Washington, DC, USA, 2005.
30. Wang, J.; Meier, S.; Soininen, J.; Casamayor, E.O.; Pan, F.; Tang, X.; Yang, X.; Zhang, Y.; Wu, Q.; Zhou, J.; et al. Regional and global elevational patterns of microbial species richness and evenness. *Ecography* **2017**, *40*, 393–402. [CrossRef]
31. Chantigny, M.H.; Harrison-Kirk, T.; Curtin, D.; Beare, M. Temperature and duration of extraction affect the biochemical composition of soil water-extractable organic matter. *Soil Biol. Biochem.* **2014**, *75*, 161–166. [CrossRef]
32. Huang, M.; Chai, L.; Jiang, D.; Zhang, M.; Jia, W.; Huang, Y. Spatial Patterns of Soil Fungal Communities Are Driven by Dissolved Organic Matter (DOM) Quality in Semi-Arid Regions. *Microb Ecol.* **2021**, *82*, 202–214. [CrossRef] [PubMed]
33. Stedmon, C.A.; Bro, R. Characterizing dissolved organic matter fluorescence with parallel factor analysis: A tutorial. *Limnol. Oceanogr-Meth.* **2008**, *6*, 572–579. [CrossRef]
34. Bolyen, E.; Rideout, J.R.; Dillon, M.R.; Bokulich, N.A.; Abnet, C.C.; Al-Ghalith, G.A.; Alexander, H.; Alm, E.J.; Arumugam, M.; Asnicar, F.; et al. Author Correction: Reproducible, interactive, scalable and extensible microbiome data science using QIIME 2. *Nat. Biotechnol.* **2019**, *37*, 1091. [CrossRef] [PubMed]
35. Callahan, B.J.; McMurdie, P.J.; Rosen, M.J.; Han, A.W.; Johnson, A.J.; Holmes, S.P. DADA2: High-resolution sample inference from Illumina amplicon data. *Nat. Methods* **2016**, *13*, 581–583. [CrossRef]

36. Langille, M.G.; Zaneveld, J.; Caporaso, J.G.; McDonald, D.; Knights, D.; Reyes, J.A.; Clemente, J.C.; Burkepile, D.E.; Vega Thurber, R.L.; Knight, R.; et al. Predictive functional profiling of microbial communities using 16S rRNA marker gene sequences. *Nat. Biotechnol.* **2013**, *31*, 814–821. [CrossRef] [PubMed]
37. Baselga, A. Partitioning the turnover and nestedness components of beta diversity. *Global Ecol. Biogeogr.* **2010**, *19*, 134–143. [CrossRef]
38. Stegen, J.C.; Lin, X.; Konopka, A.E.; Fredrickson, J.K. Stochastic and deterministic assembly processes in subsurface microbial communities. *ISME J.* **2012**, *6*, 1653–1664. [CrossRef]
39. Stegen, J.C.; Lin, X.; Fredrickson, J.K.; Chen, X.; Kennedy, D.W.; Murray, C.J.; Rockhold, M.L.; Konopka, A. Quantifying community assembly processes and identifying features that impose them. *ISME J.* **2013**, *7*, 2069–2079. [CrossRef]
40. Zhou, J.; Deng, Y.; Luo, F.; He, Z.; Tu, Q.; Zhi, X. Functional molecular ecological networks. *MBio* **2010**, *1*, 10. [CrossRef]
41. Deng, Y.; Jiang, Y.H.; Yang, Y.; He, Z.; Luo, F.; Zhou, J. Molecular ecological network analyses. *BMC Bioinformatics* **2012**, *13*, 113. [CrossRef]
42. Hu, Y.; Bai, C.; Cai, J.; Dai, J.; Shao, K.; Tang, X.; Gao, G. Co-occurrence network reveals the higher fragmentation of the bacterial community in Kaidu River than its tributaries in Northwestern China. *Microbes Environ.* **2018**, *33*, 127–134. [CrossRef] [PubMed]
43. Kothawala, D.N.; von Wachenfeldt, E.; Koehler, B.; Tranvik, L.J. Selective loss and preservation of lake water dissolved organic matter fluorescence during long-term dark incubations. *Sci. Total Environ.* **2012**, *433*, 238–246. [CrossRef] [PubMed]
44. Osburn, C.L.; Wigdahl, C.R.; Fritz, S.C.; Saros, J.E. Dissolved organic matter composition and photoreactivity in prairie lakes of the U.S. Great Plains. *Limnol. Oceanogr.* **2011**, *56*, 2371–2390. [CrossRef]
45. Retelletti Brogi, S.; Jung, J.Y.; Ha, S.Y.; Hur, J. Seasonal differences in dissolved organic matter properties and sources in an Arctic fjord: Implications for future conditions. *Sci. Total Environ.* **2019**, *694*, 133740. [CrossRef]
46. Jørgensen, L.; Stedmon, C.A.; Kragh, T.; Markager, S.; Middelboe, M.; Søndergaard, M. Global trends in the fluorescence characteristics and distribution of marine dissolved organic matter. *Marine Chemistry* **2011**, *126*, 139–148. [CrossRef]
47. Hambly, A.C.; Arvin, E.; Pedersen, L.F.; Pedersen, P.B.; Seredynska-Sobecka, B.; Stedmon, C.A. Characterising organic matter in recirculating aquaculture systems with fluorescence EEM spectroscopy. *Water Res.* **2015**, *83*, 112–120. [CrossRef] [PubMed]
48. Fan, K.; Weisenhorn, P.; Gilbert, J.A.; Chu, H. Wheat rhizosphere harbors a less complex and more stable microbial co-occurrence pattern than bulk soil. *Soil Biol. Biochem.* **2018**, *125*, 251–260. [CrossRef]
49. Wang, M.; Chen, Y. Generation and characterization of DOM in wastewater treatment processes. *Chemosphere* **2018**, *201*, 96–109. [CrossRef]
50. Simsek, H.; Wadhawan, T.; Khan, E. Overlapping photodegradable and biodegradable organic nitrogen in wastewater effluents. *Environ. Sci. Technol.* **2013**, *47*, 7163–7170. [CrossRef]
51. Li, J.; Wang, L.; Geng, J.; Li, S.; Yu, Q.; Xu, K.; Ren, H. Distribution and removal of fluorescent dissolved organic matter in 15 municipal wastewater treatment plants in China. *Chemosphere* **2020**, *251*, 126375. [CrossRef]
52. Wang, Y.; Hu, Y.; Yang, C.; Wang, Q.; Jiang, D. Variations of DOM quantity and compositions along WWTPs-river-lake continuum: Implications for watershed environmental management. *Chemosphere* **2019**, *218*, 468–476. [CrossRef] [PubMed]
53. Maqbool, T.; Zhang, J.; Qin, Y.; Ly, Q.V.; Asif, M.B.; Zhang, X.; Zhang, Z. Seasonal occurrence of N-nitrosamines and their association with dissolved organic matter in full-scale drinking water systems: Determination by LC-MS and EEM-PARAFAC. *Water Res.* **2020**, *183*, 116096. [CrossRef] [PubMed]
54. Zhang, L.; Fang, W.; Li, X.; Gao, G.; Jiang, J. Linking bacterial community shifts with changes in the dissolved organic matter pool in a eutrophic lake. *Sci. Total Environ.* **2020**, *719*, 137387. [CrossRef]
55. Yang, X.; Zhou, Z.; Raju, M.N.; Cai, X.; Meng, F. Selective elimination of chromophoric and fluorescent dissolved organic matter in a full-scale municipal wastewater treatment plant. *J. Environ. Sci.* **2017**, *57*, 150–161. [CrossRef]
56. Perujo, N.; Freixa, A.; Vivas, Z.; Gallegos, A.M.; Butturini, A.; Romaní, A.M. Fluvial biofilms from upper and lower river reaches respond differently to wastewater treatment plant inputs. *Hydrobiologia* **2015**, *765*, 169–183. [CrossRef]
57. Pascual-Benito, M.; Balleste, E.; Monleon-Getino, T.; Urmeneta, J.; Blanch, A.R.; Garcia-Aljaro, C.; Lucena, F. Impact of treated sewage effluent on the bacterial community composition in an intermittent mediterranean stream. *Environ. Pollut.* **2020**, *266 Pt 1*, 115254. [CrossRef]
58. Zhang, W.; Lei, M.; Li, Y.; Wang, P.; Wang, C.; Gao, Y.; Wu, H.; Xu, C.; Niu, L.; Wang, L.; et al. Determination of vertical and horizontal assemblage drivers of bacterial community in a heavily polluted urban river. *Water Res.* **2019**, *161*, 98–107. [CrossRef]
59. Brislawn, C.J.; Graham, E.B.; Dana, K.; Ihardt, P.; Fansler, S.J.; Chrisler, W.B.; Cliff, J.B.; Stegen, J.C.; Moran, J.J.; Bernstein, H.C. Forfeiting the priority effect: Turnover defines biofilm community succession. *ISME J.* **2019**, *13*, 1865–1877. [CrossRef]
60. Xu, R.; Zhang, S.; Meng, F. Large-sized planktonic bioaggregates possess high biofilm formation potentials: Bacterial succession and assembly in the biofilm metacommunity. *Water Res.* **2020**, *170*, 115307. [CrossRef]
61. Li, Y.; Xu, C.; Zhang, W.; Lin, L.; Wang, L.; Niu, L.; Zhang, H.; Wang, P.; Wang, C. Response of bacterial community in composition and function to the various DOM at river confluences in the urban area. *Water Res.* **2020**, *169*, 115293. [CrossRef]
62. Neis, E.P.; Dejong, C.H.; Rensen, S.S. The role of microbial amino acid metabolism in host metabolism. *Nutrients* **2015**, *7*, 2930–2946. [CrossRef] [PubMed]
63. Cabrera, J.M.; Garcia, P.E.; Pedrozo, F.L.; Queimalinos, C.P. Dynamics of the dissolved organic matter in a stream-lake system within an extremely acid to neutral pH range: Agrio-Caviahue watershed. *Spectrochim. Acta. A. Mol. Biomol. Spectrosc.* **2020**, *235*, 118278. [CrossRef] [PubMed]

64. Zhou, L.; Zhou, Y.; Tang, X.; Zhang, Y.; Jang, K.S.; Szekely, A.J.; Jeppesen, E. Resource aromaticity affects bacterial community successions in response to different sources of dissolved organic matter. *Water Res.* **2021**, *190*, 116776. [CrossRef] [PubMed]
65. Wu, K.; Lee, T.H.; Chen, Y.L.; Wang, Y.S.; Wang, P.H.; Yu, C.P.; Chu, K.H.; Chiang, Y.R. Metabolites Involved in Aerobic Degradation of the A and B Rings of Estrogen. *Appl. Environ. Microb.* **2019**, *85*. [CrossRef]
66. Zhang, P.; Cao, C.; Wang, Y.H.; Yu, K.; Liu, C.; He, C.; Shi, Q.; Wang, J.J. Chemodiversity of water-extractable organic matter in sediment columns of a polluted urban river in South China. *Sci. Total Environ.* **2021**, *777*, 146127. [CrossRef]
67. Lu, Y.; Zhang, W.; Li, Y.; Zhang, C.; Wang, L.; Niu, L.; Zhang, H. Microbial community shift via black carbon: Insight into biological nitrogen removal from microbial assemblage and functional patterns. *Environ. Res.* **2021**, *192*, 110266. [CrossRef]
68. Morrien, E.; Hannula, S.E.; Snoek, L.B.; Helmsing, N.R.; Zweers, H.; de Hollander, M.; Soto, R.L.; Bouffaud, M.L.; Buee, M.; Dimmers, W.; et al. Soil networks become more connected and take up more carbon as nature restoration progresses. *Nat. Commun.* **2017**, *8*, 14349. [CrossRef]
69. Yang, W.; Jing, X.; Guan, Y.; Zhai, C.; Wang, T.; Shi, D.; Sun, W.; Gu, S. Response of Fungal Communities and Co-occurrence Network Patterns to Compost Amendment in Black Soil of Northeast China. *Front. Microbiol.* **2019**, *10*, 1562. [CrossRef]
70. Zhu, P.; Li, Y.; Gao, Y.; Yin, M.; Wu, Y.; Liu, L.; Du, N.; Liu, J.; Yu, X.; Wang, L. Insight into the effect of nitrogen-rich substrates on the community structure and the co-occurrence network of thermophiles during lignocellulose-based composting. *Bioresour. Technol.* **2021**, *319*, 124111. [CrossRef]
71. Newton, R.J.; Jones, S.E.; Eiler, A.; McMahon, K.D.; Bertilsson, S. A guide to the natural history of freshwater lake bacteria. *Microbiol. Mol. Biol. Rev.* **2011**, *75*, 14–49. [CrossRef]
72. Kirchman, D.L. The ecology of Cytophaga-Flavobacteria in aquatic environments. *FEMS Microbiol. Ecol.* **2002**, *39*, 91–100. [CrossRef]
73. Yang, J.; Jiang, H.; Liu, W.; Huang, L.; Huang, J.; Wang, B.; Dong, H.; Chu, R.K.; Tolic, N. Potential utilization of terrestrially derived dissolved organic matter by aquatic microbial communities in saline lakes. *ISME J.* **2020**, *14*, 2313–2324. [CrossRef] [PubMed]
74. Schellenberger, S.; Kolb, S.; Drake, H.L. Metabolic responses of novel cellulolytic and saccharolytic agricultural soil Bacteria to oxygen. *Environ. Microbiol.* **2009**, *12*, 845–861. [CrossRef] [PubMed]
75. Joung, Y.; Kim, H.; Kang, H.; Lee, B.I.; Ahn, T.S.; Joh, K. Lacihabitans soyangensis gen. nov.; sp. nov.; a new member of the family Cytophagaceae, isolated from a freshwater reservoir. *Int. J. Syst. Evol. Microbiol.* **2014**, *64 Pt 9*, 3188–3194. [CrossRef]
76. Zhou, J.; Deng, Y.; Luo, F.; He, Z.; Yang, Y. Phylogenetic molecular ecological network of soil microbial communities in response to elevated CO_2. *mBio* **2011**, *2*, e00122-11. [CrossRef]
77. Dedysh, S.N.; Haupt, E.S.; Dunfield, P.F. Emended description of the family Beijerinckiaceae and transfer of the genera Chelatococcus and Camelimonas to the family Chelatococcaceae fam. nov. *Int. J. Syst. Evol. Microbiol.* **2016**, *66*, 3177–3182. [CrossRef]
78. Janda, J.M.; Abbott, S.L. The Genus Aeromonas: Taxonomy, Pathogenicity, and Infection. *Clin. Microbiol. Rev.* **2010**, *23*, 35–73. [CrossRef] [PubMed]
79. Zepp, R.G.; Sheldon, W.M.; Moran, M.A. Dissolved organic fluorophores in southeastern US coastal waters: correction method for eliminating Rayleigh and Raman scattering peaks in excitation–emission matrices. *Mar. Chem.* **2004**, *89*, 15–36.
80. Lee, M.H.; Lee, Y.K.; Derrien, M.; Choi, K.; Shin, K.H.; Jang, K.S.; Hur, J. Evaluating the contributions of different organic matter sources to urban river water during a storm event via optical indices and molecular composition. *Water Res.* **2019**, *165*, 115006.
81. Ye, Q.H.; Wang, Y.H.; Zhang, Z.T.; Huang, W.L.; Li, L.P.; Li, J.T.; Liu, J.S.; Zheng, Y.; Mo, J.M.; Zhang, W.; et al. Dissolved organic matter characteristics in soils of tropical legume and non-legume tree plantations. *Soil Biol. Biochem.* **2020**, *148*, 107880.

International Journal of
Environmental Research and Public Health

MDPI

Article

Phytoplankton Composition and Their Related Factors in Five Different Lakes in China: Implications for Lake Management

Junmei Jia [1,*], Qiuwen Chen [2,3,*], Haidong Ren [1,4], Renjie Lu [5], Hui He [1] and Peiwen Gu [1]

1 State Key Laboratory of Plateau Ecology and Agriculture, Qinghai University, Xining 810016, China; mingqiangren@163.com (H.R.); hehui1113@sina.com (H.H.); gpw214206@163.com (P.G.)
2 State Key Laboratory of Hydrology-Water Resources and Hydraulic Engineering, Nanjing 210029, China
3 CEER, Nanjing Hydraulic Research Institute, Nanjing 210029, China
4 Key Laboratory of the Northern Qinghai-Tibet Plateau Geological Processes and Mineral Resources, Qinghai Geological Survey Institute, Xining 810012, China
5 Jiangsu Suzhou Environmental Monitoring Center, Suzhou 215000, China; lurenjie990@163.com
* Correspondence: jiadao_mei@126.com (J.J.); qwchen@nhri.cn (Q.C.)

Abstract: In this paper, two trophic lakes: Lake Taihu and Lake Yanghe, and three alpine lakes: Lake Qinghai, Lake Keluke, and Lake Tuosu, were investigated to discover the connections between environmental factors and the phytoplankton community in lakes with differences in trophic levels and climatic conditions. Three seasonal data, including water quality and phytoplankton, were collected from the five lakes. The results demonstrated clear differences in water parameters and phytoplankton compositions in different lakes. The phytoplankton was dominated by Bacillariophyta, followed by Cyanobacteria and Chlorophyta in Lake Qinghai, Lake Keluke, and Lake Tuosu. It was dominated by Cyanobacteria (followed by Chlorophyta and Bacillariophyta in Lake Yanghe) and Cyanobacteria (followed by Chlorophyta and Cryptophyta in Lake Taihu). The temperature was an essential factor favoring the growth of Cyanobacteria, Chlorophyta, and Bacillariophyta, especially Cyanobacteria and Chlorophyta. The pH had significantly negative relationships with Cyanobacteria, Chlorophyta, and Bacillariophyta. Particularly, a high pH might be a strong and negative factor for phytoplankton growth in alpine lakes. A high salinity was also an adverse factor for phytoplankton. Those results could provide fundamental information about the phytoplankton community and their correlated factors in the alpine lakes of the Tibetan Plateau, contributing to the protection and management of alpine lakes.

Keywords: phytoplankton; environmental factors; connections; eutrophic lakes; alpine lakes

Citation: Jia, J.; Chen, Q.; Ren, H.; Lu, R.; He, H.; Gu, P. Phytoplankton Composition and Their Related Factors in Five Different Lakes in China: Implications for Lake Management. *Int. J. Environ. Res. Public Health* **2022**, *19*, 3135. https://doi.org/10.3390/ijerph19053135

Academic Editors: Piotr Rzymski and Kenneth W. Cummins

Received: 20 December 2021
Accepted: 2 March 2022
Published: 7 March 2022

1. Introduction

The phytoplankton community, as a crucial primary producer, has profound influences on the geochemical cycling and the function of aquatic ecosystems [1]. In many eutrophic aquatic systems around the world, the phytoplankton community is dominated by several bloom-forming species, and the blooming of phytoplankton threatens those aquatic systems. Previous studies have suggested that the phytoplankton, dominated by cyanobacteria, affect the zooplankton structure and weaken the zooplankton biodiversity [2]. Additionally, the toxic species have adverse effects on other aquatic organisms, and the toxins accumulate in their body [3]. Thus, the exploration of environmental factors influencing the structure and dynamic of phytoplankton in eutrophic lakes has caught the attention of scientists worldwide [4–6]. Lakes located in Tibetan Plateau, such as Lake Qinghai, are neglected since they have low phytoplankton abundance and present no phytoplankton blooms. However, endemic fish and rare birds live in those lakes, and the alteration of the phytoplankton may affect the survival of rare fish and birds. Climate change may lead to a severe situation in those alpine lakes, as they are more susceptible to climate change. Therefore, it is critical to reveal phytoplankton structures and their driving factors in both trophic lakes and alpine lakes.

The factors affecting the growth of phytoplankton have been deeply explored in temperate and subtropical lakes. Temperature and nutrients are considered the most important factors regulating the growth of phytoplankton [1,7–9]. High temperature favors phytoplankton by increasing the growth rate and shifting the phytoplankton with higher optimum temperature species, such as cyanobacteria [10–12]. The effect of nutrients on cyanobacteria has been investigated in the long term. P has been identified as a limited nutrient factor in freshwater systems [13–16]. However, some buoyant species, including *Microcystis*, *Cylindrospermopsis*, *Anabaena*, *Aphanizomenon*, and *Gloeotrichia*, are not likely limited by P, because the buoyant species can vertically migrate, consume excess phosphorus at the sediment–water interface, and then rise to the water surface to form blooms [17–21]. N is also a critical factor for the growth of cyanobacteria. N loading may promote *Microcystis* blooms by not only the enhancement of growth [22–25] but also the synthesis of protease inhibitors, discouraging zooplankton grazing [26,27]. Additionally, the N_2-fixing cyanobacteria, such as *Anabaena*, *Aphanizomenon*, *Aphanothece*, *Cylindrospermopsis*, and *Gloeotrichia*, exhibit great flexibility in the N sources to form blooms and N fixation, making N sufficient to allow biomass to be continuously produced [9,17,28,29]. Temperature and nutrients are also essential factors for phytoplankton growth in high-latitude lakes [30,31]. A nine-year study of a mountain lake in Austria demonstrated that long-term phytoplankton changes were mainly attributed to the increasing temperature, while nutrients acted as modulating factors regulating the short-term phytoplankton changes [30]. A study of Lake Qinghai suggested that the increase in P load under climate change and overgrazing favored the growth of P-limited phytoplankton [31]. Nevertheless, some environmental factors such as salinity may have been neglected in the previous study. Considering that phytoplankton blooms are seldom reported in brackish lakes, the bloom-forming cyanobacterium is restricted in saltwater. However, freshwater blooms are found in many coastal areas [32,33]. Thus, salinity may not be a decisive factor in bloom formation. It is urgent to explore the potential factors influencing the phytoplankton community in both fresh and brackish lakes under climate change.

Lake Taihu and Lake Yanghe are eutrophic lakes located in subtropical and temperate areas, respectively. Lake Qinghai, Lake Keluke, and Lake Tuosu are alpine lakes located in the Tibetan Plateau. These five lakes have different climate conditions, presenting many differences in both environmental factors and the phytoplankton community. The comparisons between them could reveal the potential factors regulating the phytoplankton community. The exploration of the phytoplankton structure and the correlated factors is essential for understanding phytoplankton succession in different lakes, especially in the alpine lakes in the Tibetan Plateau. This study aimed to reveal the differences in phytoplankton communities of different lakes and explore the correlated factors regulating the growth of different phytoplankton phyla.

2. Materials and Methods

2.1. Study Area and Sampling Locations

Lake Taihu is located at the center of the Yangtze River Delta in Eastern China (30°56′–31°33′ N, 119°56′–120°54′ E) (Figure 1), with an annual average air temperature of 15–17 °C. It has a surface area of approximately 2338 km^2, a mean water depth of 1.9 m, and a maximum depth of 2.6 m [34]. Lake Taihu serves as a critical water resource for drinking, irrigation, aquaculture, and many industries, as well as recreation and tourism. Nonetheless, Lake Taihu has suffered from high nutrient loads and eutrophication in recent decades [35].

Lake Yanghe is located in Qinhuangdao City in Northern China (39°59′–40°12′ N, 119°10′–119°15′ E) (Figure 1), with an annual average air temperature of 11 °C. It has a drainage area of 755 km^2, a mean water depth of 5.7 m, and a maximum depth of over 40 m (Li et al. 2020). Lake Yanghe is a hydraulic project on the Yanghe River, supplies industrial and domestic water to Qinhuangdao City, and provides flood control for Funing Country and Beidaihe downstream. In recent years, nutrients from surface runoffs have greatly increased due to industrial and economic development in the area, leading to the continued outbreak of cyanobacterial blooms in the summer. This has negatively influenced its function as a drinking water source.

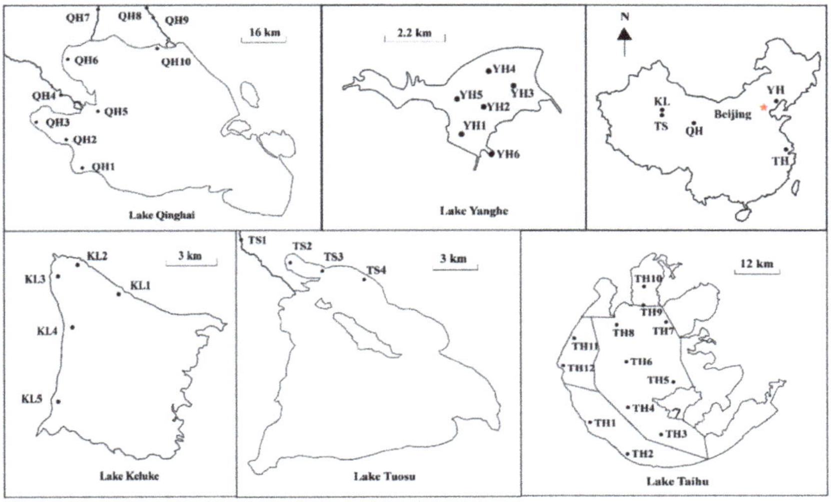

Figure 1. Sampling sites of Lake Qinghai, Lake Keluke, Lake Tuosu, Lake Taihu, and Lake Yanghe in China.

Lake Qinghai is located in Qinghai Province in Northwest China (36°32′–37°15′ N, 99°36′–100°47′ E) (Figure 1), with an area of about 4260 km^2, a maximum depth of 30 m, and a mean surface elevation of 3194 m [36,37]. This area is characterized by an alpine and continental climate, with an annual average temperature of 1.2 °C [37]. Lake Keluke (37°15′–37°20′ N, 96°51′–96°58′) and Lake Tuosu (37°04′–37°13′ N, 96°50′–97°03′E) are located in Qinghai Province in Northwest China (Figure 1), with an annual average air temperature of 11 °C [38]. Lake Keluke has an area of about 56.7 km^2, a mean water depth of 4 m, and a maximum depth of 13.3 m. Lake Tuosu has an area of about 145 km^2 and a maximum depth of 25 m [38].

2.2. Sample Collection and Processing

Water samples of five different lakes were taken in the spring, summer, and autumn. Water samples of Lake Taihu and Lake Yanghe were collected in the spring, summer, and autumn of 2014. Lake Taihu samples were collected from 12 locations (Figure 1) on 16 April, 22 July, and 24 September 2014. Lake Yanghe samples were collected from 6 locations (Figure 1) on 15 May, 21 August, and 14 October 2014. Water samples of Lake Qinghai, Lake Keluke, and Lake Tuosu were collected in the spring, summer, and autumn of 2018. Water samples of Lake Qinghai were collected on 28 and 29 May, on 20 and 21 August, and on 23 and 24 October from 10 sites (Figure 1). Water samples of Lake Keluke were collected on 1 June, on 18 August, and on 22 October from 5 sites (Figure 1). Water samples of Lake Tuosu were collected on 2 June, on 19 August, and on 22 October from 4 sites (Figure 1). Surface water samples (2 L) were taken at a 0.5-m depth if the water depth was less than 2 m and were taken at 0.5 m, 1 m, 2 m, 3 m, and 4 m if the water depth was less than 5 m [39]. All the samples were stored in polyethylene barrels at 4 °C in the dark before laboratory analysis. Water temperature, pH, and salinity were measured in situ at each site using a YSI 6600 multi-probe sonde (YSI, Yellow Springs, OH, USA). An appropriate amount of water samples was filtered through GF/C glass fiber filters (Whatman, Kent,

UK). Meanwhile, total nitrogen (TN) and total phosphorus (TP) were measured using the Chinese standard methods of HJ/T 636-2012 and GB/T 11983-1989, respectively.

Water samples (100 mL) taken from each sampling site were treated with Lugol's iodine solution to fix phytoplankton species. Phytoplankton was settled in 20 mm × 20 mm chambers and identified/enumerated by light microscopy following commonly used monographs on phytoplankton [40,41]. The Shannon–Wiener index (H) based on the number was calculated for phytoplankton [42]. The biodiversity of phytoplankton in the five lakes was calculated according to Bacillariophyta, Cyanobacteria, Chlorophyta, Cryptophyta, Pyrrophyta, Euglenophyta, and Chrysophyta.

2.3. Statistical Analysis

All data were statistically processed using SPSS 17.0 (SPSS, Inc., Chicago, IL, USA, 2008). The frequency of the water parameters and phytoplankton abundance approximated normal distributions. Pearson correlations were adopted to assess the relationships between environmental factors and phytoplankton. Additionally, the regression models were fitted to the water parameters and phytoplankton with significant relationships.

3. Results

3.1. Water Parameters and Nutrients in the Five Lakes

Table 1 exhibited the seasonal changes of the water parameters and nutrient levels. Lake Taihu had the highest average temperature of the three seasons, followed by Lake Yanghe, Lake Tuosu, and Lake Keluke, and lowest in Lake Qinghai ($p < 0.05$). The autumn had lower water temperatures than in the spring in the northern lakes, including Lake Qinghai, Lake Keluke, Lake Tuosu, and Lake Yanghe, with a higher temperature than that in the spring for Lake Taihu. Lake Qinghai and Lake Tuosu had the highest pH in the five lakes ($p < 0.05$), followed by Lake Keluke and Lake Taihu ($p < 0.05$). Lake Yanghe had the lowest pH in the five lakes ($p < 0.05$). There were significant differences in the salinity in the five lakes. Lake Tuosu had the highest salinity in the five lakes, followed by Lake Qinghai and Lake Keluke. Lake Taihu and Lake Yanghe had the lowest salinity in the five lakes. The inflow rivers had a lower temperature, pH, and salinity compared to Lake Qinghai. This is also true in Lake Tuosu.

There was a clear seasonal pattern in TN in Lake Yanghe and Lake Taihu. The mean TN in Lake Yanghe were 2.7 ± 0.02 mg/L, 0.1 ± 0.01 mg/L, and 0.5 ± 0.1 mg/L in the spring, summer, and autumn, respectively. The mean TN in Lake Taihu were 2.8 ± 0.7 mg/L, 0.6 ± 0.5 mg/L, and 0.5 ± 0.4 mg/L in the spring, summer, and autumn, respectively. The TP concentrations of the five lakes were all lower than 0.1 mg/L. Lake Taihu, Lake Qinghai, and Lake Keluke had higher TP compared to Lake Yanghe and Lake Tuosu. There were not many differences in TP concentrations of Lake Keluke and Lake Tuosu in different seasons. There were clear seasonal differences in Lake Qinghai, Lake Taihu, and Lake Yanghe. The TN was higher in the inflow rivers than the main lakes of both Lake Qinghai and Lake Tuosu, while it had an opposite pattern about TP in the inflow rivers and the main lakes in the two lakes.

3.2. Phytoplankton Composition and Biodiversity in the Five Lakes

The composition of phytoplankton differed among the five lakes (Figure 2). The dominant phylum in Lake Qinghai, Lake Keluke, and Lake Tuosu was Bacillariophyta in the spring, summer, and autumn. In Lake Qinghai, the phytoplankton abundance was 3.7×10^5 cells/L, 2.0×10^5 cells/L, and 1.5×10^5 cells/L. In the spring, summer, and autumn, Bacillariophyta accounted for 65%, 59%, and 39%, and Cyanobacteria accounted for 21%, 9%, and 37%, respectively. In Lake Keluke, it was 1.2×10^5 cells/L, 4.2×10^5 cells/L, and 2.0×10^7 cells/L. Bacillariophyta accounted for 56%, 40%, and 50% of phytoplankton, and Cyanobacteria accounted for 24%, 31%, and 28% in the spring, summer, and autumn, respectively. In Lake Tuosu, it was 2.9×10^4 cells/L, 6.9×10^4 cells/L, and 5.6×10^4 cells/L. Bacillariophyta accounted for 88%, 47%, and 82% of phytoplankton, and Cyanobacteria accounted for 9%, 22%, and 4% in the spring, summer, and autumn, respectively. The phytoplankton abundance of Lake Yanghe was 5.2 × 10^6 cells/L, 2.3 × 10^8 cells/L, and 3.1 × 10^8 cells/L in the spring, summer, and autumn, respectively. The dominant phylum of Lake Yanghe was Cyanobacteria in all three seasons, except for the spring. Chlorophyta accounted for 52% of phytoplankton in Lake Yanghe in the spring, and Cyanobacteria accounted for 89% and 98% of phytoplankton in Lake Yanghe in the summer and autumn, respectively. The dominant phylum of Lake Taihu was Cyanobacteria, which accounted for 35%, 44%, and 43% of phytoplankton in the spring (3.4 × 10^8 cells/L), summer (2.6 × 10^8 cells/L), and autumn (1.5 × 10^8 cells/L), respectively. The dominant species was *Microcystis* sp. in Lakes Taihu and Yanghe and was *Synedra* sp. In Lake Qinghai, Lake Keluke, and Lake Tuosu. The Bacillariophyta species (*Navicula* sp. And *Cyclotella* sp.); the Chlorophyta species (*Monoraphidium* sp., *Schroederia* sp., *Oocystis* sp., *Scenedesmus* sp., *Chlamydomonas* sp., and *Closterium* sp.); and the Cyanobacteria species (*Anabeana* sp. and *Oscillatoria* sp.) were discovered in all five lakes.

Table 1. Means and standard deviations of several water parameters of lakes and inflow rivers of five lakes in China.

Lakes and Rivers (n)	Seasons	Water Parameters				
		Temperature (°C)	pH	Salinity (ppt)	TN (mg/L)	TP (mg/L)
Qinghai (6)	Spring	17.9 ± 2.3	9.2 ± 0.09	10.5 ± 0.7	2.7 ± 1.2	0.06 ± 0.05
	Summer	17.5 ± 0.1	9.2 ± 0.01	9.5 ± 1.1	1.8 ± 0.8	0.01 ± 0.004
	Autumn	7.0 ± 3.5	9.4 ± 0.2	8.6 ± 2.9	3.5 ± 3.0	0.04 ± 0.04
	Mean	**14.6 ± 5.5**	**9.3 ± 0.1**	**9.6 ± 1.8**	**2.6 ± 1.8**	**0.04 ± 0.04**
Qinghai (IR) (4)	Spring	12.3 ± 3.4	8.5 ± 0.3	0.26 ± 0.1	7.0 ± 9.6	0.004 ± 0.003
	Summer	16.5 ± 0.8	8.6 ± 0.3	0.28 ± 0.07	5.8 ± 4.9	0.02 ± 0.02
	Autumn	8.1 ± 0.9	8.6 ± 0.1	0.24 ± 0.09	7.0 ± 3.6	0.02 ± 0.01
	Mean	**12.5 ± 4.0**	**8.6 ± 0.2**	**0.26 ± 0.09**	**6.6 ± 6.6**	**0.01 ± 0.01**
Keluke (5)	Spring	20.6 ± 4.0	8.5 ± 0.2	10.2 ± 20.4	5.1 ± 8.0	0.05 ± 0.05
	Summer	21.2 ± 0.8	8.4 ± 0.3	3.0 ± 3.8	3.9 ± 1.1	0.05 ± 0.03
	Autumn	6.0 ± 0.6	8.2 ± 0.4	7.7 ± 11.7	7.5 ± 7.3	0.06 ± 0.05
	Mean	**15.9 ± 7.6**	**8.4 ± 0.3**	**6.9 ± 13.1**	**5.5 ± 6.0**	**0.05 ± 0.04**
Tuosu (3)	Spring	20.3 ± 0.9	9.2 ± 0.03	22.2 ± 0.4	0.3 ± 0.4	0.02 ± 0.002
	Summer	23.1 ± 1.2	8.8 ± 0.3	8.5 ± 7.4	2.0 ± 0.4	0.02 ± 0.003
	Autumn	10.1 ± 0.3	9.1 ± 0.02	13.9 ± 6.7	1.6 ± 0.3	0.02 ± 0.01
	Mean	**17.8 ± 5.9**	**9.0 ± 0.2**	**14.9 ± 7.8**	**1.3 ± 0.8**	**0.02 ± 0.008**
Tuosu (IR)(1)	**Mean**	**16.1 ± 6.8**	**8.6 ± 0.2**	**0.77 ± 0.1**	**2.2 ± 0.05**	**0.01 ± 0.004**
Yanghe (6)	Spring	17.1 ± 0.7	5.7 ± 0.8	0.17 ± 0.08	2.7 ± 0.02	0.01 ± 0.001
	Summer	26.9 ± 0.7	8.4 ± 0.2	0.16 ± 0.005	0.1 ± 0.01	0.01 ± 0.002
	Autumn	15.5 ± 0.1	6.5 ± 0.6	0.18 ± 0.001	0.5 ± 0.1	0.03 ± 0.01
	Mean	**19.8 ± 5.2**	**6.9 ± 1.3**	**0.17 ± 0.04**	**1.1 ± 1.2**	**0.02 ± 0.01**
Taihu (12)	Spring	17.4 ± 0.2	6.8 ± 0.3	0.30 ± 0.04	2.8 ± 0.7	0.09 ± 0.11
	Summer	30.6 ± 0.4	7.4 ± 0.4	0.24 ± 0.03	0.6 ± 0.5	0.01 ± 0.003
	Autumn	23 ± 0.4	6.9 ± 0.3	0.22 ± 0.02	0.5 ± 0.4	0.04 ± 0.03
	Mean	**23.7 ± 5.5**	**7.0 ± 0.4**	**0.25 ± 0.04**	**1.3 ± 1.2**	**0.05 ± 0.07**

n indicates the number of sampling sites; IR indicates inflow rivers.

There were clear differences in phytoplankton biodiversity of the five lakes. In Lake Qinghai, the phytoplankton biodiversity was 0.19 ± 0.30, 0.38 ± 0.37, and 0.39 ± 0.46 in the spring, summer, and autumn, respectively. In Lake Keluke, the phytoplankton biodiversity was 0.87 ± 0.56, 0.75 ± 0.67, and 0.72 ± 0.45, respectively. In Lake Tuosu, the phytoplankton biodiversity was 0.41 ± 0.39, 1.25 ± 0.19, and 0.66 ± 0.30 in the spring, summer, and autumn, respectively. In Lake Yanghe, the phytoplankton biodiversity was 0.91 ± 0.18, 0.10 ± 0.07, and 0.34 ± 0.43 in the spring, summer, and autumn, respectively. In Lake Taihu, the phytoplankton biodiversity was 0.38 ± 0.33, 0.51 ± 0.37, and 0.41 ± 0.39, respectively.

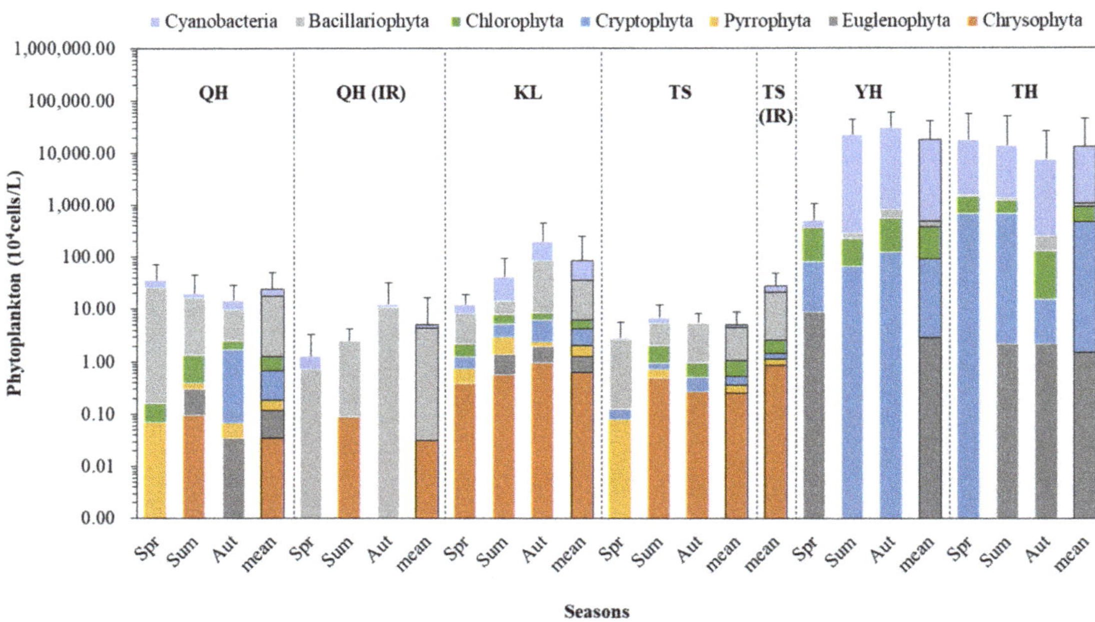

Figure 2. Phytoplankton composition based on the abundance in different seasons in Lake Qinghai (QH), Lake Keluke (KL), Lake Tuosu (TS), Lake Yanghe (YH), Lake Taihu (TH), and the inflow rivers (IR) of Lake Qinghai and Lake Tuosu. The error bars indicate the standard deviations of phytoplankton abundance in each lake.

3.3. Correlations between Water Parameters and Phytoplankton in the Five Lakes

The phytoplankton had significant relationships with temperature based on the annual average data from Lake Qinghai, Lake Keluke, Lake Tuosu, Lake Yanghe, and Lake Taihu (Figure 3). Bacillariophyta, Cyanobacteria, Chlorophyta, and phytoplankton abundance had significantly positive relationships with temperature under R^2 = 0.59, 0.70, 0.71, and 0.74 ($p < 0.05$). Additionally, the ratio of Bacillariophyta/phytoplankton had a significantly negative relationship with temperature ($p < 0.01$), and the ratio of Cyanobacteria/phytoplankton had a significantly positive relationship with temperature ($p < 0.01$).

The phytoplankton had a significant relationship with pH based on the annual average data from the five lakes in China (Figure 4). Bacillariophyta, Cyanobacteria, Chlorophyta, and phytoplankton abundance had significantly negative relationships with pH under R^2 = 0.58, 0.82, 0.80, and 0.81 ($p < 0.01$). Additionally, the ratio of Bacillariophyta/phytoplankton had a significantly positive relationship with the pH ($p < 0.01$). The ratio of Cyanobacteria/phytoplankton had a significantly negative relationship with the pH ($p < 0.01$). Additionally, phytoplankton had a significant relationship with salinity based on the annual average data from the five lakes in China (Figure 5). Bacillariophyta, Cyanobacteria, Chlorophyta, and phytoplankton abundance had significant negative relationships with salinity ($p < 0.05$) (Figure 5).

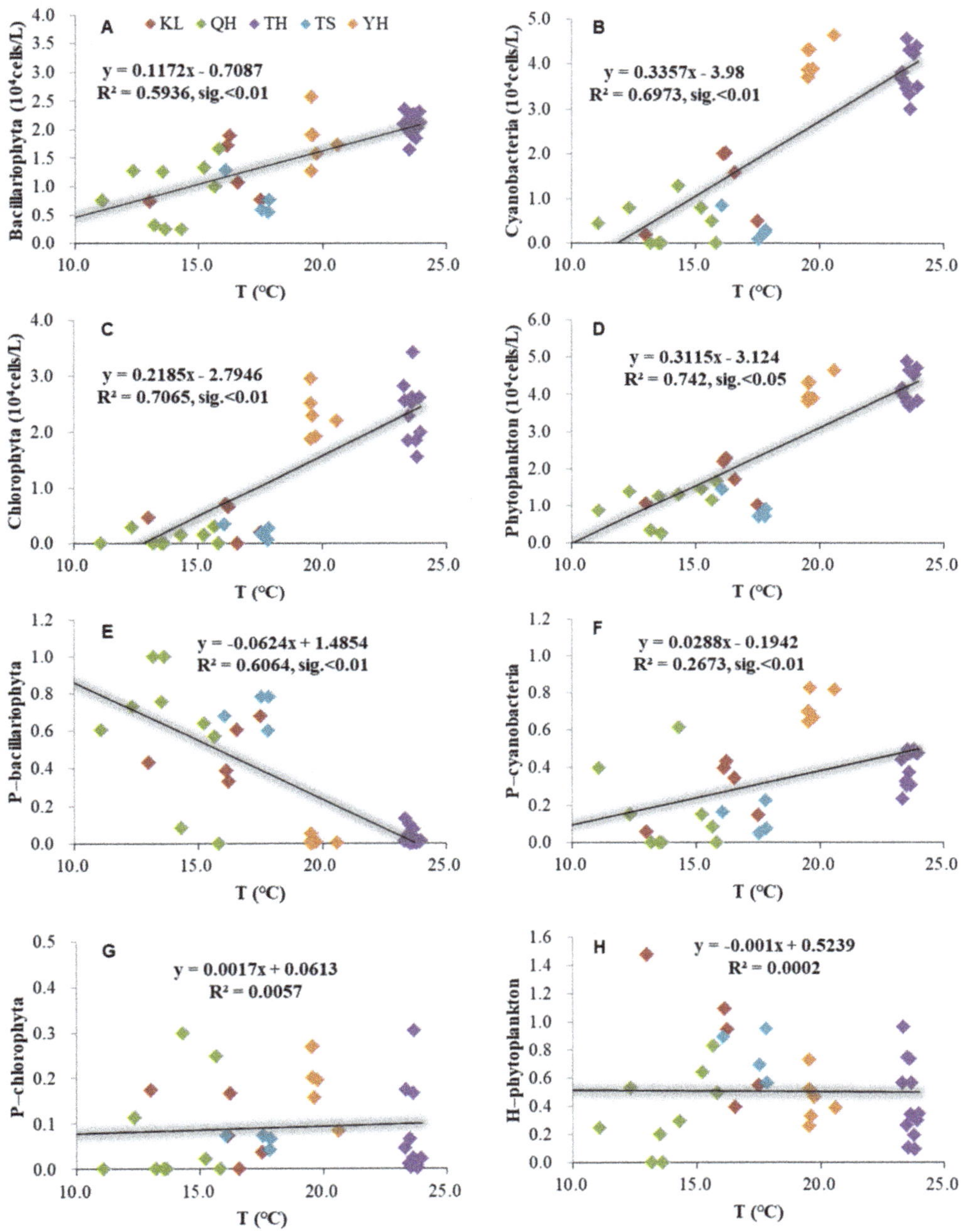

Figure 3. Relationships between temperature (T) and phytoplankton based on the annual average data of each site of Lake Qinghai, Lake Keluke, Lake Tuosu, Lake Yanghe, and Lake Taihu. (**A–D**) showed the relationships between T and phytoplankton, and (**E–H**) showed the relationships between T and phytoplankton composition. Bacillariophyta, Chlorophyta, Cyanobacteria, and phytoplankton abundance were transformed by Log (X + 1). P-i indicates the i/phytoplankton ratio based on abundance. H-phytoplankton represents the biodiversity of phytoplankton.

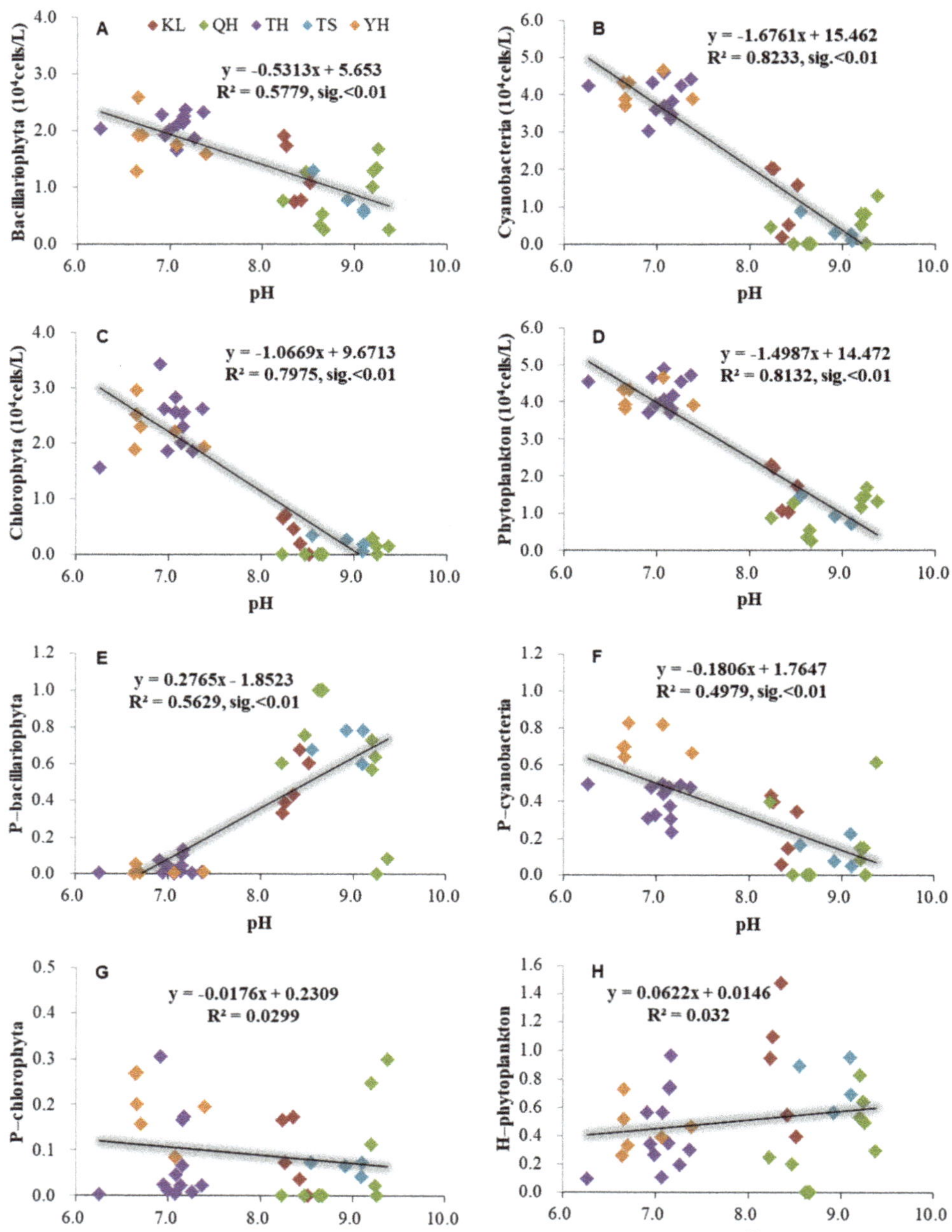

Figure 4. Relationships between pH and phytoplankton based on the annual average data of each site of Lake Qinghai, Lake Keluke, Lake Tuosu, Lake Yanghe, and Lake Taihu. (**A–D**) showed the relationships between pH and phytoplankton, and (**E–H**) showed the relationships between pH and phytoplankton composition. Bacillariophyta, Chlorophyta, Cyanobacteria, and phytoplankton abundance were transformed by Log (X + 1). P-i indicates the i/phytoplankton ratio based on abundance. H-phytoplankton represents the biodiversity of phytoplankton.

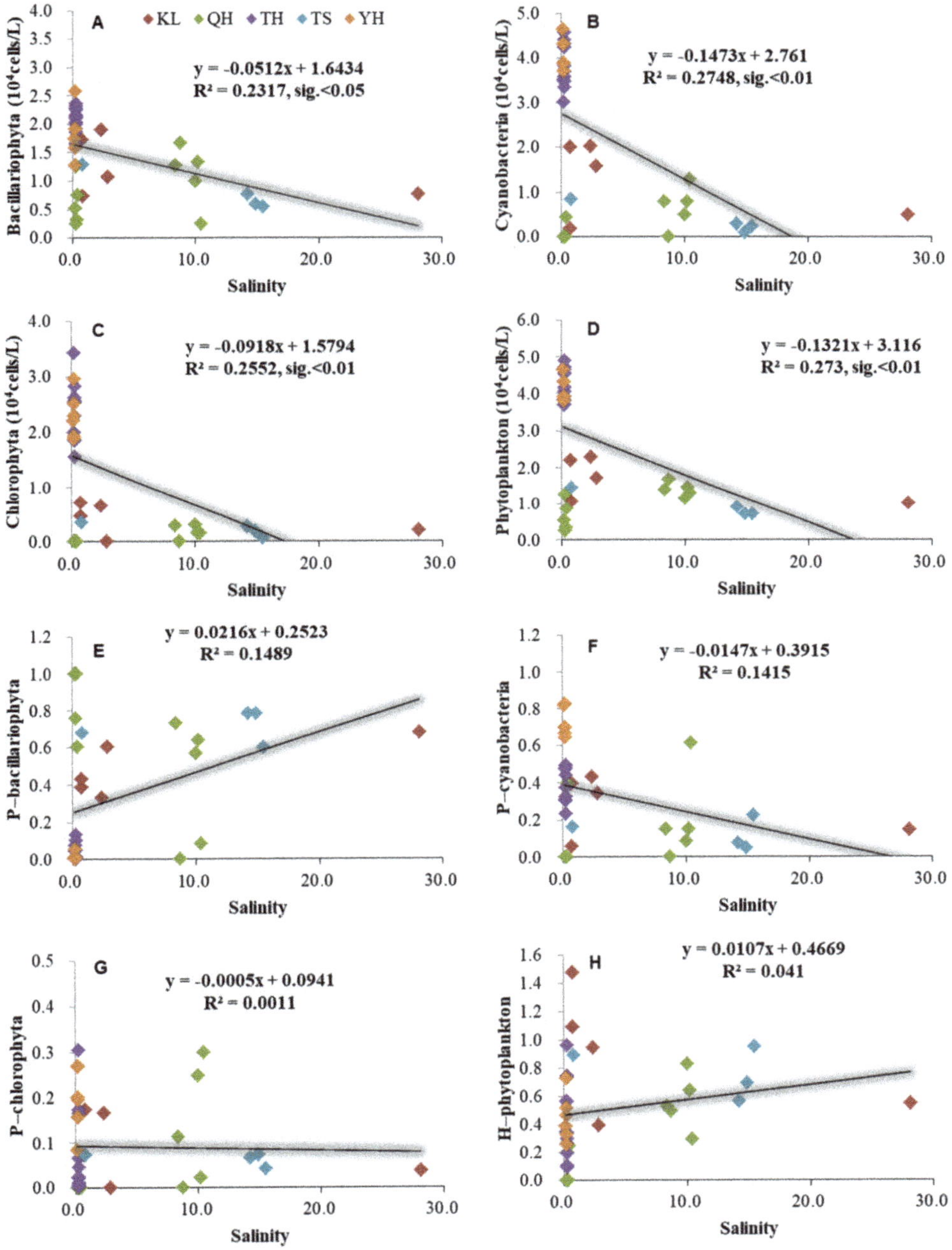

Figure 5. Relationships between salinity and phytoplankton based on the annual average data of each site of Lake Qinghai, Lake Keluke, Lake Tuosu, Lake Yanghe, and Lake Taihu. (**A–D**) showed the relationships between salinity and phytoplankton, and (**E–H**) showed the relationships between salinity and phytoplankton composition. Bacillariophyta, Chlorophyta, Cyanobacteria, and phytoplankton abundance were transformed by Log (X + 1). P-i indicates the i/phytoplankton ratio based on abundance. H-phytoplankton represents the biodiversity of phytoplankton.

4. Discussion

The temperature was one of the most essential factors regulating the growth of phytoplankton [1,7]. In the present study, phytoplankton abundance had a significantly positive relationship with water temperature. Additionally, the lakes with higher temperatures generally possessed higher phytoplankton abundance, such as Lake Taihu and Lake Yanghe. Thus, the temperature was a crucial factor promoting the growth of phytoplankton in the five lakes. Previous studies demonstrated that higher temperatures favored the growth of Cyanobacteria [10–12]. This is consistent with the result in the present study that the lakes with higher temperatures generally had higher Cyanobacteria abundance and a Cyanobacteria/phytoplankton ratio. Lake Taihu and Lake Yanghe had higher average water temperatures compared to Lake Qinghai, Lake Keluke, and Lake Tuosu. Hence, Lake Taihu and Lake Yanghe had higher average Cyanobacteria abundance and a higher Cyanobacteria/phytoplankton ratio than Lake Qinghai, Lake Keluke, and Lake Tuosu. Additionally, the Cyanobacteria/phytoplankton ratio exhibited a significantly positive relationship with temperature.

The nutrients were another vital factor regulating the growth of phytoplankton. In the present study, TN had a high concentration in the five lakes, and there were not many differences in TP concentration. A previous study revealed that the growth of the dominant phytoplankton was not nutrient-limited under P enrichment ≥ 0.20 mg·L^{-1} (P) and N enrichment ≥ 0.80 mg·L^{-1} (N) [8]. Therefore, the growth of the dominant phytoplankton in all the lakes was not N-limited. In other words, TN does not shape the differences in phytoplankton abundance in the five lakes. Phosphorus is one of the limiting factors for phytoplankton growth in all five lakes, since the TP was much lower than 0.20 mg·L^{-1} in those lakes. Thus, TP does not shape the differences in phytoplankton abundance in the five lakes. However, the high nutrient level in Lake Qinghai, Lake Keluke, and Lake Tuosu, located in the Tibetan Plateau, should be stressed under climate change. In recent years, the lakes in the Tibetan Plateau are undergoing an increase in temperature and precipitation. The rising temperature may result in a blooming of phytoplankton, especially Cyanobacteria, in those lakes with enough nutrients. The increase in precipitation may induce more nutrient input, especially TP, in the lakes of the Tibetan Plateau. This has been confirmed in Lake Qinghai that the TP increased but TN decreased in the inflowing rivers, contributing to the alleviation of the P deficiency in Lake Qinghai and the promotion of the growth of phytoplankton [31].

In the present study, the pH had significantly negative relationships with Bacillariophyta, Cyanobacteria, Chlorophyta, and phytoplankton. This may be caused by a relatively high phytoplankton abundance and low pH in Lake Taihu and Lake Yanghe, as well as relatively low phytoplankton abundance and high pH in other lakes. A previous study suggested that the natural phytoplankton biomass decreased in high pH (9.5) incubation, while the phytoplankton biomass increased in pH 8–9 incubation [43]. In Lake Taihu and Lake Yanghe, the mean pH was low (pH = 7.0 and 6.9, respectively), and the phytoplankton could grow well if the pH increased from 7.0 to 9.0, since the blooming of the phytoplankton was associated with an increase in pH [44]. This would be explained by the blooming of the phytoplankton, which depletes the dissolved CO_2 concentration and, therefore, results in an increase in pH [45–47]. This is consistent with previous studies that the pH increased with increasing the cell densities of Cyanobacteria, and the Cyanobacteria blooms were associated with high pH [44,48,49]. In Lake Yanghe, the Cyanobacteria accounted for 98% of the total phytoplankton when the Cyanobacteria abundance reached 2.2×10^8 cells/L, and the pH reached 8.7 in the summer. However, the high pH may not favor the growth of phytoplankton. Additionally, *Microcystis aeruginosa* and *Scenedesmus quadricauda* were at the stationary phase under high pH values (10.0), and both algae resumed growing when the pH was decreased using HCl [50]. Additionally, Touloupakis et al. [51] confirmed that the light conversion efficiency of Cyanobacteria decreased linearly with the increase in pH at a range of 7.5–11. Therefore, a high pH (high OH^-) could cause an adverse effect on the phytoplankton growth [52], and the phytoplankton cannot grow well in Lake Qinghai and Lake Tuosu at high pH values (both mean pH $\geq$ 9.0).

High salinity was an adverse factor for phytoplankton growth, since it can cause oxidative stress to algal cells, resulting in cell death [53,54]. This is consistent with the results in the present study. The Bacillariophyta, Cyanobacteria, Chlorophyta, and phytoplankton all had significantly negative relationships with salinity. However, the low phytoplankton abundance in some brackish lakes should not be simply attributed to high salinity. Recently, more and more studies have revealed that freshwater phytoplankton blooms occur in brackish waters [32,33]. Additionally, indoor cultivation has implied that the freshwater strain *Microcystis aeruginosa*, a typical blooming species, acclimated to a salinity gradient that could reach 7.5 [52]. In the present study, Lake Keluke had a salinity lower than the maximum tolerance salinity based on indoor cultivation but a low phytoplankton abundance. Hence, the low phytoplankton abundance of Lake Keluke should not be simply attributed to high salinity. In Lake Tuosu, nevertheless, the salinity reached 11.3 $\pm$ 9.2 and could have an adverse effect on many phytoplankton species.

The factors elaborated above are imperative to phytoplankton, especially for phytoplankton of the alpine lakes in the Tibetan Plateau. The lakes in the Tibetan Plateau experience a warm and wet climate [55,56], leading to changes in the water temperature, nutrients, and salinity and eventually facilitating the growth of phytoplankton [57]. Particularly, phytoplankton blooms may occur in those brackish lakes with enough nutrients under climate change. The brackish lakes in the Tibetan Plateau should receive priority for management, since some rare fish and birds live in those lakes. Their simple and fragile food chains could be altered under climate change. Buffer zones may be an effective approach for those alpine lakes to alleviate diffusive pollutions from agriculture and livestock.

5. Conclusions

The phytoplankton community was dominated by Bacillariophyta, followed by Cyanobacteria and Chlorophyta in the alpine lakes, including Lake Qinghai, Lake Keluke, and Lake Tuosu. It was dominated by Cyanobacteria, followed by Chlorophyta and Bacillariophyta in Lake Yanghe. Additionally, it was dominated by Cyanobacteria, followed by Chlorophyta and Cryptophyta in Lake Taihu. The temperature was a crucial factor influencing the growth of Cyanobacteria, Chlorophyta, and Bacillariophyta, especially Cyanobacteria and Chlorophyta. The pH had significantly negative relationships with Cyanobacteria, Chlorophyta, and Bacillariophyta. Moreover, it could be a strongly negative factor for phytoplankton growth in alpine lakes. Salinity had significantly negative relationships with Cyanobacteria, Chlorophyta, and Bacillariophyta. It was also an adverse factor for phytoplankton when it was very high.

Author Contributions: J.J.: conceptualization, methodology, investigation, writing—original draft preparation, writing—review and editing, and funding acquisition; Q.C.: conceptualization and writing—review and editing resources; H.R. investigation, writing—original draft preparation, and writing—review and editing; R.L. writing—original draft preparation, writing—review and editing, and funding acquisition, H.H.: investigation and writing—original draft preparation; and P.G.: investigation and writing—original draft preparation. All authors have read and agreed to the published version of the manuscript.

Funding: This research was funded by the Project of Qinghai Science & Technology Department (2018-ZJ-958Q) and the National Natural Science Foundation of China (No. 42167012), as well as the Suzhou Minsheng Science and Technology Development Program Project (No. SS201838).

Data Availability Statement: The data presented in this study are available on request from the corresponding author. The data are not publicly available due to privacy.

Conflicts of Interest: The authors declare no conflict of interest.

References

1. Cao, J.; Hou, Z.; Li, Z.; Chu, Z.; Yang, P.; Zheng, B. Succession of phytoplankton functional groups and their driving factors in a subtropical plateau lake. *Sci. Total Environ.* **2018**, *631–632*, 1127–1137. [CrossRef] [PubMed]
2. Jia, J.; Shi, W.; Chen, Q.; Lauridsen, T.L. Spatial and temporal variations reveal the response of zooplankton to cyanobacteria. *Harmful Algae* **2017**, *64*, 63–73. [CrossRef]
3. Jia, J.; Luo, W.; Lu, Y.; Giesy, J.P. Bioaccumulation of microcystins (MCs) in four fish species from Lake Taihu, China: Assessment of risks to humans. *Sci. Total Environ.* **2014**, *487*, 224–232. [CrossRef] [PubMed]
4. Su, X.; Steinman, A.D.; Oudsema, M.; Hassett, M.; Xie, L. The influence of nutrients limitation on phytoplankton growth and microcystins production in Spring Lake, USA. *Chemosphere* **2019**, *234*, 34–42. [CrossRef] [PubMed]
5. Yang, Y.; Pan, J.; Han, B.-P.; Naselli-Flores, L. The effects of absolute and relative nutrient concentrations (N/P) on phytoplankton in a subtropical reservoir. *Ecol. Indic.* **2020**, *115*, 106466. [CrossRef]
6. Derot, J.; Jamoneau, A.; Teichert, N.; Rosebery, J.; Morin, S.; Laplace-Treyture, C. Response of phytoplankton traits to environmental variables in French lakes: New perspectives for bioindication. *Ecol. Indic.* **2020**, *108*, 105659. [CrossRef]
7. Li, J.; Ma, R.; Xue, K.; Loiselle, S. Drivers to spatial and temporal dynamics of column integrated phytoplankton biomass in the shallow lake of Chaohu, China. *Ecol. Indic.* **2020**, *109*, 105812. [CrossRef]
8. Xu, H.; Paerl, H.W.; Qin, B.; Zhu, G.; Gao, G. Nitrogen and phosphorus inputs control phytoplankton growth in eutrophic Lake Taihu, China. *Limnol. Oceanogr.* **2010**, *55*, 420–432. [CrossRef]
9. Vrede, T.; Ballantyne, A.; Mille-Lindblom, C.; Algesten, G.; Gudasz, C.; Lindahl, S.; Brunberg, A.K. Effects of N:P loading ratios on phytoplankton community composition, primary production and N fixation in a eutrophic lake. *Freshwater Biol.* **2009**, *54*, 331–344. [CrossRef]
10. Johnk, K.D.; Huisman, J.; Sharples, J.; Sommeijer, B.; Visser, P.M.; Stroom, J.M. Summer heatwaves promote blooms of harmful cyanobacteria. *Glob. Chang. Biol.* **2008**, *14*, 495–512. [CrossRef]
11. Zhang, M.; Qin, B.; Yu, Y.; Yang, Z.; Shi, X.; Kong, F. Effects of temperature fluctuation on the development of cyanobacterial dominance in spring: Implication of future climate change. *Hydrobiologia* **2016**, *763*, 135–146. [CrossRef]
12. Kosten, S.; Huszar, V.L.M.; Bécares, E.; Costa, L.S.; Donk, E.; Hansson, L.-A.; Jeppesen, E.; Kruk, C.; Lacerot, G.; Mazzeo, N.; et al. Warmer climates boost cyanobacterial dominance in shallow lakes. *Glob. Chang. Biol.* **2012**, *18*, 118–126. [CrossRef]
13. Schindler, D.W. Eutrophication and Recovery in Experimental Lakes-Implications for Lake Management. *Science* **1974**, *184*, 897–899. [CrossRef] [PubMed]
14. Hecky, R.E.; Kilham, P. Nutrient limitation of phytoplankton in fresh-water and marine environments—A review of recent-evidence on the effects of enrichment. *Limnol. Oceanogr.* **1988**, *33*, 796–822.
15. Nixon, S.W. Coastal marine eutrophication—A definition, social causes, and future concerns. *Ophelia* **1995**, *41*, 199–219. [CrossRef]
16. Lv, J.; Wu, H.; Chen, M. Effects of nitrogen and phosphorus on phytoplankton composition and biomass in 15 subtropical, urban shallow lakes in Wuhan, China. *Limnologica* **2011**, *41*, 48–56. [CrossRef]
17. Dokulil, M.T.; Teubner, K. Cyanobacterial dominance in lakes. *Hydrobiologia* **2000**, *438*, 1–12. [CrossRef]
18. Conley, D.J.; Paerl, H.W.; Howarth, R.W.; Boesch, D.F.; Seitzinger, S.P.; Havens, K.E.; Lancelot, C.; Likens, G.E. ECOLOGY Controlling Eutrophication: Nitrogen and Phosphorus. *Science* **2009**, *323*, 1014–1015. [CrossRef]
19. Paerl, H.W.; Fulton, R.S. *Ecology of Harmful Marine Algae*; Springer: Berlin, Germany, 2006; pp. 95–107.
20. Paerl, H.W.; Huisman, J. Climate-Blooms like it hot. *Science* **2008**, *320*, 57–58. [CrossRef]
21. Fernandez, C.; Estrada, V.; Parodi, E.R. Factors Triggering Cyanobacteria Dominance and Succession During Blooms in a Hypereutrophic Drinking Water Supply Reservoir. *Water Air Soil Pollut.* **2015**, *226*, 1–13. [CrossRef]
22. Gobler, C.J.; Davis, T.W.; Coyne, K.J.; Boyer, G.L. Interactive influences of nutrient loading, zooplankton grazing, and microcystin synthetase gene expression on cyanobacterial bloom dynamics in a eutrophic New York lake. *Harmful Algae* **2007**, *6*, 119–133. [CrossRef]
23. Sunda, W.G.; Graneli, E.; Gobler, C.J. Positive feedback and the development and persistence of ecosystem disruptive algal blooms. *J. Phycol.* **2006**, *42*, 963–974. [CrossRef]
24. Harke, M.J.; Davis, T.W.; Watson, S.B.; Gobler, C.J. Nutrient-Controlled Niche Differentiation of Western Lake Erie Cyanobacterial Populations Revealed via Metatranscriptomic Surveys. *Environ. Sci. Technol.* **2016**, *50*, 604–615. [CrossRef]
25. Davis, T.W.; Harke, M.J.; Marcoval, M.A.; Goleski, J.; Orano-Dawson, C.; Berry, D.L.; Gobler, C.J. Effects of nitrogenous compounds and phosphorus on the growth of toxic and non-toxic strains of Microcystis during cyanobacterial blooms. *Aquat. Microb. Ecol.* **2010**, *61*, 149–162. [CrossRef]
26. Agrawal, M.K.; Bagchi, D.; Bagchi, S.N. Acute inhibition of protease and suppression of growth in zooplankter, Moina macrocopa, by Microcystis blooms collected in Central India. *Hydrobiologia* **2001**, *464*, 37–44. [CrossRef]
27. Agrawal, M.K.; Zitt, A.; Bagchi, D.; Weckesser, J.; Bagchi, S.N.; von Elert, E. Characterization of proteases in guts of Daphnia magna and their inhibition by Microcystis aeruginosa PCC 7806. *Environ. Toxicol.* **2005**, *20*, 314–322. [CrossRef] [PubMed]
28. O'Neil, J.M.; Davis, T.W.; Burford, M.A.; Gobler, C.J. The rise of harmful cyanobacteria blooms: The potential roles of eutrophication and climate change. *Harmful Algae* **2012**, *14*, 313–334. [CrossRef]
29. Schindler, D.W.; Hecky, R.E.; Findlay, D.L.; Stainton, M.P.; Parker, B.R.; Paterson, M.J.; Beaty, K.G.; Lyng, M.; Kasian, S.E.M. Eutrophication of lakes cannot be controlled by reducing nitrogen input: Results of a 37-year whole-ecosystem experiment. *Proc. Natl. Acad. Sci. USA* **2008**, *105*, 11254–11258. [CrossRef]

30. Tolotti, M.; Thies, H.; Nickus, U.; Psenner, R. Temperature modulated effects of nutrients on phytoplankton changes in a mountain lake. *Hydrobiologia* **2012**, *698*, 61–75. [CrossRef]
31. Ren, Z.; Niu, D.; Ma, P.; Wang, Y.; Fu, H.; Elser, J.J. Cascading influences of grassland degradation on nutrient limitation in a high mountain lake and its inflow streams. *Ecology* **2019**, *100*, e02755. [CrossRef]
32. Preece, E.P.; Hardy, F.J.; Moore, B.C.; Bryan, M. A review of microcystin detections in Estuarine and Marine waters: Environmental implications and human health risk. *Harmful Algae* **2017**, *61*, 31–45. [CrossRef]
33. Von Alvensleben, N.; Magnusson, M.; Heimann, K. Salinity tolerance of four freshwater microalgal species and the effects of salinity and nutrient limitation on biochemical profiles. *J. Appl. Phycol.* **2015**, *28*, 861–876. [CrossRef]
34. Zhang, D.W.; Xie, P.; Liu, Y.Q.; Qiu, T. Transfer, distribution and bioaccumulation of microcystins in the aquatic food web in Lake Taihu, China, with potential risks to human health. *Sci. Total Environ.* **2009**, *407*, 2191–2199. [CrossRef]
35. Jin, X.; Hu, X. A comprehensive plan for treating the major polluted regions of Lake Taihu, China. *Lakes Reserv. Res. Manag.* **2003**, *8*, 217–230. [CrossRef]
36. Zhang, X.; Lin, C.; Zhou, X.; Lei, K.; Guo, B.; Cao, Y.; Lu, S.; Liu, X.; He, M. Concentrations, fluxes, and potential sources of nitrogen and phosphorus species in atmospheric wet deposition of the Lake Qinghai Watershed, China. *Sci. Total Environ.* **2019**, *682*, 523–531. [CrossRef] [PubMed]
37. Cao, Y.; Lin, C.; Zhang, X.; Liu, X.; He, M.; Ouyang, W. Distribution, source, and ecological risks of polycyclic aromatic hydrocarbons in Lake Qinghai, China. *Environ. Pollut.* **2020**, *266 Pt 1*, 115401. [CrossRef]
38. Ding, Z.; Zhang, J.; Yang, P.; Zhou, S.; Zhang, Y. Comparison of sediment proxies of cores and their environmental significance at different locations of Lake Toson in Qaidam Basin. *J. Lake Sci.* **2020**, *32*, 259–270.
39. National Environmental Protection Bureau. *Standard Methods for the Examination of Water and Wastewater (Version 4)*; China Environmental Science Publish Press: Beijing, China, 2002. (In Chinese)
40. Watanabe, M.M.; Kaya, K.; Takamura, N. Fate of the toxic cyclic heptapeptides, the microcystins, from blooms of Microcystis (cyanobacteria) in a hypertrophic lake. *J. Phycol.* **1992**, *28*, 761–767. [CrossRef]
41. Song, L.R.; Chen, W.; Peng, L.; Wan, N.; Gan, N.Q.; Zhang, X.M. Distribution and bioaccumulation of microcystins in water columns: A systematic investigation into the environmental fate and the risks associated with microcystins in Meiliang Bay, Lake Taihu. *Water Res.* **2007**, *41*, 2853–2864. [CrossRef]
42. Jeppesen, E.; Jensen, J.P.; Sondergaard, M.; Lauridsen, T.; Landkildehus, F. Trophic structure, species richness and biodiversity in Danish lakes: Changes along a phosphorus gradient. *Freshwater Biol.* **2000**, *45*, 201–218. [CrossRef]
43. Ji, X.; Verspagen, J.M.H.; Stomp, M.; Huisman, J. Competition between cyanobacteria and green algae at low versus elevated CO2: Who will win, and why? *J. Exp. Bot.* **2017**, *68*, 3815–3828. [CrossRef] [PubMed]
44. Kranz, S.A.; Gladrow, D.W.; Nehrke, G.; Langer, G.; Rosta, B. Calcium carbonate precipitation induced by the growth of the marine cyanobacteriaTrichodesmium. *Limnol. Oceanogr.* **2010**, *55*, 2563–2569. [CrossRef]
45. Talling, J.F. The Depletion of Carbon Dioxide from Lake Water by Phytoplankton. *J. Ecol.* **1976**, *64*, 79–121. [CrossRef]
46. Balmer, M.B.; Downing, J.A. Carbon dioxide concentrations in eutrophic lakes: Undersaturation implies atmospheric uptake. *Inland Waters* **2011**, *1*, 125–132. [CrossRef]
47. Verschoor, A.M.; Van Dijk, M.A.; Huisman, J.; Van Donk, E. Elevated CO2 concentrations affect the elemental stoichiometry and species composition of an experimental phytoplankton community. *Freshwater Biol.* **2013**, *58*, 597–611. [CrossRef]
48. Liu, X.; Lu, X.H.; Chen, Y.W. The effects of temperature and nutrient ratios on Microcystis blooms in Lake Taihu, China: An 11-year investigation. *Harmful Algae* **2011**, *10*, 337–343. [CrossRef]
49. Hörnström, E. Phytoplankton in 63 limed lakes in comparison with the distribution in 500 untreated lakes with varying pH. *Hydrobiologia* **2002**, *470*, 115–126. [CrossRef]
50. Ma, Z.; Chu, Z.; Hu, X.; Jin, X.; Zhang, G. The Effect of pH on Competition of M. aeruginosa and S. quadricauda at Different Phosphorus Mass Concentration. *Res. Environ. Sci.* **2005**, *18*, 30–33.
51. Touloupakis, E.; Cicchi, B.; Silva Benavides, A.M.; Torzillo, G. Effect of high pH on growth of Synechocystis sp PCC 6803 cultures and their contamination by golden algae (*Poterioochromonas* sp.). *Appl. Microbiol. Biotechnol.* **2016**, *100*, 1333–1341. [CrossRef]
52. Georges des Aulnois, M.; Roux, P.; Caruana, A.; Réveillon, D.; Briand, E.; Hervé, F.; Savar, V.; Bormans, M.; Amzil, Z. Physiological and Metabolic Responses of Freshwater and Brackish-Water Strains of Microcystis aeruginosa Acclimated to a Salinity Gradient: Insight into Salt Tolerance. *Appl. Environ. Microbiol.* **2019**, *85*, e01614-19. [CrossRef]
53. Ross, C.; Warhurst, B.C.; Brown, A.; Huff, C.; Ochrietor, J.D. Mesohaline conditions represent the threshold for oxidative stress, cell death and toxin release in the cyanobacterium Microcystis aeruginosa. *Aquat. Toxicol.* **2019**, *206*, 203–211. [CrossRef] [PubMed]
54. Affenzeller, M.J.; Darehshouri, A.; Andosch, A.; Lutz, C.; Lutz-Meindl, U. Salt stress-induced cell death in the unicellular green alga Micrasterias denticulata. *J. Exp. Bot.* **2009**, *60*, 939–954. [CrossRef] [PubMed]
55. Yao, T.; Yu, W. Recent Glacial Retreat and Its Impact on Hydrological Processes on the Tibetan Plateau, China, and Surrounding Regions. *Arct. Antarct. Alp. Res.* **2007**, *39*, 642–650. [CrossRef]
56. Lin, Q.; Xu, L.; Hou, J.; Liu, Z.; Jeppesen, E.; Han, B.P. Responses of trophic structure and zooplankton community to salinity and temperature in Tibetan lakes: Implication for the effect of climate warming. *Water Res.* **2017**, *124*, 618–629. [CrossRef]
57. Jeppesen, E.; Meerhoff, M.; Holmgren, K.; González-Bergonzoni, I.; Teixeira-de Mello, F.; Declerck, S.A.J.; De Meester, L.; Søndergaard, M.; Lauridsen, T.L.; Bjerring, R.; et al. Impacts of climate warming on lake fish community structure and potential effects on ecosystem function. *Hydrobiologia* **2010**, *646*, 73–90. [CrossRef]

International Journal of
Environmental Research and Public Health

MDPI

Article

Comparative Study of Algal Responses and Adaptation Capability to Ultraviolet Radiation with Different Nutrient Regimes

Lingxiao Ren [1,*], Jing Huang [2], Keqiang Ding [1], Yi Wang [1], Yangyang Yang [3], Lijuan Zhang [1] and Haoyu Wu [1]

1 School of Environmental Engineering, Nanjing Institute of Technology, Nanjing 211167, China; dingkq@njit.edu.cn (K.D.); yiwang@njit.edu.cn (Y.W.); lijuan_zhang@njit.edu.cn (L.Z.); why1871739065@163.com (H.W.)
2 Three Gorges Beijing Enterprises Nanjing Water Group Co., Ltd., Nanjing 210000, China; huangjing@bewg.net.cn
3 School of Environmental Engineering, Xuzhou University of Technology, Xuzhou 221018, China; yangyy7075@126.com
* Correspondence: rlxjht@gmail.com; Tel.: +86-158-5066-2170

Abstract: Frequent outbreaks of harmful algal blooms (HABs) represent one of the most serious outcomes of eutrophication, and light radiation plays a critical role in the succession of species. Therefore, a better understanding of the impact of light radiation is essential for mitigating HABs. In this study, *Chlorella pyrenoidosa* and non-toxic and toxic *Microcystis aeruginosa* were mono-cultured and co-cultured to explore algal responses under different nutrient regimes. Comparisons were made according to photosynthetically active radiation (PAR), UV-B radiation exerted oxidative stresses, and negative effects on the photosynthesis and growth of three species under normal growth conditions, and algal adaptive responses included extracellular polymeric substance (EPS) production, the regulation of superoxide dismutase (SOD) activity, photosynthetic pigments synthesis, etc. Three species had strain-specific responses to UV-B radiation and toxic *M. aeruginosa* was more tolerant and showed a higher adaptation capability to UV-B in the mono-cultures, including the lower sensitivity and better self-repair efficiency. In addition to stable μ_{max} in PAR ad UV-B treatments, higher EPS production and enhanced production of photosynthetic pigments under UV-B radiation, toxic *M. aeruginosa* showed a better recovery of its photosynthetic efficiency. Nutrient enrichment alleviated the negative effects of UV-B radiation on three species, and the growth of toxic *M. aeruginosa* was comparable between PAR and UV-B treatment. In the co-cultures with nutrient enrichment, *M. aeruginosa* gradually outcompeted *C. pyrenoidosa* in the PAR treatment and UV-B treatment enhanced the growth advantages of *M. aeruginosa*, when toxic *M. aeruginosa* showed a greater competitiveness. Overall, our study indicated the adaptation of typical algal species to ambient UV-B radiation and the stronger competitive ability of toxic *M. aeruginosa* in the UV-radiated waters with severer eutrophication.

Keywords: *Microcystis aeruginosa*; *Chlorella pyrenoidosa*; ultraviolet B radiation; photosynthetic efficiency; adaptation capability; nutrient enrichment

Citation: Ren, L.; Huang, J.; Ding, K.; Wang, Y.; Yang, Y.; Zhang, L.; Wu, H. Comparative Study of Algal Responses and Adaptation Capability to Ultraviolet Radiation with Different Nutrient Regimes. *Int. J. Environ. Res. Public Health* **2022**, *19*, 5485. https://doi.org/10.3390/ijerph19095485

Academic Editor: Nansheng Chen

Received: 24 February 2022
Accepted: 28 April 2022
Published: 30 April 2022

1. Introduction

With the rapid economic development and pollutant discharge, eutrophication has seriously affected aquatic ecosystems over the last several decades [1,2]. Frequent outbreaks of harmful algal blooms (HABs) represent one of the most serious outcomes of eutrophication [3,4] and many studies have investigated the effects of environmental factors, such as temperature, light, and nutrients, on the growth of typical species and the development of HABs [5,6]. These factors could partly explain the underlying mechanism of HAB formation and the seasonal succession of species. For a long time, cyanobacteria gained much attention from environmental degradation and human health perspectives [7–9]. Especially, many scholars have focused on *Microcystis* in recent years, which is a dominant cyanobacterial genus in many eutrophic waters and often exhibits a greater threat to the microcystins produced by toxic species [10–12].

Light could directly affect the photosynthesis and growth of cyanobacteria [13,14], which were closely associated with light intensity, exposure time, and light wavelength. Especially, other than necessary photosynthetically active radiation (PAR; 400–700 nm), enhanced ultraviolet (UV) radiation is reported for many aquatic ecosystems throughout the world due to serious stratospheric ozone depletion [15,16]. Therefore, the effects of UV radiation on typical cyanobacterial species have received considerable attention in recent years, and most studies have used *Microcystis* as a model species [17–19]. Freshwater ecosystems in the middle and lower reaches of Yangtze River are susceptible to enhanced UV radiation due to the lack of depth refuge [20], and *Microcystis* often occur as the surface blooms that encounter higher irradiance [21,22]. It is assumed that *Microcystis* should be more threatened and suffered greater UV-induced damage. However, the frequency and intensity of the dominance of *Microcystis* continue to increase in typical eutrophic lakes in China, such as Lake Taihu. Hence, it is crucial to investigate and compare the responses of *Microcystis* and other algal species to UV radiation.

The composition of HABs in freshwater ecosystems is varied and often includes cyanobacteria and green microalgae as the major components [21,23]. For example, *Microcystis* and *Chlorella* were the most dominant species in eutrophic lakes in China, although their cell densities fluctuated wildly during different seasons [12]. Meanwhile, *Microcystis* blooms were often formed by mixed species when the seasonal succession and competition between the non-toxic and toxic species have been widely studied [24,25]. For example, the toxic *Microcystis aeruginosa* was determined to be more harmful to *Chlorella vulgaris* than the non-toxic species at higher temperatures [19]. Although numerous studies have focused on the effects of UV on algae in recent years, relatively few studies have deeply examined and compared the adaptive strategies to the ambient UV radiation of non-toxic and toxic *M. aeruginosa* and other species [16,26]. Some scholars investigated the effects of nutrient enrichment on algal responses to UV radiation and the results are varied. For example, Li et al. [27] reported that the effects of UV-B on phytoplankton productivity might be underestimated in iron-deficient ecosystems, and Yang et al. [28] reported that the negative impact was most pronounced when UV-B exposure and P limitation were combined. Meanwhile, Zheng et al. [29] reported that impacts of solar UV radiation on algal growth differed significantly at different N concentrations. However, the influence mechanisms of nutrient enrichment on algal adaptation and biotic interactions to UV radiation also remain unclear. In addition, many studies have investigated the effects of UV radiation on algal growth in the pure mono-culture systems [13,30], and it remains unclear how the coexistence of algae was affected by UV radiation, despite the fact that algal species coexist in the natural ecosystems. In this regard, the co-cultures with different nutrient conditions may provide useful information to address cyanobacterial blooms and algal competition in the natural waters and to better explain the synergistic effects of eutrophication and irradiation in mixed communities.

In this study, we selected *C. pyrenoidosa* and non-toxic and toxic *M. aeruginosa* to investigate their various physiological responses with ambient irradiation treatment under different nutrient regimes. The main goals were to: (i) analyze the effects and mechanisms of ambient UV-B radiation on three species, (ii) compare and explore the responses of the adaptation capability of three species to UV radiation, and (iii) study the effects of nutrient enrichment on algal growth and competition.

2. Materials and Methods

2.1. Algal Culture

C. pyrenoidosa (FACHB 5), non-toxic *M. aeruginosa* (FACHB 469), and toxic *M. aeruginosa* (FACHB 905) were obtained from the Freshwater Algae Culture Collection of the Institute of Hydrobiology, Chinese Academy of Sciences (FACHB). For the three algal species, *M. aeruginosa* is a dominant genus during the outbreaks of HABs, and *C. pyrenoidosa* was selected because of its common distribution and frequent co-existence with cyanobacteria in many Chinese eutrophic ecosystems [14]. All strains were pre-cultured separately and exponential growth was maintained by transferring 5 mL of growing cultures to fresh standard BG_{11} medium in Erlenmeyer flasks every 8–10 days for enlargement [31]. Pre-culture was performed under sterile conditions and the flasks were placed at 25 °C under 40 μmol photons m^{-2} s^{-1} PAR with cool white fluorescent lamps (light/dark regime of 12 h:12 h) in the illuminated incubator (GZX-250BS-II). All flasks were shaken three times per day to prevent the cells from adhering to inner walls, and the position of flasks was exchanged randomly to ensure uniform light exposure.

2.2. Experimental Setup

After pre-culture and enlargement–cultivation, the exponentially growing algal cells were collected and suspended in phosphate buffer solution (PBS, pH = 7.4) for washing and reservation prior to running our formal experiments. After 3–4 days, algal cells were collected again and inoculated into 500-mL flasks containing 300–400 mL of modified BG_{11} medium for experiments in the mono-cultures and co-cultures. In the first scheme of modified BG_{11} medium, the composition was as shown in Table S1, and concentrations of nitrogen (N), phosphorus (P), and iron were comparable to those in the natural waters, representing normal growth conditions. In the second scheme of modified BG_{11} medium, the initial N, P, and iron concentrations were appropriately increased (Table S1), representing nutrient enrichment conditions. The initial cell density of three species was 1.0×10^6 cells mL^{-1}, which approximated the cell number at the beginning of HABs in most eutrophic lakes in China [32]. Meanwhile, co-cultures were conducted to simulate natural conditions and to explore the characteristics of algal competition. To this end, *C. pyrenoidosa* was co-cultured with non-toxic and toxic *M. aeruginosa*, when the inoculation ratio was 1:1 and the cell density of each strain was 1.0×10^6 cells mL^{-1}.

On each day, algal cultures in the flasks were transferred into sterilized petri dishes with quartz glass covers (20 cm in diameter) for PAR or UV-B exposure after slightly shaking the flasks, representing PAR treatment and UV-B treatment, and the two treatments both lasted for 4 h (9:00–13:00). In the PAR treatment, petri dishes were maintained in another illumination incubator and continuously irradiated with 50 μmol photons m^{-2} s^{-1} PAR. In the UV-B treatment, petri dishes were stored in clean chambers and subjected to high-pressure mercury UV-B lamps with the dominant wavelength of 313 nm (TL20W/01RS, Philips, Eindhoven, Netherlands, Figure S1). In the UV-B treatment, light exposure was restricted to UV-B and no photosynthetically active wavelengths were given to algal cells. The effective irradiation intensity of PAR and UV-B in our study was 70 and 0.8 W m^{-2}, respectively. After the irradiation treatment for 4 h, algal cultures were returned to flasks and incubated under the conditions as described in the pre-cultures for the rest time on each day (dark during 0:00–6:00 and 18:00–24:00, 40 μmol photons m^{-2} s^{-1} PAR during 6:00–9:00 and 13:00–18:00). The incubation lasted for 14 days in our study and a schematic diagram of the experiment is shown in Figure S2.

Based on field monitoring, the adopted PAR and UV-B in different treatments was in accordance with the natural conditions at noon in the middle and lower reaches of the Yangtze River [26]. The vertical sides of petri dishes were all covered with aluminum foil to ensure vertical radiation and the irradiance was measured using a miniature fiber optic spectrometer (FLA4000A+, Flight, Hangzhou, China).

2.3. Analytical Methods of Parameters

2.3.1. Cell Density and Photosynthetic Efficiency

Subsamples were regularly taken for determining cell density in the mono-cultures and co-cultures. Cells were both enumerated by using a flow cytometer (CytoFLEX S, Beckman Coulter, Fullerton, CA, USA), when *M. aeruginosa* and *C. pyrenoidosa* could be clearly differentiated by autofluorescence in the co-cultures [33]. Then, the algal growth rate was determined as follows: $\mu = (\ln N_2 - \ln N_1)/(t_2 - t_1)$, where N_1 and N_2 was the cell density on days t_1 and t_2, respectively. The maximum μ during the whole incubation period was defined as μ_{max}, which is an important index to indicate algal growth potential.

A Phyto-PAM fluorometer (Hein Walz, Effeltrich, Germany) was adopted to determine the effective quantum yield (F_v/F_m) of algal species. The Phyto-PAM fluorometer has been increasingly used in laboratory and in situ experiments, and F_v/F_m can effectively indicate the efficiency of algal photosynthesis apparatus [34–36].

2.3.2. Release of K^+ by Algal Cells

As K^+ is absorbed into the vacuole of algae cells and is mainly stored as an enzyme activator, the algal release of K^+ can be manifested for cell membrane damage [37,38]. During the incubation process, 10-mL supernatant samples of three species were regularly taken and immediately filtered through 0.2-μm mixed cellulose ester filters (Whatman, Little Chalfont, Buckinghamshire, UK) after daily irradiation. Then, the solution was acidified to pH = 2 with HNO_3 and K^+ content was determined by the inductively coupled plasma mass spectrometry (IC-PMS) (XII series, Thermo, Waltham, MA, USA). Afterwards, the release rate of K^+ by algal cells was calculated as a percentage and ultrasonic disrupted samples were adopted to make a comparison.

2.3.3. Characterization of Extracellular Polymeric Substance (EPS)

The extraction of EPS was conducted according to the methods described by Gao et al. [39] and Yang et al. [40]. Firstly, samples of the algal cultures were sonicated with 100 W ultrasound treatment for 5 min to obtain a uniform distribution, then filtered through 0.45-μm filters (Whatman, Little Chalfont, Buckinghamshire, UK) in order to separate soluble EPS (SEPS) from cell pellets [41]. The supernatant was collected and stored at 4 °C in the dark. Then, the harvested cells were washed in ultra-pure water, re-suspended in 0.05% NaCl solution, and centrifuged at 16,000× *g* for 20 min. The resulting supernatant was collected as bounded EPS (BEPS) and also stored at 4 °C in the dark.

On Day 1, the filtered SEPS and BEPS fractions were taken without dilution and the excitation emission matrix (EEM) spectra were determined by using a fluorescence spectrometer (F-7000, Hitachi, Tokyo, Japan). The excitation wavelengths were increased from 200 to 400 nm in 5-nm steps and the emission spectra were recorded from 250 to 500 nm in 1-nm increments. The increments were all set at 5 nm, and a scan speed of 2400 nm min^{-1} was applied. The blank scans were performed using modified BG_{11} medium, in which no fluorescence substance was present.

In addition, the contents of SEPS and BEPS were quantified spectrophotometrically (UV-2700, Shimadzu, Kyoto, Japan) during the whole incubation process by the anthrone sulfuric acid method and the values were normalized to cell density [42].

2.3.4. Reactive Oxygen Species (ROS) in Cells and Superoxide Dismutase (SOD) Activity

Before and during the incubation process in the mono-cultures, subsamples were regularly taken for the determination of ROS in algal cells of three species and activity of SOD. The details can be seen in the Supplementary Materials (SM).

2.3.5. Cell Adsorption Spectra and Contents of Photosynthetic Pigments

At the beginning of incubation, a scanning spectrophotometer (Beckman Coulter, Fullerton, CA, USA) was used to measure the whole-cell absorption spectra of three species between 400 and 750 nm. Cell cultures with the optical density value at 680 nm (OD_{680}) of 0.10 were used for measurement and the absorption peaks could indicate the existence of photosynthetic pigments in algal cells [43].

At different stages of the incubation (Day 1 and 8) in the mono-cultures, a subsample of algal cultures was collected and filtrated through 0.2-mm mixed cellulose ester filters (Whatman) to determine the contents of photosynthetic pigments (pg $cell^{-1}$) in single cells, including chlorophyll a (Chl-a), carotenoid (CAR), and phycocyanin (PC). The details can be seen in SM.

2.4. Statistical Analysis

All experiments were conducted in triplicate and means ± standard deviations of three replicates were calculated. The parametric three-way repeated-measures analysis of variance (RM-ANOVA) was used to determine the effects of irradiation treatments (PAR and UV-B), species (*C. pyrenoidosa*, non-toxic and toxic *M. aeruginosa*), and sampling time on the cell density, growth rates, EPS contents, and other parameters. Data were tested for normality and the variance assumptions of parametric ANOVA, and no data transformation was needed. If the interaction factor was significant at $p < 0.05$, a one-way ANOVA followed by Tukey's test was adopted to determine where differences lie. Meanwhile, the student's *t*-test was adopted to test the differences in the algal cell density of different species on a specific day in the co-cultures. All statistical analyses were performed using SPSS 22.0 (Chicago, IL, USA).

3. Results

3.1. Algal Growth in the Mono-Cultures under Normal Growth Conditions

3.1.1. Cell Density and Algal Photosynthetic Efficiency

In the PAR treatment, three species could all persistently grow and reached the maximum cell density on Day 14 (Figure 1). Meanwhile, the maximum cell density of *C. pyrenoidosa* was higher compared to the other two species, and toxic *M. aeruginosa* propagated more slowly. By comparison, the significant inhibitive effects ($p < 0.05$) of UV-B radiation were observed on algal growth, including cell density on a specific day, the maximum cell density during the incubation, and the duration of exponential growth. More specifically, three species both grow slowly and their cell densities began to decrease on Day-10 in the UV-B treatment. For algal photosynthetic efficiency, F_v/F_m of three species in the PAR treatment gradually increased before Day-10 and decreased afterwards. The variation patterns of algal F_v/F_m were different in the UV-B treatment, which decreased from the beginning, increased during Day 2–10, and declined afterwards. Moreover, algal F_v/F_m on a specific day in the UV-B treatment was always lower ($p < 0.05$) than that in the PAR treatment.

For *C. pyrenoidosa* and non-toxic *M. aeruginosa* under normal growth conditions, μ_{max} in the UV-B treatment was significantly lower ($p < 0.05$) compared with those in the PAR treatment (Figure 2), which indicated the negative effects of UV-B on the intrinsic growth potential of two species. Although μ_{max} of toxic *M. aeruginosa* was lower in the PAR treatment compared with the other two species, it did not change significantly ($p > 0.05$) with daily UV-B radiation in our study.

3.1.2. Diurnal Changes of Algal F_v/F_m

The diurnal changes of F_v/F_m were similar for three species in the PAR and UV-B treatments (Figure 3). During the incubation, diurnal F_v/F_m did not change significantly after 4 h of PAR treatment ($p > 0.05$), irrespective of species. In comparison, algal diurnal F_v/F_m decreased after 4 h of UV-B radiation ($p < 0.05$), which could then increase after the withdrawal of UV-B radiation on each day. However, the decline degree and recovery of F_v/F_m was dependent on the incubation time and species.

For three species, their F_v/F_m decreased to 15.2–40.6% of the initial values after UV-B radiation on Day 2 and they recovered to 75.2–83.3% of the initial values within 20 h, when the decline was lower for toxic *M. aeruginosa* ($p < 0.05$). On Days 6 and 8, the decline of algal F_v/F_m worsened after UV-B radiation and algal F_v/F_m could all recover to the initial values within 20 h. Meanwhile, the recovery rate increased with the development of incubation, and the recovery rate was highest for toxic *M. aeruginosa* ($p < 0.05$). In comparison, the inhibition of UV-B radiation on F_v/F_m was maximum for *C. pyrenoidosa* on each day and the recovery rate of its F_v/F_m was lower.

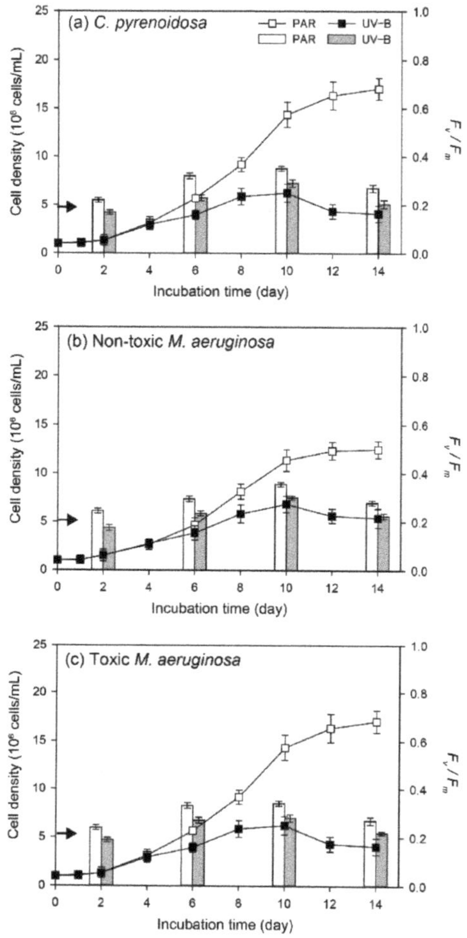

Figure 1. Cell density (line and scatter) and F_v/F_m (vertical bar) of three species in the PAR and UV-B treatments under normal growth conditions (the arrow indicates the initial value of F_v/F_m).

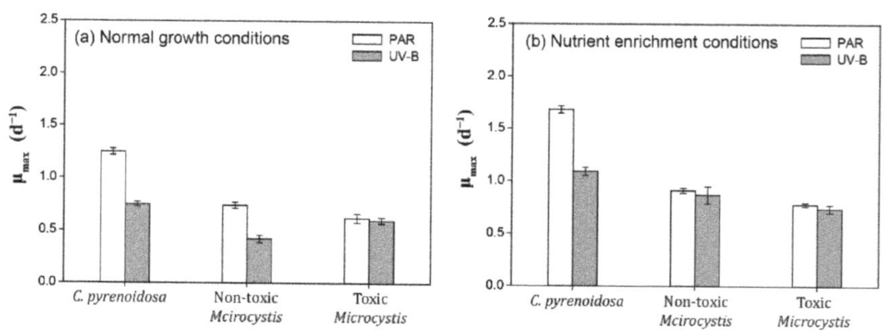

Figure 2. Maximum growth rates of three species in the PAR and UV-B treatment under normal growth conditions and nutrient enrichment conditions.

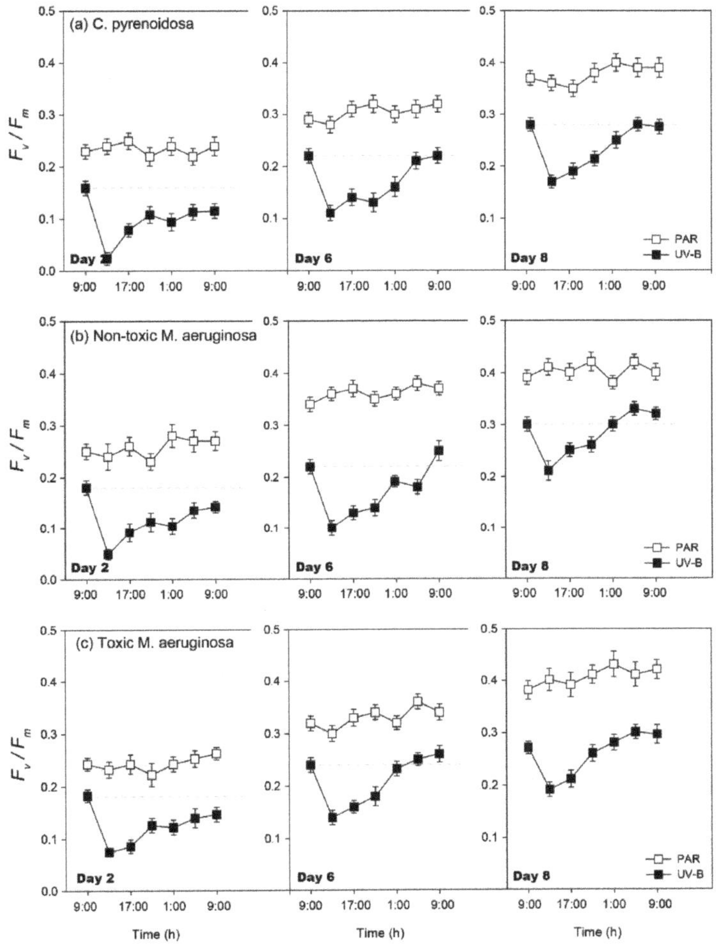

Figure 3. Diurnal changes of F_v/F_m of three species in the PAR and UV-B treatments under normal growth conditions on Day 2, Day 6, and Day 8.

3.2. K^+ Contents in the Algal Cultures

The release rates of K^+ by three species were all less than 5% before Day 6 in the PAR treatment (Figure 4), which could indicate the integrity of cells. Moreover, algal release rates of K^+ did not differ significantly between PAR and UV-B treatments during this period ($p > 0.05$), when the death and propagation of algal cells might be in a state of balance. After reaching the exponential growth stage in the PAR treatment, algal metabolism was enhanced with higher cell density, leading to the gradual increase of K^+ in the cultures (9.38–10.22% on Day 12). In comparison, the cell rupture of three species and algal release rates of K^+ were significantly promoted in the UV-B treatment ($p < 0.05$), indicating the greater damage of UV-B radiation on algal cells during this period.

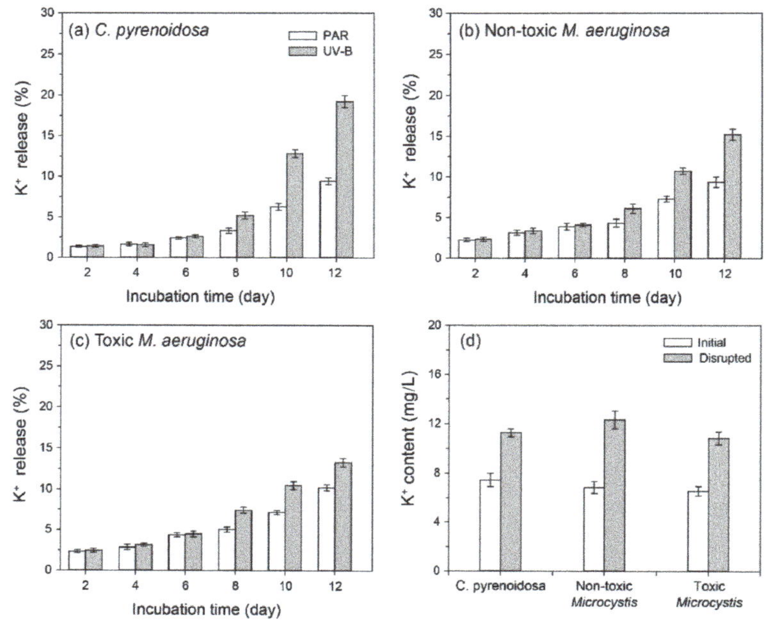

Figure 4. (**a–c**) The release rates of K^+ by three species in the PAR and UV-B treatments in the mono-cultures. (**d**) A comparison of the initial and ultrasonic disrupted cells of three species.

3.3. EPS Determination of Algal Cells

At the beginning of incubation, 3-D EEM spectra of algal EPS were determined and EEM contours were depicted (Figure S3). Results showed that EEM contours of BEPS and SEPS were similar for *C. pyrenoidosa* and non-toxic and toxic *M. aeruginosa*, indicating the similar metabolism patterns of three species. For BEPS, two peaks were presented near Ex/Em of 225/325 nm (peak T_2) and 280/325 nm (peak T_1), which belonged to the low-molecular aromatic protein and soluble microbial by-product like protein (such as tyrosine and tryptophan-like substances), respectively [44]. In contrast, three peaks were presented near Ex/Em of 280/325 nm (peak T_1), 340/430 nm (peak C, humic-acid like substances), and 275/435 nm (peak A, fulvic-acid like substances) in SEPS.

During the incubation, EPS contents increased for all three species in the PAR and UV-B treatments, and SEPS and BEPS had distinct changing trends during the incubation (Figure 5). For two *M. aeruginosa* species in the PAR treatment, SEPS content gradually increased and then remained constant, but BEPS content both increased in the early stage and decreased with increasing cell density. Meanwhile, the production of BEPS and SEPS was stronger for toxic *M. aeruginosa* ($p < 0.05$) during the incubation. In comparison, algal production of BEPS and SEPS by *M. aeruginosa* species was enhanced ($p < 0.05$) before Day 8 in the UV-B treatment. With the decline of cell densities, BEPS contents of two *M. aeruginosa* species decreased and their SEPS contents increased greatly after Day 10. In contrast, EPS production by *C. pyrenoidosa* was weaker in the PAR treatment ($p < 0.05$) and UV-B radiation did not significantly promote BEPS production before Day 8.

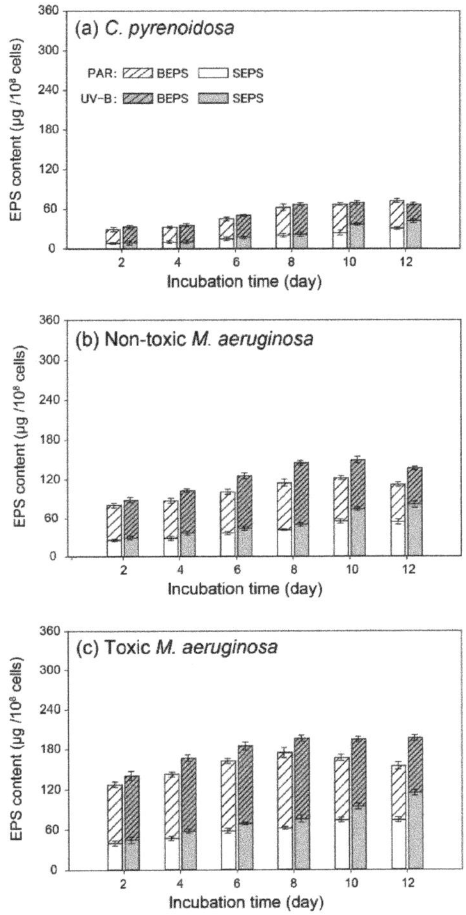

Figure 5. Contents of BEPS and SEPS produced by (**a**) *C. pyrenoidosa*, (**b**) non-toxic *M. aeruginosa* and (**c**) toxic *M. aeruginosa* cells in the PAR and UV-B treatments.

3.4. Antioxidant Responses of Algal Species under Normal Growth Conditions

3.4.1. ROS in Algal Cells and SOD Activity

The variation patterns of ROS in algal cells and algal SOD activity were similar for three species (Figure 6). In the PAR treatment, ROS and algal SOD activities were constant ($p > 0.05$) before Day 4 compared to the initial values, indicating that PAR treatment did not cause evident oxidative stresses on three species. However, ROS gradually increased after Day 6 in the PAR treatment, and algal SOD activity was elevated for all three species. Compared with PAR treatment, ROS of two *M. aeruginosa* species were comparable in the UV-B treatment on Day 2 and they were higher ($p < 0.05$) after Day 6. In comparison, ROS of *C. pyrenoidosa* was higher in the UV-B treatment on Day 2 and it was further promoted after Day 6, which was higher than that in the cells of two *M. aeruginosa* species. Irrespective of species, algal SOD activities were significantly higher ($p < 0.05$) in the UV-B treatment before Day 8, which could provide effective antioxidant protection. However, as the incubation progressed, algal SOD activity decreased sharply and leveled off until the end of incubation.

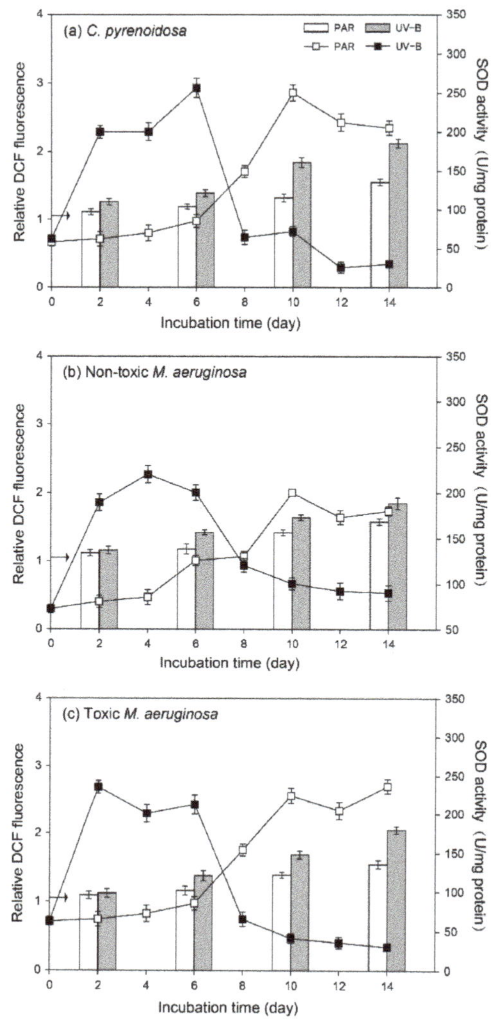

Figure 6. ROS in the cells of three species (vertical bar) and algal SOD activity (line and scatter) during the incubation in the PAR and UV-B treatments under normal growth conditions (the arrow indicates the initial value of ROS contents).

3.4.2. Contents of Photosynthetic Pigments

For all three species, the whole-cell absorption spectra indicated that they have Chl-a (two absorption peaks in the blue and red parts of the spectra at around 440 and 680 nm) and CAR with an absorption peak at around 495 nm (Figure S4). Moreover, non-toxic and toxic *M. aeruginosa* had an extra absorption peak at around 620 nm, which was regarded to phycocyanin in cyanobacterial cells [45].

Similar Chl-a contents of algal single cells were observed in the PAR and UV-B treatments on Day 1 ($p > 0.05$, Table 1). However, CAR of three species and PC of two *Microcystis* species were significantly promoted ($p < 0.05$) in the UV-B treatment on Day 1. Moreover, algal CAR/Chl-a and PC/Chl-a ratios in the UV-B treatment were higher ($p < 0.05$) than those in the PAR treatment at this moment. As incubation progressed on Day 8, Chl-a in algal cells were lower ($p < 0.05$) in the UV-B treatment compared to those in the PAR treatment, which could indicate the damage to chlorophyll synthesis. Meanwhile, algal

CAR and PC contents in the UV-B treatment decreased greatly ($p < 0.05$) compared with Day 1, which were lower ($p < 0.05$) than those in the PAR treatment. Compared with PAR treatment, the CAR/Chl-a ratio of *C. pyrenoidosa* on Day 8 was lower in the UV-B treatment ($p < 0.05$), whereas CAR/Chl-a and PC/Chl-a ratios of two *M. aeruginosa* species were significantly higher on Day 8 in the UV-B treatment ($p < 0.05$).

Table 1. Contents of photosynthetic pigments of three species in the PAR and UV-B treatments on Day 1 and Day 8 under normal growth conditions).

Contents of Piments (pg/cell)		*C. pyrenoidosa*		**Non-Toxic** *M. aeruginosa*		**Toxic** *M. aeruginosa*	
		PAR	**UV-B**	**PAR**	**UV-B**	**PAR**	**UV-B**
Day 1	Chl-a	0.14 ± 0.02	0.15 ± 0.02	0.18 ± 0.03	0.16 ± 0.01	0.17 ± 0.02	0.17 ± 0.03
	CAR	0.12 ± 0.02	**0.19 ± 0.02** *	0.07 ± 0.01	**0.11 ± 0.01** *	0.06 ± 0.02	**0.12 ± 0.02** *
	PC	\	\	0.62 ± 0.07	**0.84 ± 0.06** *	0.55 ± 0.03	**0.75 ± 0.04** *
	CAR/Chl-a	0.86 ± 0.02	**1.25 ± 0.06** *	0.41 ± 0.02	**0.68 ± 0.05** *	0.37 ± 0.05	**0.72 ± 0.01** *
	PC/Chl-a	\	\	3.46 ± 0.15	**5.15 ± 0.35** *	3.26 ± 0.16	**4.64 ± 0.28** *
Day 8	Chl-a	0.36 ± 0.04	**0.24 ± 0.03** *	0.27 ± 0.01	**0.11 ± 0.01** *	0.25 ± 0.01	**0.12 ± 0.02** *
	CAR	0.18 ± 0.01	**0.09 ± 0.01** *	0.11 ± 0.01	**0.07 ± 0.01** *	0.10 ± 0.02	**0.05 ± 0.01**
	PC	\	\	0.70 ± 0.01	**0.54 ± 0.04** *	0.67 ± 0.03	**0.46 ± 0.03** *
	CAR/Chl-a	0.50 ± 0.02	**0.39 ± 0.03** *	0.42 ± 0.05	**0.68 ± 0.02** *	0.42 ± 0.06	**0.63 ± 0.03** *
	PC/Chl-a	\	\	2.61 ± 0.03	**5.07 ± 0.38** *	2.68 ± 0.05	**4.12 ± 0.24** *

* Bold values with * indicated significant higher contents in the UV-B treatment compared with PAR treatment at $p < 0.05$, while those bold and underlined values with * indicated significant lower contents in the UV-B treatment compared with PAR treatment at $p < 0.05$.

3.5. *Algal Growth in the Mono-Cultures under Nutrient Enrichment Conditions*

3.5.1. Cell Density and Algal Photosynthetic Efficiency

Compared to normal growth conditions, three species all grew steadily under nutrient enrichment conditions and their cell densities did not decrease at the later stage (Figure 7). For *C. pyrenoidosa* and non-toxic *M. aeruginosa*, their cell densities on a specific day and the maximum cell density in the UV-B treatment were significantly lower ($p < 0.05$) than those in the PAR treatment. However, no significant difference in the cell density of toxic *M. aeruginosa* ($p > 0.05$) was observed between PAR and UV-B treatments during the whole incubation process. F_v/F_m of three species gradually increased and decreased afterwards in both PAR and UV-B treatments. Although algal F_v/F_m in the UV-B treatment were lower ($p < 0.05$) than that in the PAR treatment before Day 6, the difference became smaller at the later stage of incubation. For example, F_v/F_m of non-toxic and toxic *M. aeruginosa* were both comparable ($p > 0.05$) in the UV-B and PAR treatments on Day 10 and Day 14.

As expected, μ_{max} of three species increased with nutrient enrichment compared with those under normal growth conditions in the PAR and UV-B treatments (Figure 2). μ_{max} of non-toxic and toxic- *M. aeruginosa* were lower compared with *C. pyrenoidosa*, but they were both comparable between PAR and UV-B treatments ($p > 0.05$).

3.5.2. Diurnal Changes of Algal F_v/F_m

Compared with normal growth conditions, diurnal changes of algal F_v/F_m were similar under nutrient enrichment conditions (Figure 8). However, the decline degrees of algal F_v/F_m after UV-B radiation were lower (17.5–50.8%), and the recovery efficiency of F_v/F_m was better with nutrient enrichment. For example, F_v/F_m of two *M. aeruginosa* species could both totally recover to the initial values after UV-B radiation on Day 2, and F_v/F_m of *C. pyrenoidosa* on Day 6 and Day 8 totally recovered to the initial values within 16 h and 8 h after UV-B radiation, respectively. For three species in the UV-B treatment, the decline degree of F_v/F_m was also lower for toxic *M. aeruginosa* and it exhibited a faster recovery rate. This result was consistent with that under normal growth conditions.

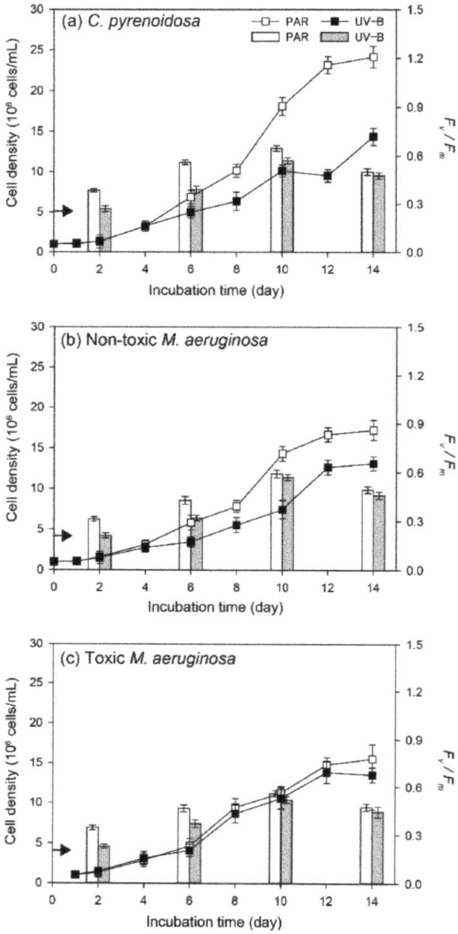

Figure 7. Cell density (line and scatter) and F_v/F_m (vertical bar) of three species in the PAR and UV-B treatments under nutrient enrichment conditions (the arrow indicates the initial value of F_v/F_m).

3.6. Antioxidant Responses of Algal Species under Nutrient Enrichment Conditions

3.6.1. ROS in Algal Cells and SOD Activity

Under nutrient enrichment conditions, the variation patters of ROS in algal cells and algal SOD activity were also similar for three species (Figure 9). More specifically, PAR treatment did not cause great oxidative stresses on algae, but ROS and algal SOD activity gradually increased at the later stage of incubation. In the UV-B treatment, ROS in algal cells also increased gradually, and they only showed higher values ($p < 0.05$) than those in the PAR treatment after Day 10. For algal SOD activity in the UV-B treatment, they all exhibited a sharp increase and decreased gradually to maintain a stable value. For both PAR and UV-B treatments, ROS in algal cells were lower ($p < 0.05$) than those under normal growth conditions on a specific day.

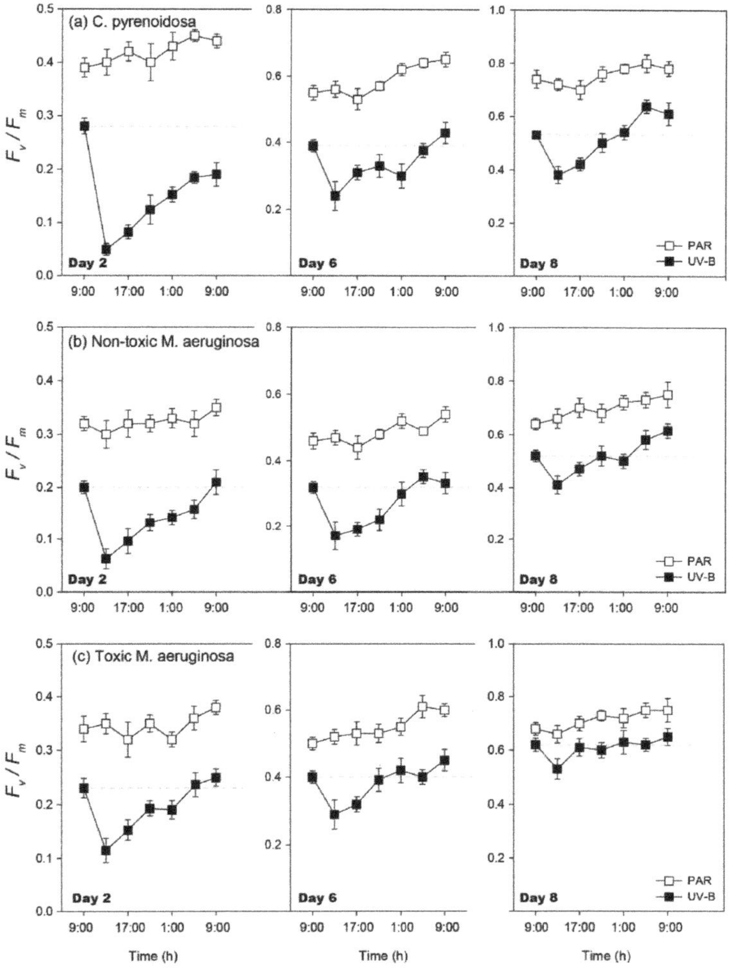

Figure 8. Diurnal changes of F_v/F_m of three species in the PAR and UV-B treatments under nutrient enrichment conditions on Day 2, Day 6, and Day 8.

3.6.2. Contents of Photosynthetic Pigments

As shown in Table 2, similar patterns were overserved for the algal synthesis of photosynthetic pigments on Day 1. Compared to the initial values, Chl-a contents of algal single cells were comparable in the PAR and UV-B treatment ($p > 0.05$), but CAR and PC in single cells increased greatly ($p < 0.05$) in the UV-B treatment, resulting in the higher CAR/Chl-a and PC/Chl-a ratios of three species on Day 1. Moreover, CAR and PC contents, CAR/Chl-a and PC/Chl-a ratios were all higher ($p < 0.05$) with nutrient enrichment compared to those under normal growth conditions.

On Day 8, despite the fact that the Chl-a contents of algal single cells were lower ($p < 0.05$) in the UV-B treatments, they showed an increasing trend compared with those on Day 1. This was consistent with the patterns of cell density. In addition, CAR and PC in single cells were also higher ($p < 0.05$) in the UV-B treatment at this moment, and CAR/Chl-a and PC/Chl-a ratios were promoted with UV-B radiation. This pattern was remarkably different from that under normal growth conditions.

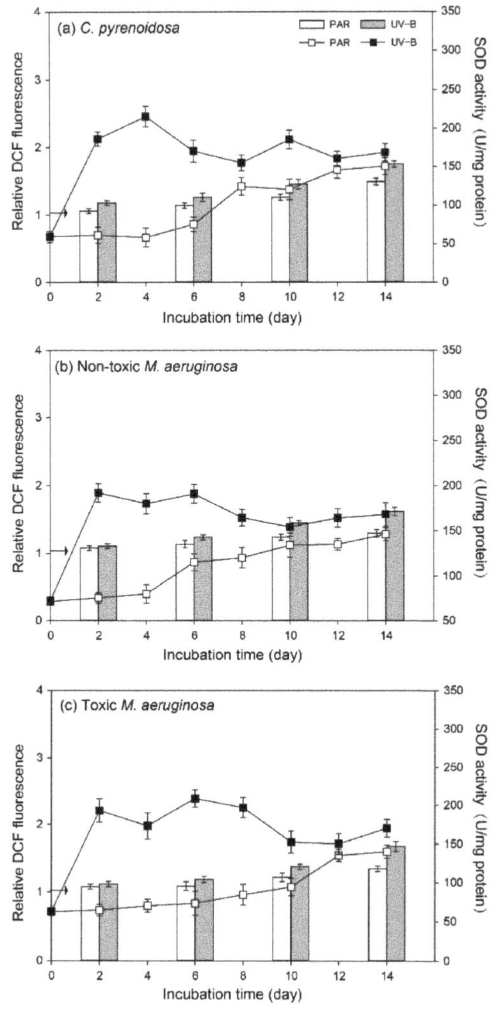

Figure 9. ROS in the cells of three species (vertical bar) and algal SOD activity (line and scatter) during the incubation in the PAR and UV-B treatments under nutrient enrichment conditions (the arrow indicates the initial value of ROS contents).

3.7. Interspecific Competition in the Co-Cultures

Algal growth patterns were comparable in the PAR treatment (Figure 10), i.e., *C. pyrenoidosa* grew rapidly after the lag period and soon outcompeted non-toxic or toxic *M. aeruginosa*, while the cell density of *C. pyrenoidosa* decreased when that of *M. aeruginosa* started to increase. Compared with mono-cultures, the maximum cell densities of three species were all lower ($p < 0.05$) in the co-cultures (Table S2). However, the maximum cell density of *C. pyrenoidosa* seemed to decrease more in the PAR treatment, and the decline was great in the co-cultures of *C. pyrenoidosa* with toxic *M. aeruginosa*. In comparison, two *M. aeruginosa* species gained the obvious dominance and maintained competitive advantages from the beginning in the UV-B treatment, and the growth of *C. pyrenoidosa* was also markedly inhibited, which achieved 42.1% and 31.4% of the maximum cell density in the mono-cultures. Meanwhile, despite the faster growth of *C. pyrenoidosa* in the mono-cultures, μ_{max} of *C. pyrenoidosa* decreased greatly when it was co-cultured with *M. aeruginosa* in the PAR and UV-B treatments. However, μ_{max} of non-toxic *M. aeruginosa* only decreased slightly and μ_{max} of toxic *M. aeruginosa* even increased slightly.

Table 2. Contents of photosynthetic pigments of three species in the PAR and UV-B treatments on Day 1 and Day 8 under nutrient enrichment conditions.

Contents of Piments (pg/cell)		*C. pyrenoidosa*		Non-Toxic *M. aeruginosa*		Toxic *M. aeruginosa*	
		PAR	UV-B	PAR	UV-B	PAR	UV-B
Day 1	Chl-a	0.14 ± 0.01	0.15 ± 0.02	0.18 ± 0.02	0.16 ± 0.02	0.17 ± 0.02	0.17 ± 0.02
	CAR	0.12 ± 0.02	**0.23 ± 0.02 ***	0.09 ± 0.01	**0.14 ± 0.01 ***	0.08 ± 0.01	**0.15 ± 0.01 ***
	PC	\	\	0.67 ± 0.02	**0.97 ± 0.04 ***	0.58 ± 0.02	**0.94 ± 0.02 ***
	CAR/Chl-a	0.84 ± 0.05	**1.52 ± 0.08 ***	0.41 ± 0.01	**0.83 ± 0.02 ***	0.37 ± 0.05	**0.88 ± 0.09 ***
	PC/Chl-a	\	\	3.57 ± 0.22	**6.01 ± 0.48 ***	3.17 ± 0.18	**5.58 ± 0.42 ***
Day 8	Chl-a	0.50 ± 0.02	**0.35 ± 0.04 ***	0.41 ± 0.02	**0.30 ± 0.02 ***	0.39 ± 0.01	**0.29 ± 0.01 ***
	CAR	0.23 ± 0.02	**0.27 ± 0.01 ***	0.18 ± 0.02	**0.26 ± 0.01 ***	0.15 ± 0.01	**0.19 ± 0.01**
	PC	\	\	0.70 ± 0.02	**0.83 ± 0.03 ***	0.69 ± 0.01	**0.79 ± 0.03 ***
	CAR/Chl-a	0.46 ± 0.03	**0.78 ± 0.06 ***	0.45 ± 0.05	**0.86 ± 0.06 ***	0.38 ± 0.02	**0.68 ± 0.02 ***
	PC/Chl-a	\	\	1.72 ± 0.02	**3.83 ± 0.21 ***	1.78 ± 0.09	**2.84 ± 0.25 ***

* Bold values with * indicated significant higher contents in the UV-B treatment compared with PAR treatment at $p < 0.05$, while those bold and underlined values with * indicated significant lower contents in the UV-B treatment compared with PAR treatment at $p < 0.05$.

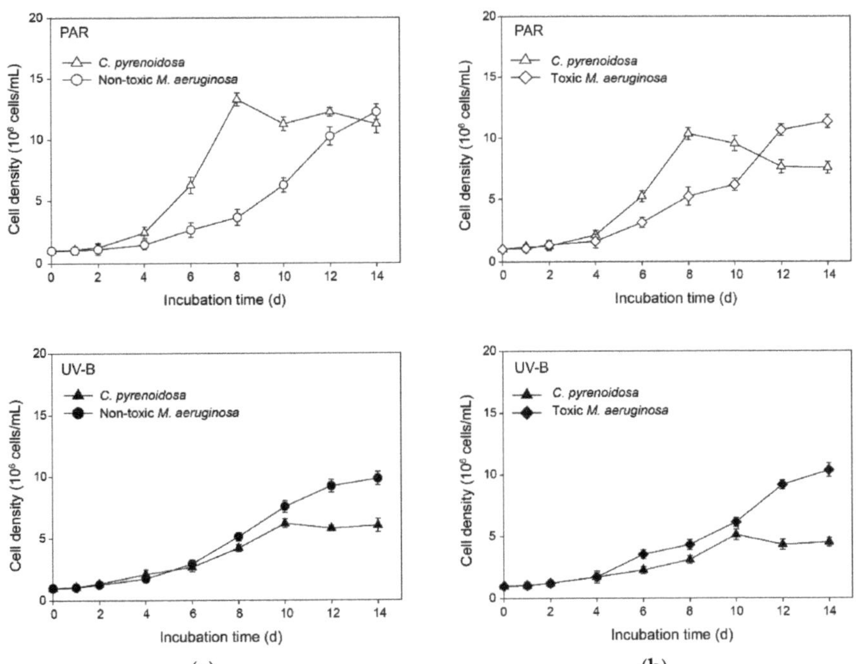

Figure 10. Cell densities of algal species in the co-cultures of (**a**) *C. pyrenoidosa* & non-toxic *M. aeruginosa*, and of (**b**) *C. pyrenoidosa* & toxic *M. aeruginosa* in the PAR and UV-B treatments under nutrient enrichment conditions.

4. Discussion

4.1. Effects of UV-B Radiation and Algal Responses

Although UV irradiance usually constitutes a few percent of solar radiation (5.85–8.51% in China), many studies have analyzed the effects of UV radiation on algal growth and negative effects were often reported [46,47]. Based on field and laboratory experiments, the main influencing mechanisms of UV-B radiation on algae include cell vitality impairment, ROS production, DNA damages, changes of nutrient utilization, etc. [26,27,48,49]. Our results are consistent with these findings, namely in that ambient UV-B radiation could exert negative effects on typical algal species in freshwater ecosystems, which also showed adaptative responses to UV-B radiation.

Since the growth and vitality of photosynthetic organisms are mainly governed by photosynthetic activity and the photosynthetic apparatus is an important damage target of UV-B radiation [50,51], daily F_v/F_m of all three species in the UV-B treatment were often lower compared to those in the PAR treatment. This result indicated that ambient UV-B might cause damages to D1 or D2 protein in algal photosystems [52] and 50 μmol m^{-2} s^{-1} of PAR did not have similar effects. However, different from many other studies using high-dose UV-B radiation whereby algal photosynthetic systems were greatly damaged [53,54], algal F_v/F_m gradually increased during Day 2–10 under two different growth conditions. Based on algal Chl-a contents on Day 1 and the release rates of K^+, the adopted UV-B treatment in our study did not have direct lethal effects on algae, and changes of algal F_v/F_m and growth could be the balance between the light-induced effects and adaptive physiological processes of cells. This was confirmed by the diurnal changes of algal F_v/F_m, as F_v/F_m of three species recovered with different rates after UV-B exposure, which could have resulted from processes, such as oxidation resistance, nucleotide resynthesis, ATP supply, or the repair of damaged proteins [55,56].

UV radiation could cause the overexcitation of substances and produce excess ROS in algal cells or in the cultures [57,58], leading to the impairment of algal photosynthetic systems and normal growth. Consistently, ROS contents in UV-B radiated algal cells were higher during the incubation under normal growth conditions. The increase of ROS in the later period in the PAR treatment was consistent with the work of Latifi et al. [59] in that some environmental factors, such as nutrient deficiency and light limitation, could indirectly generate ROS at multiple sites of the photosynthetic electron transport chain in algal cells. However, as mentioned above, the oxidative stresses and resulting damages could be mitigated with algal adaptive strategies. In our study, EPS production, the up-regulation of SOD activity, CAR and PC synthesis, and recovery of F_v/F_m by three species, could act as their effective adaptation mechanisms, resulting in decreased sensitivity to UV-B exposure and increased self-repair efficiency. For example, algal EPS consisted of polysaccharides, proteins, lipids, and humic substances and often appeared as a structureless slimy layer around cells, which was helpful to algal aggregation and its resistance to environmental stresses [38,42,60]. Meanwhile, higher CAR and PC in cells could adsorb UV-B light and quench ROS to alleviate damage to algal photosynthetic systems and DNA [50,61]. Moreover, higher CAR in the cells could increase the algal utilization efficiency of light and promote its generation of ATP and other substances [50,62], such as antioxidant enzymes, nucleotides, and proteins, to repair damaged apparatus in algal cells [18,30]. In the UV-B treatment, the high SOD activity and enhanced production of CAR and PC by algae in the early stage could partly explain the gradual increase of algal F_v/F_m and cell density. However, these adaptive responses of three species might be not enough to remove UV-induced oxidative stresses at the later stage under normal growth conditions, and inhibition on algal growth could occur. Our previous study indicated that algal adaptation to UV-B radiation required energy and essential nutrient substances [26], and this could be the possible reason for algal decay at the later stage in the UV-B treatment. Especially, cumulative ROS might damage the antioxidant systems of UV-radiated algae after Day 8 under normal growth conditions and result in low SOD activity and algal death.

4.2. Comparison of Algal Adaptation to UV-B Radiation

In previous studies, scholars have often investigated the strategies of cyanobacteria to alleviate the harmful effects of UV-B radiation, such as the production of UV-absorbing compounds (UVCs) to mitigate photo-induced damages, vertical migration of cells to decrease the irradiation stress, enhanced self-repair, etc. [46,51,63]. In this study, three species exhibited strain-specific responses to UV-B radiation, when toxic *M. aeruginosa* was more tolerant and showed a higher adaptation capability, including lower sensitivity to UV-B radiation and better self-repair efficiency.

Firstly, *C. pyrenoidosa* grew faster, whereas toxic *M. aeruginosa* had similar μ_{max} in the PAR and UV-B treatments, which might indicate the stronger plasticity of toxic *M. aeruginosa* to maintain a stable growth potential. The lower growth rate of toxic *M. aeruginosa* was probably caused by the excess energy cost for microcystin production [64]. Secondly, EPS production by toxic *M. aeruginosa* could provide a better adaptation to UV-B radiation. In this study, toxic *M. aeruginosa* produced more BEPS in the early stage in the UV-B treatment, when tryptophan-like substances in BEPS could absorb UV-B radiation and play the role of precursor to UV-absorbing metabolites [65]. The decrease of BEPS and increase of SEPS at the later stage could be explained as some UV-absorbing compounds were degraded, and this contributed to the decreased adaptation of algae to UV radiation under normal growth conditions [49,57]. Meanwhile, toxic *M. aeruginosa* excreted more SEPS, and organic matter in SEPS had a positive effect on algal aggregation [41,66]. The aggregated morphology of algal cells could be beneficial to reduce photo-induced damage by shading [67,68] and this was regarded as one kind of defense against UV-B in the natural waters. Furthermore, the high iron availability for algal cells could decrease UV-induced damages [27] and higher EPS could serve as an important iron reservoir that helps toxic *M. aeruginosa* to better cope with UV-B radiation [11,60].

Moreover, toxic *M. aeruginosa* exhibited a better antioxidant response in the UV-B treatment. In our study, two *M. aeruginosa* species could promote the synthesis of CAR and PC in the UV-B treatment, which was further enhanced with nutrient enrichment. As mentioned above, the beneficial effects of CAR and PC included the alleviation of photo-induced damage and the promotion of self-repair [15,69]. Moreover, microcystin synthesis by toxic *M. aeruginosa* could contribute to a higher fitness of cells under UV-B irradiation through a covalent interaction with the cysteine residue of proteins [70]. Consequently, F_v/F_m decline was lower and the recovery rate was faster for toxic *M. aeruginosa* under two different conditions. Xu et al. [18] also indicated that toxic *M. aeruginosa* had a competitive advantage relative to non-toxic strain in a changing light environment via stronger antioxidant capacity (higher SOD activity and the synthesis of microcystin) and quicker PSII recovery capacity The decrease of CAR and PC on Day 8 under normal growth conditions was related to the photooxidation and photodegradation of pigments, when the biological resources in the cultures might be not enough for the algal resynthesis of pigments and other efficient metabolic processes [71]. Compared to PAR treatment, the higher CAR/Chl-a and PC/Chl-a ratios of two *M. aeruginosa* species under normal growth conditions and higher CAR/Chl-a and PC/Chl-a ratios of all three species under nutrient enrichment conditions probably indicated their increased acclimation to prolonged UV-B exposure [13,72]. This was consistent with results obtained by Jiang et al. [50]. Although increased cell density partially reduced UV-B radiation at the later stage of incubation, our results could be mainly ascribed to the adaptation capability of algae to UV radiation in the 2-cm depth dishes.

4.3. *Effects of Nutrient Enrichment and Algal Competition Characteristics*

Whereas scholars have often studied the influences and mechanisms of UV radiation on algae, fewer studies have focused on the effects of nutrient enrichment. Meanwhile, the role of UV-B radiation in determining interspecific competition has not been clearly elucidated.

Combining the diurnal changes of algal F_v/F_m and algal growth patterns under different growth conditions, nutrient enrichment alleviated the negative effects of UV-B radiation on three species in our study. This was in accordance with our previous findings that higher P availability could enhance algal adaptation to UV radiation [26]. Zheng et al. [29] also reported that UV-induced inhibition of algal growth and photosynthetic production changed in accordance with the changes of the chemical environment in the water. In our study, the beneficial effects of nutrient enrichment also included decreasing algal sensitivity to UV-B radiation and increasing its self-repair efficiency. For example, higher contents of CAR and PC in cells with nutrient enrichment could help algae to counteract UV-induced damages [50,73], which resulted in lower ROS in cells and lower decline degrees of algal F_v/F_m on each day. Meanwhile, three species did not require a great deal of energy and biological resources to deal with UV-B radiation, and they could better promote their growth after self-repair with more nutrients in the medium, such as the photo-reactivation of DNA or resynthesis of D1 proteins [18,22]. Therefore, algal μ_{max} values were higher and three species persistently grew in the UV-B treatment under nutrient enrichment conditions. Since toxic *M. aeruginosa* exhibited a higher adaptation capability to UV-B radiation, as previously discussed, the beneficial effects of nutrient enrichment were best for toxic *M. aeruginosa*, and its growth was comparable between PAR and UV-B treatment during the whole incubation.

Wind-induced mixing of water and sediment resuspension could cause pulse fluctuations of irradiation conditions and high nutrient availability in the water, where different algal species coexist. Thus, the co-cultures under nutrient enrichment conditions might partly explain the competitive advantages of typical species in the field. Different from the mono-cultures, *C. pyrenoidosa* was not always the fastest-growing species in the co-cultures, and exposure to UV-B radiation could enhance the growth advantages of *M. aeruginosa*. Our previous study indicated that the augmentation of algal P quota could alleviate or eliminate the negative effects of UV radiation on algae [26]. Considering that *M. aeruginosa* had a faster and better P accumulation ability compared to other typical species in freshwater ecosystems [74], *M. aeruginosa* might have a stronger adaptation capability to UV-B radiation and a stronger competitive advantage in the co-cultures. However, since nutrients were not limited under nutrient enrichment conditions, allelopathy effects between species might have a more important role in the co-cultures [19,75]. In our study, two *M. aeruginosa* species demonstrated a greater inhibition effect on *C. pyrenoidosa* growth compared with the negative effects of *C. pyrenoidosa* on *M. aeruginosa*. For example, when the secondary metabolites of green algae showed declining inhibitory effects as incubation progressed, the extracts of cyanobacteria and microcystins were often more effective to inhibit the growth of other species [24,76]. Therefore, μ_{max} of *C. pyrenoidosa* decreased greatly and two *M. aeruginosa* species outcompeted *C. pyrenoidosa* at the later stage in the PAR treatment. Meanwhile, toxic *M. aeruginosa* showed a greater competitiveness to maintain high μ_{max} and inhibit *C. pyrenoidosa* growth in the co-cultures. As mentioned above, the higher EPS contents and microcystin of *M. aeruginosa* cells were conducive to the adaptation of *Microcystis* to UV-B radiation [11,70]. Furthermore, the aggregation of *Microcystis* might prevent *C. pyrenoidosa* to utilize PAR for self-repair or recovery after UV-B radiation [77]. Consequently, non-toxic and toxic *M. aeruginosa* were dominant from the beginning in the UV-B treatment, and toxic *M. aeruginosa* also had a greater impact in depressing the growth of *C. pyrenoidosa*. In this sense, the dominance of cyanobacteria and advantages of toxic *M. aeruginosa* could be enhanced in UV-radiated waters with severer eutrophication. However, the complexities and likely influence of coexisting yet unexamined factors deserve a further in situ study in the future.

5. Conclusions

(1) Compared with PAR, 4 h of ambient UV-B radiation could exert oxidative stresses and negative effects on the photosynthesis and growth of three algal species under normal growth conditions. The adopted UV- B treatment did not cause lethal effects on algae, and three species could grow with adaptive responses, including EPS production, regulation of SOD activity, synthesis of photosynthetic pigments, and F_v/F_m recovery.

(2) Three species exhibited strain-specific responses to ambient UV-B radiation in the mono-cultures, when toxic *M. aeruginosa* was more tolerant and showed a higher adaptation capability to UV-B, including lower sensitivity and better self-repair efficiency.

In addition to stable μ_{max} in two treatments, higher production of EPS, and enhanced production of CAR and PC under UV-B radiation, toxic *M. aeruginosa* showed a better recovery of its photosynthetic efficiency.

(3) Nutrient enrichment could alleviate the negative effects of UV-B radiation on algae, and the growth of toxic *M. aeruginosa* was comparable between PAR and UV-B treatment. In the co-cultures with nutrient enrichment, *M. aeruginosa* gradually outcompeted *C. pyrenoidosa* in the PAR treatment, and UV-B treatment enhanced the growth advantages of *M. aeruginosa*, when toxic *M. aeruginosa* showed a greater competitiveness to maintain high μ_{max} and inhibit the growth of *C. pyrenoidosa*.

Supplementary Materials: The following supporting information can be downloaded at: https://www.mdpi.com/article/10.3390/ijerph19095485/s1, Figure S1. The spectral power distribution and weighted UV radiation of UV- B lamps (TL20W/01RS, Philips) used in the irradiation experiment. Figure S2. A schematic diagram of the irradiation experiments. Figure S3. Fluorescence EEM spectra for EPS produced by three species. Figure S4. Whole-cell absorption spectra of algal cultures at the beginning of mono-cultures. Cell cultures with OD_{680} of 0.10 were used for measurement and adsorption values were normalized to the optical density at OD_{680}. Table S1. Composition of the modified BG_{11} medium under different growth conditions in our experiment. Table S2. The maximum growth rate (μ_{max}, d^{-1}) and maximum cell density (10^6 cells/mL) of three species in the mono-cultures and co-cultures under nutrient enrichment conditions and the percentage change of maximum cell density showing in parentheses. References [57,78–80] are cited in the supplementary materials.

Author Contributions: Conceptualization, J.H. and L.R.; methodology, K.D.; data curation, Y.W.; writing—original draft preparation, L.R.; writing—review and editing, Y.Y. and L.Z.; investigation, H.W. All authors have read and agreed to the published version of the manuscript.

Funding: This work was financially supported by the Natural Science Foundation of Jiangsu Province (BK20210933) and National Natural Science Foundation of China (22006066).

Institutional Review Board Statement: Not applicable.

Informed Consent Statement: Not applicable.

Data Availability Statement: The data that support the findings of this study are available from the corresponding author, upon reasonable request.

Conflicts of Interest: The authors declare no conflict of interest.

References

1. Paerl, H.W.; Huisman, J. Blooms like it hot. *Science* **2008**, *320*, 57–58. [CrossRef] [PubMed]
2. Li, Y.; Shang, J.H.; Zhang, C.; Zhang, W.L.; Niu, L.H.; Wang, L.F.; Zhang, H.J. The role of freshwater eutrophication in greenhouse gas emissions: A review. *Sci. Total Environ.* **2021**, *768*, 144582. [CrossRef]
3. Paerl, H.W. Mitigating harmful cyanobacterial blooms in a human- and climatically-impacted world. *Life* **2014**, *4*, 988–1012. [CrossRef]
4. Rigosi, A.; Carey, C.C.; Ibelings, B.W.; Brookes, J.D. The interaction between climate warming and eutrophication to promote cyanobacteria is dependent on trophic state and varies among taxa. *Limnol. Oceanogr.* **2014**, *59*, 99–114. [CrossRef]
5. Carey, C.C.; Ibelings, B.W.; Hoffmann, E.P.; Hamilton, D.P.; Brookes, J.D. Eco-physiological adaptations that favour freshwater cyanobacteria in a changing climate. *Water Res.* **2012**, *46*, 1394–1407. [CrossRef]
6. Lürling, M.; Eshetu, F.; Faassen, E.J.; Kosten, S.; Huszar, V.L.M. Comparison of cyanobacterial and green algal growth rates at different temperatures. *Freshw. Biol.* **2013**, *58*, 552–559. [CrossRef]
7. Paerl, H.W.; Xu, H.; McCarthy, M.J.; Zhu, G.; Qin, B.; Li, Y.; Gardner, W.S. Controlling harmful cyanobacterial blooms in a hyper-eutrophic lake (Lake Taihu, China): The need for a dual nutrient (N & P) management strategy. *Water Res.* **2011**, *45*, 1973–1983. [PubMed]
8. Huisman, J.; Codd, G.A.; Paerl, H.W.; Ibelings, B.W.; Verspagen, J.M.H.; Visser, P.M. Cyanobacterial blooms. *Nat. Rev. Microbiol.* **2018**, *16*, 471–483. [CrossRef] [PubMed]
9. Qin, B.Q.; Paerl, H.W.; Brookes, J.D.; Liu, J.G.; Jeppesen, E. Why Lake Taihu continues to be plagued with cyanobacterial blooms through 10 years (2007–2017) efforts. *Sci. Bull.* **2019**, *64*, 7–9. [CrossRef]
10. Davis, T.W.; Berry, D.L.; Boyer, G.L.; Gobler, C.J. The effects of temperature and nutrients on the growth and dynamics of toxic and non-toxic strains of *Microcystis* during cyanobacteria blooms. *Harmful Algae* **2009**, *8*, 715–725. [CrossRef]

11. Wang, X.Q.; Jiang, H.B.; Qiu, B.S. Effects of iron availability on competition between Microcystis and Pseudanabaena or Chlorella species. *Eur. J. Phycol.* **2015**, *50*, 260–270. [CrossRef]
12. Ren, L.X.; Huang, J.; Wang, B.; Wang, H.Y.; Gong, R.; Hu, Z.X. Effects of temperature on the growth and competition between *Microcystis aeruginosa* and *Chlorella pyrenoidosa* with different phosphorus availabilities. *Desalination Water Treat.* **2021**, *241*, 87–111. [CrossRef]
13. Chen, Z.; Jiang, H.B.; Gao, K.S.; Qiu, B.S. Acclimation to low ultraviolet-B radiation increases photosystem I abundance and cyclic electron transfer with enhanced photosynthesis and growth in the cyanobacterium *Nostoc sphaeroides*. *Environ. Microbiol.* **2020**, *22*, 183–197. [CrossRef] [PubMed]
14. Tan, X.; Zhang, D.F.; Duan, Z.P.; Parajuli, K.; Hu, J.Y. Effects of light color on interspecific competition between *Microcystis aeruginosa* and *Chlorella pyrenoidosa* in batch experiment. *Environ. Sci. Pollut. Res.* **2019**, *27*, 344–352. [CrossRef] [PubMed]
15. Häder, D.P.; Williamson, C.E.; Wängberg, S.Å.; Rautio, M.; Rose, K.C.; Gao, K.S.; Helbling, E.W.; Sinha, R.P.; Worrest, R. Effects of UV radiation on aquatic ecosystems and interactions with other environmental factors. *Photochem. Photobiol. Sci. B Biol.* **2015**, *14*, 108–126. [CrossRef]
16. Carrillo, P.; Medina-Sánchez, J.M.; Herrera, G.; Durán, C.; Segovia, M.; Cortés, D.; Salles, S.; Korbee, N.; Figueroa, F.L.; Mercado, J.M. Interactive effect of UVR and phosphorus on the coastal phytoplankton community of the Western Mediterranean Sea: Unravelling eco-physiological mechanisms. *PLoS ONE* **2015**, *10*, e0142987. [CrossRef] [PubMed]
17. Beardall, J.; Stojkovic, S.; Gao, K.S. Interactive effects of nutrient supply and other environmental factors on the sensitivity of marine primary producers to ultraviolet radiation: Implications for the impacts of global change. *Aquat. Biol.* **2014**, *22*, 5–23. [CrossRef]
18. Xu, Z.G.; Gao, G.; Tu, B.; Qiao, H.J.; Ge, H.M.; Wu, H.Y. Physiological response of the toxic and non-toxic strains of a bloom-forming cyanobacterium *Microcystis aeruginosa* to changing ultraviolet radiation regimes. *Hydrobiologia* **2019**, *833*, 143–156. [CrossRef]
19. Ma, Z.L.; Fang, T.X.; Thring, R.W.; Li, Y.B.; Yu, H.G.; Zhou, Q.; Zhao, M. Toxic and non-toxic strains of *Microcystis aeruginosa* induce temperature dependent allelopathy toward growth and photosynthesis of *Chlorella vulgaris*. *Harmful Algae* **2015**, *48*, 21–29. [CrossRef]
20. Zhang, Y.L.; Yin, Y.; Zhang, E.L.; Zhu, G.W.; Liu, M.L.; Feng, L.Q.; Qin, B.Q.; Liu, X.H. Spectral attenuation of ultraviolet and visible radiation in lakes in the Yunnan Plateau, and the middle and lower reaches of the Yangtze River, China. *Photochem. Photobiol. Sci.* **2011**, *10*, 469–482. [CrossRef]
21. Cai, L.L.; Zhu, G.W.; Zhu, M.Y.; Yang, G.J. Succession of phytoplankton structure and its relationship with algae bloom in littoral zone of Meiliang Bay, Taihu Lake. *Ecol. Sci.* **2012**, *31*, 345–351. (In Chinese)
22. Xu, H.; McCarthy, M.J.; Paerl, H.W.; Brookes, J.D.; Zhu, G.; Hall, N.S.; Qin, B.; Zhang, Y.; Zhu, M.; Hampel, J.J.; et al. Contributions of external nutrient loading and internal cycling to cyanobacterial bloom dynamics in Lake Taihu, China: Implications for nutrient management. *Limnol. Oceanogr.* **2021**, *9999*, 1492–1509. [CrossRef]
23. Chen, Y.W.; Qin, B.Q.; Teubner, K.; Dokulil, M.T. Long-term dynamics of phytoplankton assemblages: Microcystis-domination in Lake Taihu, a large shallow lake in China. *J. Plankton Res.* **2003**, *25*, 445–453. [CrossRef]
24. Bittencourt-Oliveira, M.C.; Chia, M.A.; Oliveira, H.S.B.; Araujo, M.K.C.; Molica, R.J.R.; Dias, C.T.S. Allelopathic interactions between microcystin-producing and non-microcystin-producing cyanobacteria and green microalgae: Implications for microcystins production. *J. Appl. Phycol.* **2015**, *27*, 275–284. [CrossRef]
25. Yang, J.; Deng, X.; Xian, Q.; Qian, X.; Li, A. Allelopathic effect of *Microcystis aeruginosa* on *Microcystis wesenbergii: Microcystin*-LR as a potential allelochemical. *Hydrobiologia* **2014**, *727*, 65–73. [CrossRef]
26. Ren, L.X.; Wang, P.F.; Wang, C.; Paerl, H.W.; Wang, H.Y. Effects of phosphorus availability and the phosphorus utilization behavior by *Microcystis aeruginosa* on its adaptation capability to ultraviolet radiation. *Environ. Pollut.* **2020**, *256*, 113441. [CrossRef]
27. Li, Z.K.; Dai, G.Z.; Juneau, P.; Qiu, B.S.; Posewitz, M. Different physiological responses of cyanobacteria to ultraviolet-b radiation under iron-replete and iron-deficient conditions: Implications for underestimating the negative effects of UV-B radiation. *J. Phycol.* **2017**, *53*, 425–436. [CrossRef]
28. Yang, Z.; Kong, F.X.; Shi, X.L.; Yu, Y.; Zhang, M. UV-B radiation and phosphorus limitation interact to affect the growth, pigment content, and photosynthesis of the toxic cyanobacterium *Microcystis aeruginosa*. *J. Appl. Phycol.* **2014**, *26*, 1669–1674. [CrossRef]
29. Zheng, Y.Q.; Gao, K.S. Impacts of solar UV radiation on the photosynthesis, growth, and UV-absorbing compounds in *Gracilaria lemaneiformis* (Rhodophyta) grown at different nitrate concentrations. *J. Phycol.* **2009**, *45*, 314–323. [CrossRef]
30. Jiang, H.B.; Qiu, B.S. Inhibition of photosynthesis by UV-B exposure and its repair in the bloom-forming cyanobacterium *Microcystis aeruginosa*. *J. Appl. Phycol.* **2011**, *23*, 691–696. [CrossRef]
31. Stanier, R.Y.; Kunisawa, R.; Mandel, M.; Cohenbazire, G. Purification and properties of unicellular blue-green algae (order *Chroococcales*). *Bacteriol. Rev.* **1971**, *35*, 171–205. [CrossRef] [PubMed]
32. Ye, W.J.; Liu, X.L.; Tan, J.; Li, D.T.; Yang, H. Diversity and dynamics of microcystin-producing cyanobacteria in China's third largest lake, Lake Taihu. *Harmful Algae* **2009**, *8*, 637–644. [CrossRef]
33. Duan, Z.P.; Tan, X.; Li, N.G. Ultrasonic selectivity on depressing photosynthesis of cyanobacteria and green algae probed by chlorophyll-a fluorescence transient. *Water Sci. Technol.* **2017**, *76*, 2085–2094. [CrossRef] [PubMed]
34. Shen, H.; Song, L.R. Comparative studies on physiological responses to phosphorus in two phenotypes of bloom-forming *Microcystis*. *Hydrobiologia* **2007**, *592*, 475–486. [CrossRef]

35. Helbling, E.W.; Carrillo, P.; Medinasánchez, J.M.; Durán, C.; Herrera, G.; Villarargaiz, M.; Villafa, V.E. Interactive effects of vertical mixing, nutrients and ultraviolet radiation: In situ photosynthetic responses of phytoplankton from high mountain lakes in Southern Europe. *Biogeosciences* **2013**, *10*, 1037–1050. [CrossRef]
36. Wang, Z.C.; Li, D.H.; Cao, X.Y.; Song, C.L.; Zhou, Y.Y. Photosynthesis regulates succession of toxic and nontoxic strains in blooms of Microcystis (Cyanobacteria). *Phycologia* **2015**, *54*, 640–648. [CrossRef]
37. Chen, J.J.; Yeh, H.H. The mechanisms of potassium permanganate on algae removal. *Water Res.* **2005**, *39*, 4420–4428. [CrossRef]
38. Hou, J.; Yang, Z.J.; Wang, P.F.; Wang, C.; Yang, Y.; Wang, X. Changes in *Microcystis aeruginosa* cell integrity and variation in microcystin-LR and proteins during Tanfloc flocculation and floc storage. *Sci. Total Environ.* **2018**, *626*, 264–273. [CrossRef]
39. Gao, L.; Pan, X.L.; Zhang, D.Y.; Mu, S.Y.; Lee, D.J.; Halik, U. Extracellular polymeric substances buffer against the biocidal effect of H_2O_2 on the bloom-forming cyanobacterium *Microcystis aeruginosa*. *Water Res.* **2015**, *69*, 51–58. [CrossRef]
40. Yang, Y.Y.; Hou, J.; Wang, P.F.; Wang, C.; Wang, X.; You, G.X. Influence of extracellular polymeric substances on cell-NPs heteroaggregation process and toxicity of cerium dioxide NPs to *Microcystis aeruginosa*. *Environ. Pollut.* **2018**, *242*, 1206–1216. [CrossRef]
41. Xu, H.; Cai, H.; Yu, G.; Jiang, H. Insights into extracellular polymeric substances of cyanobacterium *Microcystis aeruginosa* using fractionation procedure and parallel factor analysis. *Water Res.* **2013**, *47*, 2005–2014. [CrossRef] [PubMed]
42. Li, M.; Zhu, W.; Gao, L.; Lu, L. Changes in extracellular polysaccharide content and morphology of *Microcystis aeruginosa* at different specific growth rates. *J. Appl. Phycol.* **2013**, *25*, 1023–1030. [CrossRef]
43. Cai, X.; Hutchins, D.A.; Fu, F.X.; Gao, K.S. Effects of ultraviolet radiation on photosynthetic performance and N_2 fixation in Trichodesmium erythraeum IMS 101. *Biogeosciences Discuss.* **2017**, *14*, 4455–4466. [CrossRef]
44. Qu, F.; Liang, H.; He, J.; Ma, J.; Wang, Z.; Yu, H.; Li, G. Characterization of dissolved extracellular organic matter (dEOM) and bound extracellular organic matter (bEOM) of Microcystis aeruginosa and their impacts on UF membrane fouling. *Water Res.* **2012**, *46*, 2881–2890. [CrossRef]
45. Six, C.; Thomas, J.C.; Garczarek, L.; Ostrowski, M.; Dufresne, A.; Blot, N.; Scanlan, D.J.; Partensky, F. Diversity and evolution of phycobilisomes in marine *Synechococcus* spp.: A comparative genomics study. *Genome Biol.* **2007**, *8*, R259. [CrossRef]
46. Häder, D.P.; Kumar, H.D.; Smith, R.C.; Worrest, R.C. Effects of solar UV radiation on aquatic ecosystems and interactions with climate change. *Photochem. Photobiol. Sci.* **2007**, *6*, 267–285. [CrossRef]
47. Hessen, D.O.; Frigstad, H.; Færøvig, P.J.; Wojewodzic, M.W.; Leu, E. UV radiation and its effects on P-uptake in arctic diatoms. *J. Exp. Mar. Biol. Ecol.* **2012**, *411*, 45–51. [CrossRef]
48. Wang, G.H.; Chen, K.; Chen, L.Z.; Hu, C.X.; Zhang, D.L.; Liu, Y.D. The involvement of the antioxidant system in protection of desert cyanobacterium *Nostoc* sp. against UV-B radiation and the effects of exogenous antioxidants. *Ecotoxicol. Environ. Saf.* **2008**, *69*, 150–157. [CrossRef]
49. Zhu, Z.; Fu, F.X.; Qu, P.P.; Mak, E.W.K.; Jiang, H.B.; Zhang, R.F.; Zhu, Z.Y.; Gao, K.S.; Hutchins, D.A. Interactions between ultraviolet radiation exposure and phosphorus limitation in the marine nitrogen-fixing cyanobacteria *Trichodesmium* and *Crocosphaera*. *Limnol. Oceanogr.* **2020**, *65*, 363–376. [CrossRef]
50. Jiang, H.B.; Qiu, H.S. Photosynthetic adaption of a bloom-forming cyanobacterium *Microcystis aeruginosa* (Cyanophyceae) to prolonged UV-B exposure. *J. Phycol.* **2005**, *41*, 983–992. [CrossRef]
51. Helbling, E.W.; Buma, A.G.J.; Boelen, P.; Strate, H.J.; Giordanino, M.V.F.; Villafane, V.E. Increase in Rubisco activity and gene expression due to elevated temperature partially counteracts ultraviolet radiation–induced photoinhibition in the marine diatom *Thalassiosira weissflogii*. *Limnol. Oceanogr.* **2011**, *56*, 1330–1342. [CrossRef]
52. Xu, J.T.; Gao, K.S. UV-A enhanced growth and UV-B induced positive effects in the recovery of photochemical yield in Gracilaria lemaneiformis (Rhodophyta). *J. Photochem. Photobiol. B* **2010**, *100*, 117–122. [CrossRef] [PubMed]
53. Sakai, H.; Oguma, K.; Katayama, H.; Ohgaki, S. Effects of low- or medium-pressure ultraviolet lamp irradiation on *Microcystis aeruginosa* and *Anabaena variabilis*. *Water Res.* **2007**, *41*, 11–18. [CrossRef] [PubMed]
54. Li, P.; Liu, W.H.; Gao, K.S. Effects of temperature, pH, and UV radiation on alkaline phosphatase activity in the terrestrial cyanobacterium Nostoc flagelliforme. *J. Appl. Phycol.* **2013**, *25*, 1031–1038. [CrossRef]
55. Shelly, K.; Heraud, P.; Beardall, J. Interactive effect of PAR and UV-B radiation on PSII electron transport in the marine alga Dunaliella tertiolecta (Chlorophyceae). *J. Phycol.* **2010**, *39*, 509–512. [CrossRef]
56. Villafañe, V.E.; Cabrerizo, M.J.; Erzinger, G.S.; Bermejo, P.; Strauch, S.M.; Valiñas, M.S.; Helbling, E.W. Photosynthesis and growth of temperate and sub-tropical estuarine phytoplankton in a scenario of nutrient enrichment under solar ultraviolet radiation exposure. *Estuaries Coasts* **2017**, *40*, 842–855. [CrossRef]
57. Zhang, Y.; Jiang, H.B.; Qiu, B.S. Effects of UVB Radiation on competition between the bloom-forming cyanobacterium *Microcystis aeruginosa* and the chlorophyceae *Chlamydomonas microsphaera*. *J. Phycol.* **2013**, *49*, 318–328. [CrossRef]
58. Ling, F.; Hamzeh, M.; Dodard, S.; Zhao, Y.H.; Sunahara, G.I. Effects of TiO_2 nanoparticles on ROS production and growth inhibition using freshwater green algae pre-exposed to UV irradiation. *Environ. Toxicol. Pharmacol.* **2015**, *39*, 1074–1080.
59. Latifi, A.; Ruiz, M.; Zhang, C.C. Oxidative stress in cyanobacteria. *FEMS Microbiol Rev.* **2009**, *33*, 258–278. [CrossRef]
60. Li, Z.K.; Dai, G.Z.; Juneau, P.; Qiu, B.S. Capsular polysaccharides facilitate enhanced iron acquisition by the colonial cyanobacterium *Mcrocystis* sp. isolated from a freshwater lake. *J. Phycol.* **2016**, *52*, 105–115.
61. He, Y.Y.; Häder, D.P. Reactive oxygen species and UV-B: Effect on cyanobacteria. *Photochem. Photobiol. Sci.* **2002**, *1*, 729–736. [CrossRef]

62. Paerl, H.W.; Jane, T.; Bl, P.T. Carotenoid enhancement and its role in maintaining blue-green algal (*Microcystis aeruginosa*) surface blooms. *Limnol. Oceanogr.* **1983**, *28*, 847–857. [CrossRef]
63. Qin, H.J.; Li, S.S.; Li, D.H. Differential responses of different phenotypes of *Microcystis* (Cyanophyceae) to UV-B radiation. *Phycologia* **2015**, *54*, 118–129. [CrossRef]
64. Briand, E.; Bormans, M.; Quiblier, C.; Salençon, M.J.; Humbert, J.F. Evidence of the cost of the production of microcystins by *Microcystis aeruginosa* under differing light and nitrate environmental conditions. *PLoS ONE* **2012**, *7*, e29981. [CrossRef]
65. Janssen, E.M.L.; Erickson, P.R.; Mcneill, K. Dual roles of dissolved organic matter as sensitizer and quencher in the photooxidation of tryptophan. *Environ. Sci. Technol.* **2014**, *48*, 4916–4924. [CrossRef] [PubMed]
66. Liu, X.M.; Sheng, G.P.; Luo, H.W.; Zhang, F.; Yuan, S.J.; Juan, X.U.; Zeng, R.J.; Wu, J.G.; Yu, H.Q. Contribution of extracellular polymeric substances (EPS) to the sludge aggregation. *Environ. Sci. Technol.* **2010**, *44*, 4355–4360. [CrossRef]
67. Zhang, M.; Kong, F.X.; Tan, X.; Yang, Z.; Cao, H.S.; Xing, P. Biochemical, morphological, and genetic variations in *Microcystis aeruginosa* due to colony disaggregation. *World J. Microbiol. Biotechnol.* **2007**, *23*, 663–670. [CrossRef]
68. Sun, Y.; Chen, Y.; Wei, J.; Zhang, X.; Zhang, L.; Yang, Z.; Huang, Y. Ultraviolet-B radiation stress alters the competitive outcome of algae: Based on analyzing population dynamics and photosynthesis. *Chemosphere* **2021**, *272*, 129645. [CrossRef]
69. García-Gómez, C.; Parages, M.L.; Jiménez, C.; Palma, A.; Mata, M.T.; Segovia, M. Cell survival after UV radiation stress in the unicellular chlorophyte Dunaliella tertiolecta is mediated by DNA repair and MAPK phosphorylation. *J. Exp. Bot.* **2012**, *63*, 5259–5274. [CrossRef]
70. Zilliges, Y.; Kehr, J.C.; Meissner, S.; Ishida, K.; Mikkat, S.; Hagemann, M.; Kaplan, A.; Börner, T.; Dittmann, E. The cyanobacterial hepatotoxin microcystin binds to proteins and increases the fitness of *Microcystis* under oxidative stress conditions. *PLoS ONE* **2011**, *6*, e17615. [CrossRef]
71. Shelly, K.; Roberts, S.; Heraud, P.; Beardall, J. Interactions between UV-B exposure and phosphorus nutrition: I. Effects on growth, phosphate uptake, and chlorophyll fluorescence. *J. Phycol.* **2005**, *41*, 1204–1211. [CrossRef]
72. Boucar, M.; Shen, L.Q.; Wang, K.; Zhang, Z.C.; Qiu, B.S. UV-B irradiation enhances the production of unique mycosporine-like amino acids and carotenoids in the subaerial cyanobacterium *Pseudanabaena* sp. CCNU1. *Eur. J. Phycol.* **2021**, *56*, 316–323. [CrossRef]
73. Häder, D.P.; Sinha, R.P. Solar ultraviolet radiation-induced DNA damage in aquatic organisms: Potential environmental impact. *Mutat. Res.* **2005**, *571*, 221–233. [CrossRef] [PubMed]
74. Ren, L.X.; Wang, P.F.; Wang, C.; Chen, J.; Hou, J.; Qian, J. Algal growth and utilization of phosphorus studied by combined mono-culture and co-culture experiments. *Environ. Pollut.* **2017**, *220*, 274–285. [CrossRef]
75. Rzymski, P.; Poniedzialek, B.; Kokocinski, M.; Jurczak, T.; Lipski, D.; Wiktorowicz, K. Interspecific allelopathy in cyanobacteria: Cylindrospermopsin and *Cylindrospermopsis raciborskii* effect on the growth and metabolism of *Microcystis aeruginosa*. *Harmful Algae* **2014**, *35*, 1–8. [CrossRef]
76. Mário, U.G.B.; Alan, E.W.; João, I.R.L.; Silvano, P.P.; Riley, P.B.; Edna, G.F.; José, C. Environmental factors associated with toxic cyanobacterial blooms across 20 drinking water reservoirs in a semi-arid region of Brazil. *Harmful Algae* **2019**, *86*, 128–137.
77. Sun, Y.; Zhang, X.; Zhang, L.; Huang, Y.; Montagnes, D. UVB radiation suppresses antigrazer morphological defense in *Scenedesmus obliquus* by inhibiting algal growth and carbohydrate-regulated gene expression. *Environ. Sci. Technol.* **2020**, *54*, 4495–4503. [CrossRef] [PubMed]
78. He, Y.Y.; Häder, D.P. Involvement of reactive oxygen species in the UV-B damage to the cyanobacterium *Anabaena* sp. *J. Photoch. Photobiol. B* **2002**, *66*, 73–80. [CrossRef]
79. Takaichi, S.; Mochimaru, M. Carotenoids and carotenogenesis in cyanobacteria: Unique ketocarotenoids and carotenoid glycosides. *Cell. Mol. Life Sci.* **2007**, *64*, 2607–2619. [CrossRef]
80. Lüder, U.H.; Knoetzel, J.; Wiencke, C. Acclimation of photosynthesis and pigments to seasonally changing light conditions in the endemic Antarctic red macroalga *Palmaria decipiens*. *Polar Biol.* **2001**, *24*, 598–603. [CrossRef]

International Journal of
Environmental Research and Public Health

MDPI

Article

Transcriptomics Analysis of the Toxicological Impact of Enrofloxacin in an Aquatic Environment on the Chinese Mitten Crab (*Eriocheir sinensis*)

Qiaona Wang [1,2], Ziling Xu [1,2], Ying Wang [1,2], Guangming Huo [2], Xing Zhang [3], Jianmei Li [3], Chun Hua [2], Shengjie Li [1,2,*] and Feng Zhou [2,*]

1 School of Life Science, Nanjing Normal University, Nanjing 210023, China
2 School of Food Science, Nanjing Xiaozhuang University, Nanjing 211171, China
3 School of Food Science and Pharmaceutical Engineering, Nanjing Normal University, Nanjing 210023, China
* Correspondence: lishengjie@njxzc.edu.cn (S.L.); zfibcas@163.com (F.Z.)

Abstract: Enrofloxacin is an important antimicrobial drug that is widely used in aquaculture. Enrofloxacin residues can have negative effects on aquatic environments and animals. The toxicological effects of different concentrations of enrofloxacin residues in cultured water on Chinese mitten crabs (*Eriocheir sinensis*) were compared. A histological analysis of the *E. sinensis* hepatopancreas demonstrated that the hepatopancreas was damaged by the different enrofloxacin residue concentrations. The hepatopancreas transcriptome results revealed that 1245 genes were upregulated and that 1298 genes were downregulated in the low-concentration enrofloxacin residue group. In the high-concentration enrofloxacin residue group, 380 genes were upregulated, and 529 genes were downregulated. The enrofloxacin residues led to differentially expressed genes related to the immune system and metabolic processes in the hepatopancreas of the Chinese mitten crab, such as the genes for alkaline phosphatase, NF-kappa B inhibitor alpha, alpha-amylase, and beta-galactosidase-like. The gene ontology terms "biological process" and "molecular function" were enriched in the carboxylic acid metabolic process, DNA replication, the synthesis of RNA primers, the transmembrane transporter activity, the hydrolase activity, and the oxidoreductase activity. A Kyoto Encyclopedia of Genes and Genomes pathway analysis determined that the immune and metabolic signal transduction pathways were significantly enriched. Furthermore, the nonspecific immune enzyme (alkaline phosphatase) and the metabolic enzyme system played a role in the enrofloxacin metabolism in the *E. sinensis* hepatopancreas. These findings helped us to further understand the basis of the toxicological effects of enrofloxacin residues on river crabs and provided valuable information for the better utilization of enrofloxacin in aquatic water environments.

Keywords: enrofloxacin; *Eriocheir sinensis*; transcriptome

Citation: Wang, Q.; Xu, Z.; Wang, Y.; Huo, G.; Zhang, X.; Li, J.; Hua, C.; Li, S.; Zhou, F. Transcriptomics Analysis of the Toxicological Impact of Enrofloxacin in an Aquatic Environment on the Chinese Mitten Crab (*Eriocheir sinensis*). *Int. J. Environ. Res. Public Health* **2023**, *20*, 1836. https://doi.org/10.3390/ijerph20031836

Academic Editors: Xun Wang, Zhiyuan Wang and Xin Zhao

Received: 19 October 2022
Revised: 7 January 2023
Accepted: 8 January 2023
Published: 19 January 2023

1. Introduction

The Chinese mitten crab (*Eriocheir sinensis*) is a commonly farmed crustacean species typically found in benthic aquatic environments [1]. The Yangtze River Delta is a primary breeding area for crabs in China [2]. Generally, farmers adopt intensive and high-density farming methods to obtain high returns and profits. This breeding practice leads to a gradual deterioration in water quality and increases the incidence of infectious diseases [3]. To prevent disease outbreaks during crab farming, large amounts of antibiotics are used and, as such, are released into the natural water environment.

Enrofloxacin is a third-generation fluoroquinolone antibacterial drug [4,5] which is often used to treat bacterial infections in crab breeding operations in China because of its wide antibacterial spectrum and high potency [6,7]. Although enrofloxacin is approved for aquaculture in China [8], when used in large quantities, the antimicrobial remains in the aquaculture water and sediment accumulates in aquatic animals. Pharmacokinetic studies on enrofloxacin in crustaceans, including the *Eriocheir sinensis* [9–11], Atlantic horseshoe crab [12], and giant mud crab [13], have been reported. Enrofloxacin can be biotransformed in vivo into its major metabolite, ciprofloxacin, in crabs. More importantly, Roca et al. reported that quinolone residues are not degraded during processing and that their presence in food poses a risk to human health [14]. The risk of enrofloxacin and ciprofloxacin residues in aquatic products and the aquatic environment has become an important issue that has attracted increasing attention.

Many recent studies on aquaculture animals have focused on the residual characteristics of enrofloxacin and ciprofloxacin; the rate of biotransformation; and the health risks of consuming residual enrofloxacin and ciprofloxacin in *Procambarus clarkia* [15], *Exopalaemon carinicauda* [16], and *Pangasianodon hypophthalmus* [17]. Su et al. reported that enrofloxacin has significant effects on the gene expression of the cytochrome P450 3 (CYP3), glutathione S-transferases (GST), and phosphoenolpyruvate carboxykinase (PEPCK) in crabs [18], which are key molecules that affect the metabolism, immunity, and antitumor activity of organisms [19–22]. However, there are still few enrofloxacin toxicology reports on Eriocheir sinensis in cultured water.

In this study, we built a microcosm that authentically simulated aquaculture conditions and used it as a research tool to determine the mechanisms underlying the toxicological effects exerted on crabs through exposure to different concentrations of enrofloxacin. In crustaceans, the hepatopancreas has multiple functions, including nutrient absorption and metabolization, the storage of minerals and energy reserves, the synthesis of lipoproteins, the detoxification of heavy metals, and the excretion of uric acid [23]. In addition, the hepatopancreas is an important organ and is the primary site for the synthesis, excretion, and regulation of immune and metabolic molecules [24]. Su et al. demonstrated that enrofloxacin regulated immunity- and metabolism-related gene expression, such as that of CYP3, GST, and PEPCK [18]. Here, we continued to explore the effect of enrofloxacin on the hepatopancreas by using RNA sequencing. Thus, we collected the hepatopancreatic tissue from *E. sinensis* following 10 days of exposure to two different concentrations of enrofloxacin. Histomorphological observations and transcriptome analyses revealed the toxicological effects of enrofloxacin as well as the mechanisms underlying these effects.

We further analyzed the toxicological effects of different concentrations of enrofloxacin on *E. sinensis* using hepatopancreas transcriptome analyses. These results aided in understanding their effect on the immune system and metabolic process disorders of *Eriocheir sinensis* and laid the foundation for further research on *Eriocheir sinensis* by developing an understanding of the defense mechanism essential for maintaining healthy mitten crabs in aquaculture.

2. Materials and Methods

2.1. Experimental Design and Sampling

To authentically simulate the culture environment of mitten crabs, three Chinese mitten crab culture drums with the same conditions were selected. Each drum had a diameter of 2 m, height of 1.2 m, and water depth of approximately 1 m. Twelve Chinese mitten crabs of similar sizes and weights were placed in each barrel. The amounts of enrofloxacin, 1.875 and 3.750 g, were determined according to the national drug standard, and the enrofloxacin was evenly sprinkled into two different Chinese mitten crab culture drums. After standing for 24 h, the concentration of enrofloxacin in the culture water was 0.63 mg/L in the low-concentration enrofloxacin residue group and 1.25 mg/L in the high-concentration enrofloxacin residue group. After 10 days of sprinkling, crabs were anesthetized on ice and then sampled from the control, low-concentration, and high-concentration enrofloxacin residue groups. The drugs in the aquaculture water were not completely degraded by the time of sampling. The remaining hepatopancreas samples were stored at −20 °C.

2.2. Histopathological Analysis of the Hepatopancreas

Hepatopancreatic tissue samples were immediately fixed in 4% paraformaldehyde for 24 h, dehydrated in gradient concentration of ethanol, and embedded in paraffin wax. Using a microtome, 4–5 μm thick sections were obtained and then stained with hematoxylin and eosin (HE). Histopathological changes were observed under a Nikon 50i optical microscope (Nikon Corporation, Tokyo, Japan).

2.3. Total RNA Extraction and Sequencing

The transcriptome sequencing and analyses were conducted by Novogene Co., Ltd. (Beijing, China). Total RNA was extracted using a TRIzol® Reagent Kit (Invitrogen, California, USA) according to the manufacturer's protocol. The RNA quality and quantity were examined using 1% Tris–acetate (TAE) agarose gel electrophoresis. Equal quantities (0.5 μg) of RNA from *E. sinensis* hepatopancreas samples were separately pooled to eliminate sample variation and to create two main samples. The samples were used for RNA-seq library construction using the NEBNext® Ultra™ RNA Library Prep Kit for Illumina® (New England Biolabs, Ipswich, MA, USA) according to the protocol. The AMPure XP system (Beckman Coulter, Beverly, MA, USA) was used to purify library fragments for selecting complementary DNA (cDNA). After the library was constructed, a Qubit2.0 (Thermo Fisher Scientific, Waltham, MA, USA) Fluorometer was used for initial quantification, and the library was diluted to 1.5 ng/μL. The Agilent 2100 BioAnalyzer (Agilent Technologies, Palo Alto, CA, USA) was used to determine the insert size of the library. For ensuring the library quality, quantitative real-time PCR (qRT-PCR) was used to accurately quantify an effective library concentration of higher than 2 nM. Mixed DNA libraries were diluted to 4–5 pM for sequencing using an Illumina NovaSeq 6000 instrument (Illumina Inc, San Diego, CA, USA).

2.4. De Novo Transcriptome Assembly

We removed low-quality adapter sequences by filtering raw reads. The resulting clean reads were assembled to produce complete reference sequences using the Trinity program (v2.4.0; min_kmer_cov:3). Longer contigs were assembled until they could not be extended to either side. The unigenes were obtained by removing redundant transcripts. Using BLASTx (2.2.28+; threshold E-value=1×10^{-5}), the assembled transcripts were aligned with the following National Center for Biotechnology Information protein databases: nonredundant (NR), nucleotide sequence (NT), protein family (PFAM), gene ontology (GO), and protein sequence (Swiss-Prot). The best hits were used for the functional annotation of the unigenes. Blast2GO (b2g4pipe_v2.5, threshold E-value=1×10^{-6}) was used to obtain and analyze GO annotations for the uniquely assembled transcripts.

2.5. Differentially Expressed Gene (DEG) Analysis

DEGs between the control and residue groups were identified using the DESeq package (http://bioconductor.org/packages/release/bioc/html/DESeq.html, accessed on 30 October 2021). An absolute log2-fold change > 1 and FDR < 0.05 were used as thresholds to define DEGs. DEGs were then subjected to GO and Kyoto Encyclopedia of Genes and Genomes (KEGG) analyses.

2.6. Enzymatic Analysis

Acid phosphatase (ACP), alkaline phosphatase (AKP), glutathione sulfotransferase (GSH-ST), and acetylcholinesterase (AchE) activities were measured using commercial kits (Nanjing Jiancheng Bioengineering Institute, Nanjing, China) according to the manufacturer's instructions.

2.7. Statistical Analyses

All data were analyzed using SPSS statistics 20 (IBM Inc., Chicago, IL, USA) through one-way analysis of variance (ANOVA), and differences between groups were analyzed using Student's *t*-test. Statistical significance was set at $p < 0.05$.

3. Results

3.1. Enrofloxacin Residues Induced Hepatopancreas Injury in E. sinensis

The degree of hepatopancreatic damage was a direct reflection of the intensity of the toxicity of the external stimuli as observed with the aid of HE staining [25]. In the hepatopancreases of the Chinese mitten crabs in the control group, the basement membranes were complete and clear, and the nuclei were arranged in an orderly manner. The absorbing cells, alveolar cells, fibroblasts, and embryonic cells were clearly distinguished (Figure 1A). However, relative to the control group, in the low-concentration enrofloxacin residue group, the hepatopancreases demonstrated an enlarged space in their lumens and deformed basement membranes; the components in the cells were loosely arranged, the number of nuclei was significantly reduced, and this was accompanied by inflammatory cell infiltration (Figure 1B). In the high-concentration enrofloxacin residue group, the damage to the basement membranes was more severe, the internal structure of the cell membranes was deformed, the internal arrangement of the cells was disordered, the cell structure was lost, secretion in the hepatopancreatic duct cavity was increased, and hepatopancreatic duct atrophy was observed (Figure 1C).

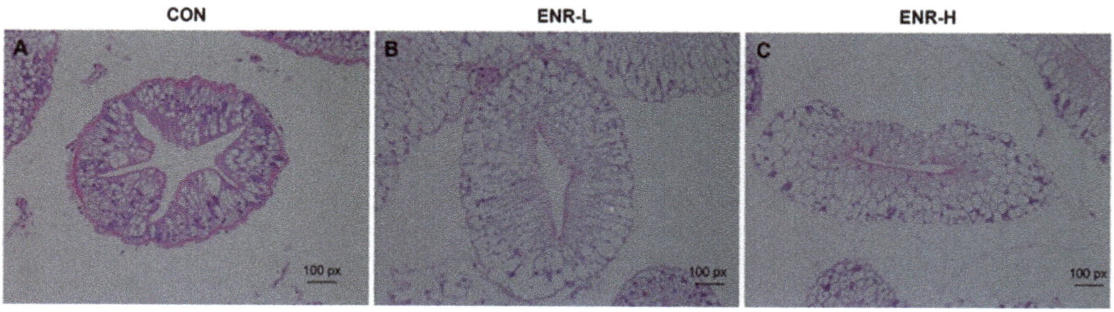

Figure 1. Hematoxylin and eosin staining of *E. sinensis* hepatopancreas specimens following enrofloxacin exposure at different doses. (**A**) CON represents the control group, (**B**) ENR-L represents the low-concentration enrofloxacin residue group, and (**C**) ENR-H represents the high-concentration enrofloxacin residue group.

3.2. Enrofloxacin Residues Led to Multiple Gene Expression Disorders in the Hepatopancreases of Crabs

To elucidate the molecular mechanism underlying the toxicological effects of enrofloxacin on crabs, a de novo assembled transcriptome analysis of the hepatopancreatic samples was performed. The data on the success rate of gene annotation were analyzed in seven databases, including NR, GO, KOG, KO, NT, SwissProt, and PFAM. The annotation success rate in NR was 24,604, accounting for 25.62%; that in GO was 26,538, accounting for 27.64%; that in KOG was 7494, accounting for 7.8%; that in KO was 9251, accounting for 9.63%; that in NT was 17,269, accounting for 17.98%; that in SwissProt was 14,294, accounting for 14.88%; and that in PFAM was 26,542, accounting for 27.64% (Table S1). Through the analysis of the transcriptomic data of the three groups of crab hepatopancreas samples, we obtained DEGs corresponding to the two doses of enrofloxacin. The DEG analysis of the RNA sequence revealed 1245 upregulated and 1298 downregulated genes in the hepatopancreases of the low-concentration enrofloxacin residue group relative to the control group (Figure 2A). Meanwhile, compared with the control group, the DEG analysis demonstrated 380 upregulated and 529 downregulated genes in the hepatopancreases of the high-concentration enrofloxacin residue group (Figure 2B).

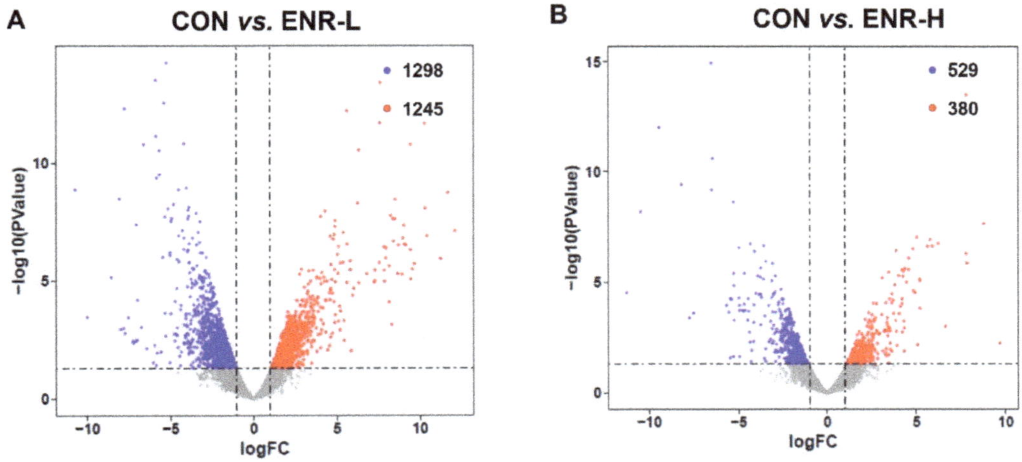

Figure 2. Volcano plot of the differences in the expression profiles of *E. sinensis* samples in the control and residue groups, respectively. CON represents control, (**A**) ENR-L represents low-concentration enrofloxacin residue group, and (**B**) ENR-H represents high-concentration enrofloxacin residue group. The x-axis represents logFC (fold change), while the y-axis represents −log10 (*p*-value). Red represents significantly upregulated genes, blue represents significantly downregulated genes, and each circle represents a single gene.

3.3. GO Analysis of DEGs Found Significant Enrichment of Biological Processes Related to Metabolism Process in Enrofloxacin Residue Groups

To investigate the function of these DEGs in the hepatopancreases of crabs exposed to enrofloxacin, the 2543 The DEGs corresponding to a low dose of enrofloxacin were entered into the GO database, and the results included the biological processes (BP), cellular components (CC), and molecular function (MF). The DEGs in the low-concentration enrofloxacin residue group, relative to the control group, demonstrated an obvious enrichment of the biological processes chiefly related to the metabolic processes, including the carbohydrate metabolic process (37 DEGs, $p = 0.000136$), tyrosine metabolic process (13 DEGs, $p = 0.001826$), and glycerolipid metabolic process (9 DEGs, $p = 0.006136$; Figure 3A; Table 1). Similarly, the 909 DEGs corresponding to a high dose of enrofloxacin were entered into the GO database, and an enrichment of the biological processes was again evident, including the carbohydrate metabolic process (17 DEGs, $p = 0.000478$), purine nucleobase metabolic process (13 DEGs, $p = 0.029346$), and tricarboxylic acid cycle (3 DEGs, $p = 0.008759$; Figure 3B; Table 1). It is worth noting that the high concentration of enrofloxacin residues led to DNA and RNA damage, and the obvious enrichment in the DEGs involved in translation, DNA-templates (12 DEGs, $p = 0.029357$), mRNA processing (8 DEGs, $p = 0.032291$), and so on (Figure 3B). Metabolic abnormalities were closely related to DNA and RNA damage as well as transcription and translation errors. DNA damage can impair metabolic organ functions and induce tissue inflammation, which disrupts the homeostasis of the systemic metabolism [26]. Moreover, the Venn diagram illustrates that, in the BP of the DEGs, the enrichment between the low- and high-concentration residue groups relative to the control group affected three common biological processes: the carboxylic acid metabolic process, carbohydrate metabolic process, and DNA replication and synthesis of RNA primers (Figure 3C). Therefore, these results illustrated that enrofloxacin residues may affect the hepatopancreatic metabolic processes of *E. sinensis*.

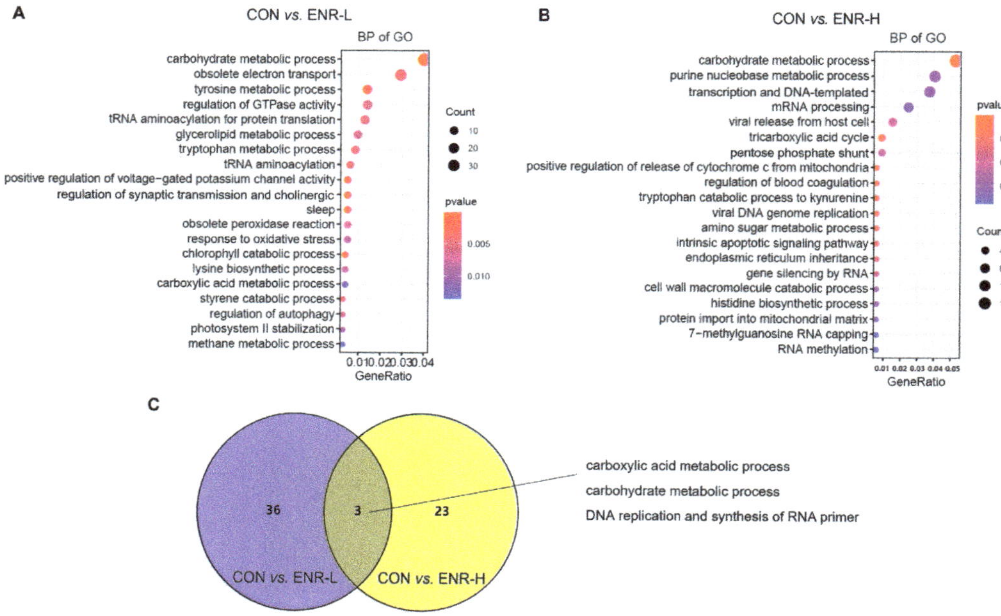

Figure 3. Gene ontology (GO) analysis of the differentially expressed genes (DEGs) in the *E. sinensis* hepatopancreases following different exposure doses of enrofloxacin. Colors represent *p*-values. The x-axis represents the gene ratio. The y-axis represents GO terms involved in biological processes. Venn diagrams showing number of biological processes in *E. sinensis* hepatopancreas samples following (**A**) low and (**B**) high enrofloxacin exposure doses. (**C**) Enrichment of DEGs of the three common biological processes affected by both low (blue) and high (yellow) enrofloxacin exposure doses.

3.4. Molecular Function in GO Analysis of DEGs

The molecular function in the GO analysis reflected the biological function that could be affected by the DEGs. Among the molecular functions, the transmembrane transporter activity was most significantly affected in the low-concentration enrofloxacin residue group relative to the control group (44 DEGs, $p = 0.000256$) followed by the hydrolase activity (34 DEGs, $p = 0.004765$), the adenosine triphosphate enzyme (ATPase) activity (27 DEGs, $p = 0.012104$), sequence-specific DNA binding (26 DEGs, $p = 0.002681$), heme binding (25 DEGs, $p = 2.19 \times 10^{-5}$), the hydrolase activity, hydrolyzing O-glycosyl compounds (22 DEGs, $p = 4.31 \times 10^{-6}$), iron ion binding (21 DEGs, $p = 0.001086$), chitin binding (18 DEGs, $p = 2.02 \times 10^{-5}$), the sulfuric ester hydrolase activity (15 DEGs, $p = 7.41 \times 10^{-9}$), and the aminoacyl-transfer RNA ligase activity (2 DEGs, $p = 0.036552$; Figure 4A). In the high-concentration enrofloxacin residue group, compared with the control, the transmembrane transporter activity was also the most significantly affected (25 DEGs, $p = 6.42 \times 10^{-6}$) followed by the hydrolase activity (19 DEGs, $p = 0.000301$), the oxidoreductase activity (19 DEGs, $p = 0.002669$), the structural molecule activity (17 DEGs, $p = 0.009975$), chitin binding (14 DEGs, $p = 2.81 \times 10^{-8}$), the catalytic activity (14 DEGs, $p = 0.022347$), the hydrolase activity and hydrolyzing O-glycosyl compounds (9 DEGs, $p = 0.048616$), and the RNA-directed 5′-3′ RNA polymerase activity (7 DEGs, $p = 0.004283$; Figure 4B). Notably, the molecular functions, including the sulfuric ester hydrolase activity, the hydrolase activity, hydrolyzing O-glycosyl compounds, chitin binding, the metallo-carboxypeptidase activity, the transmembrane transporter activity, the catalytic activity, the 1-alkyl-2-acetylglycerophosphocholine esterase activity, the oxidoreductase activity, and the phosphatase activity, occurred repeatedly in both the low- and high-concentration enrofloxacin residue groups (Figure 4C).

Table 1. The top 20 biological processes (BP) in gene ontology (GO) analysis of differentially expressed genes (DEGs).

CON [1] vs. ENR-L [2]			CON vs. ENR-H [3]		
Term	Count	p Value	Term	Count	p Value
carbohydrate metabolic process	37	0.000136285	carbohydrate metabolic process	17	0.000478166
obsolete electron transport	27	0.004245120	purine nucleobase metabolic process	13	0.029346042
tyrosine metabolic process	13	0.001825573	transcription and DNA-templated	12	0.029356640
regulation of GTPase activity	13	0.005738756	mRNA processing	8	0.032290906
tRNA aminoacylation for protein translation	12	0.005255766	viral release from host cell	5	0.020254351
glycerolipid metabolic process	9	0.006135912	tricarboxylic acid cycle	3	0.008758657
tryptophan metabolic process	8	0.004595175	pentose phosphate shunt	3	0.021326841
tRNA aminoacylation	6	0.004389519	regulation of blood coagulation	2	0.004063707
positive regulation of voltage-gated potassium channel activity	5	0.001854393	positive regulation of release of cytochrome c from mitochondria	2	0.004063707
regulation of synaptic transmission and cholinergic	5	0.001854393	tryptophan catabolic process to kynurenine	2	0.006012747
sleep	5	0.001854393	viral DNA genome replication	2	0.006012747
obsolete peroxidase reaction	5	0.006743223	amino sugar metabolic process	2	0.008303677
response to oxidative stress	5	0.008306801	intrinsic apoptotic signaling pathway	2	0.010921666
chlorophyll catabolic process	4	0.000717921	endoplasmic reticulum inheritance	2	0.013852347
lysine biosynthetic process	4	0.007702752	gene silencing by RNA	2	0.017081801
carboxylic acid metabolic process	4	0.012737180	cell wall macromolecule catabolic process	2	0.028430117
styrene catabolic process	3	0.003663577	histidine biosynthetic process	2	0.028430117
regulation of autophagy	3	0.006128731	protein import into mitochondrial matrix	2	0.032724060
photosystem II stabilization	3	0.009375217	7-methylguanosine RNA capping	2	0.037253517
methane metabolic process	3	0.013447069	RNA methylation	2	0.037253517

[1] Control group, [2] low-concentration enrofloxacin residue group, and [3] high-concentration enrofloxacin residue group.

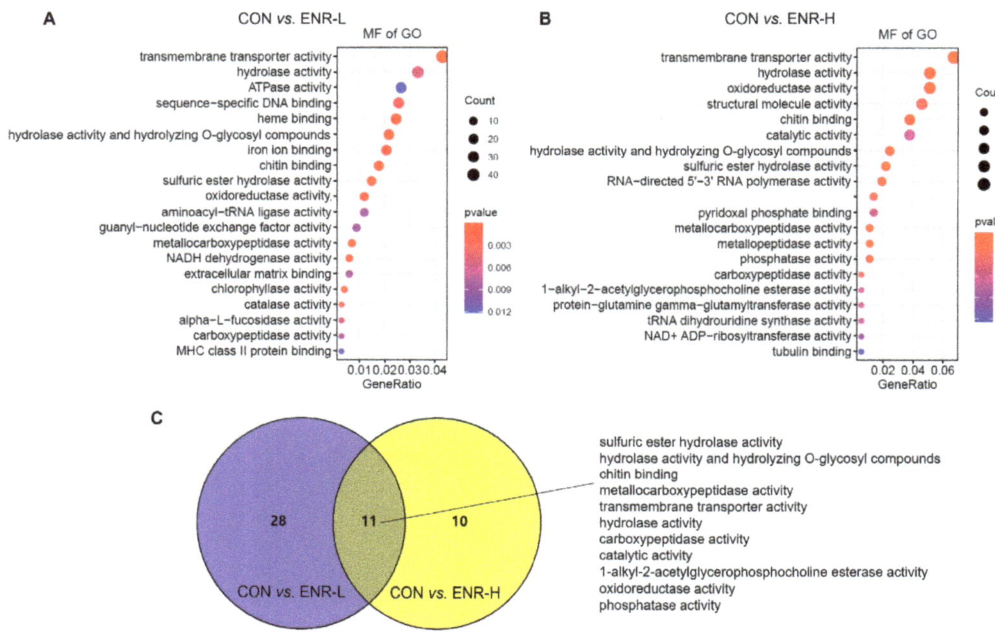

Figure 4. Gene ontology (GO) analysis of the DEGs in *E. sinensis* hepatopancreas samples following different exposure doses of enrofloxacin. Colors represent *p*-values. The x-axis represents the gene ratio. The y-axis represents GO terms involved in molecular functions. Venn diagrams showing number of molecular functions in *E. sinensis* hepatopancreas samples following (**A**) low and (**B**) high exposure doses. (**C**) Enrichment of DEGs of the common molecular functions affected by both low (blue) and high (yellow) enrofloxacin exposure doses.

3.5. Results of KEGG Analysis of DEGs

The abundant signaling pathway information in the KEGG database helped elucidate the system-level biological functions, such as the metabolic and inflammatory pathways, oxidative stress, protein modification, and cell death, among others [27]. Relative to the control group, a KEGG enrichment analysis of the DEGs in the low- and high-concentration enrofloxacin residue groups was performed. The KEGG annotation analysis demonstrated that, following the low-dose administration of enrofloxacin, the DEGs were significantly enriched in multiple basic pathways; the abundantly significant pathways were related to the starch and sucrose metabolism, the lysosome metabolism, the sphingolipid metabolism, the two-component system, other glycan degradation, the galactose metabolism, nonalcoholic fatty liver disease, folate biosynthesis, the thiamine metabolism, and the tryptophan metabolism, among others (Figure 5A). In addition, in the high-concentration enrofloxacin administration group, relative to the control group, DEGs were significantly enriched in ribosome biogenesis in eukaryotes; the starch and sucrose metabolism; carbohydrate digestion and absorption; lysosome, pantothenate, and CoA biosynthesis; folate biosynthesis; the galactose metabolism; the thiamine metabolism; the nitrogen metabolism; and the fatty acid metabolism, among others (Figure 5B). The Venn diagram illustrates that 12 pathways were enriched by KEGG following exposure to low or high doses of enrofloxacin, including the starch and sucrose metabolism, the lysosome metabolism, the sphingolipid metabolism, the two-component system, the folate metabolism, the Toll and Imd signaling pathways, the thiamine metabolism, pantothenate and CoA biosynthesis, carbohydrate digestion and absorption, aminobenzoate degradation, and protein digestion and absorption (Figure 5C). Consistent with the GO analysis, the KEGG enrichment analysis also revealed that enrofloxacin exposure may affect the metabolic processes of the crab hepatopancreas.

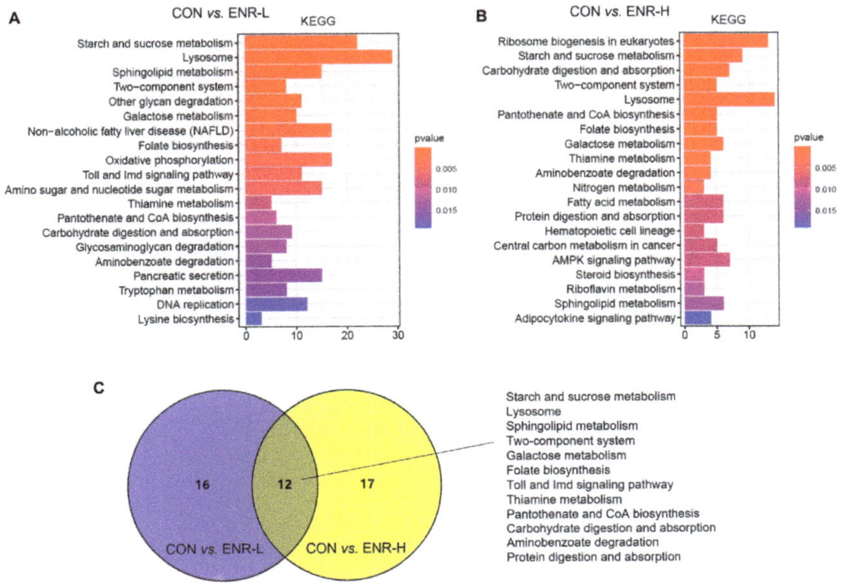

Figure 5. Kyoto Encyclopedia of Genes and Genomes (KEGG) pathway enrichment analysis of DEGs in *E. sinensis* hepatopancreas samples following different exposure doses of enrofloxacin. Colors represent *p*-values. The x-axis represents the number of genes annotated in a KEGG pathway. The y-axis represents the KEGG pathway categories. Venn diagrams showing categories of KEGG pathways in *E. sinensis* hepatopancreas samples following both (**A**) low and (**B**) high exposure doses. (**C**) Enrichment, following KEGG analysis, of DEGs of the system-level biological functions affected by both low (blue) and high (yellow) enrofloxacin exposure doses.

3.6. *Enrofloxacin Residues Led to Immune System and Metabolic Process Disorders in the Hepatopancreases of Chinese Mitten Crabs*

By analyzing the transcriptomic data of the three groups of crab hepatopancreas samples, we obtained DEGs corresponding to the two doses of enrofloxacin. The results showed that low-concentration enrofloxacin exposure produced 2543 DEGs between the control and experimental groups of which 1245 were upregulated and of which 1298 were downregulated (Figure 2A). In the meantime, high-concentration enrofloxacin exposure yielded a total of 909 DEGs between the control and experimental groups with 380 upregulated and 529 downregulated genes (Figure 2B). Compared with the control group, we found that the DEGs related to the immune system and metabolic processes showed significant changes in both the low- and high-concentration enrofloxacin residue groups. In the immune system, the DEGs, including those for alkaline phosphatase (AKP), dual oxidase 1, and nuclear factor-κB (NF-κB) inhibitor alpha, changed significantly of which NF-κB inhibitor alpha was upregulated in both the low- and high-concentration enrofloxacin residue groups, while the other DEGs were downregulated. In the metabolic processes, the DEGs changed significantly, such as those for venom phosphodiesterase 2-like, beta 1, 4-endoglucanase, alpha-amylase, arylsulfatase A-like, the ecdysteroid receptor (EcR) gene, beta-galactosidase-like, pantothenate kinase 3-like, carboxypeptidase B-like, trypsin-like serine proteinase, chitinase 3, and juvenile hormone esterase-like carboxylesterase 1, and all of them were downregulated in both the low- and high-concentration enrofloxacin residue groups (Table 2). Thus, these results demonstrated that immune system and metabolic process disorders may be key factors in *E. sinensis* hepatopancreatic damage.

Table 2. Summary of DEGs related to immune system and metabolic processes in the *E. sinensis* transcriptome.

	CON [1] vs. ENR-L [2]		CON vs. ENR-H [3]	
	logFC	*p* Value	logFC	*p* Value
Immune system				
alkaline phosphatase	−5.909012989	2.99×10^{-14}	−1.36058962	3.84×10^{-2}
dual oxidase 1	−3.065045953	3.06×10^{-3}	−2.19026803	1.42×10^{-2}
NF-kappa B inhibitor alpha	1.680036355	1.13×10^{-2}	1.668237232	2.98×10^{-2}
metabolic process				
venom phosphodiesterase 2-like	−2.154713643	1.46×10^{-4}	−1.70584579	5.71×10^{-3}
beta 1,4-endoglucanase	−4.572645684	6.59×10^{-3}	−2.81828784	1.40×10^{-2}
alpha-amylase	−1.803435554	1.97×10^{-2}	−1.50174728	1.70×10^{-2}
arylsulfatase A-like	−3.900096603	7.40×10^{-9}	−1.42018217	1.51×10^{-2}
ecdysteroid receptor (EcR) gene	−2.393264135	1.87×10^{-5}	−2.0989689	2.25×10^{-4}
beta-galactosidase-like	−2.452835189	4.92×10^{-5}	−1.93785202	8.92×10^{-4}
pantothenate kinase 3-like	−2.018241358	7.46×10^{-3}	−1.51740929	3.45×10^{-2}
carboxypeptidase B-like	−3.554127549	1.86×10^{-4}	−3.08832125	2.19×10^{-4}
trypsin-like serine proteinase	−4.213806456	4.03×10^{-3}	−3.9272743	5.16×10^{-6}
chitinase 3	−2.837019698	2.05×10^{-3}	−1.33310288	4.96×10^{-2}
juvenile hormone esterase-like carboxylesterase 1	−2.794416003	4.66×10^{-2}	−1.6048277	1.09×10^{-2}

[1] Control group, [2] low-concentration enrofloxacin residue group, and [3] high-concentration enrofloxacin residue group.

3.7. Immune Responses and Metabolic Enzymatic Activities following Exposure to Enrofloxacin Residues

Our analysis revealed that damage to the *E. sinensis* hepatopancreas was primarily associated with immune responses and metabolic processes. By analyzing DEGs, we found that enrofloxacin residues led to immune system and metabolic process disorders in the hepatopancreases of Chinese mitten crabs. The widely used indicators ACP and AKP are potential indicators for evaluating the impact of pollutants on the immune defense of biological organisms [28,29]. The other important indicators GSH-ST and AchE are indicators of the metabolic processes of the liver and hepatopancreas in animals [30,31]. Therefore, we selected those genes from the target DEGs for the qPCR analyses. The results demonstrated that the enzymatic activities of ACP and AKP in the hepatopancreases of the specimens were significantly decreased in both the low- and high-concentration enrofloxacin residue groups (Figure 6A). Notably, the results demonstrated that the enzymatic activities of GSH-ST and AchE were also significantly decreased in both the low- and high-concentration enrofloxacin residue groups (Figure 6B).

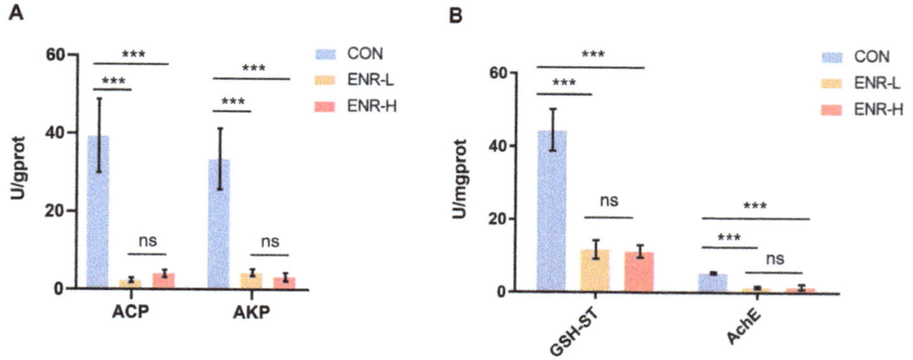

Figure 6. Effects of different enrofloxacin residues on enzyme activity in the hepatopancreases of *E. sinensis* specimens. CON represents the control group, ENR-L represents the low-concentration

enrofloxacin residue group, and ENR-H represents the high-concentration enrofloxacin residue group. *** $p < 0.001$. (**A**) Pollution impact indicators, acid phosphatase (ACP) and alkaline phosphatase (AKP), indicating the impact of low- and high-concentration enrofloxacin residues on the immune defense of *E. sinensis*. (**B**) Metabolic process indicators, glutathione sulfotransferase (GSH-ST) and acetylcholinesterase (AchE), indicating the impact of low- and high-concentration enrofloxacin residues on metabolic processes of the liver and pancreas in *E. sinensis*. The meaning of "ns" in figure is "no significance".

4. Discussion

Our results indicated that exposure to low or high doses of enrofloxacin resulted in hepatopancreatic damage in *E. sinensis*. The RNA sequencing results demonstrated that enrofloxacin-induced hepatopancreatic damage was closely related to disorders of the metabolic processes and the immune system.

Enrofloxacin, a third-generation fluoroquinolone antibacterial drug, is commonly used to treat bacterial infections in crab breeding; however, the side effects of this antimicrobial treatment deserve attention. For example, the expression level of key genes and enzymes in the hepatopancreases of the *E. sinensis* specimens treated with enrofloxacin were disturbed [18]. The enrofloxacin biotransformation product, ciprofloxacin, at low and high doses affected the expression of many genes in the hepatopancreases of the *E. sinensis* specimens, and these were primarily enriched in the metabolic processes and the immune system.

The metabolism involves a series of reactions that occur in living cells to sustain life [32]. Metabolic disorders can lead to the occurrence and development of various disorders and diseases in organisms that seriously affect the quality of life of these organisms and that may even threaten their lives. In addition, the metabolic disorders of crabs seriously affect the yield and environment of the aquaculture [33–35]. In eukaryotes, the metabolic processes are involved in various interconnected cellular pathways, molecular signaling pathways, and metabolic materials and products [36–38]. At the molecular level, metabolic changes depend on the configuration of the metabolic pathways, which are regulated by key metabolic enzymes, transcription factors, protein modifications, and the metabolite clearance status; many of these pathways are closely related to the mitochondrial and lysosomal digestive functions [39,40]. In this study, we determined that enrofloxacin exposure affected the metabolism of many substances, including carbohydrates, lipids, starch, and sucrose, as determined by the GO and KEGG enrichment analyses. Specifically, the metabolic processes, including the metabolism of carbohydrates, tyrosine, glycerolipids, tryptophan, methane, purine nucleobases, and amino sugars, were significantly altered following enrofloxacin exposure. Moreover, the functions of the mitochondria and lysosomes were also affected by enrofloxacin. Significant changes in the enrichment results relative to the biological processes (the response to oxidative stress, tricarboxylic acid cycle, and protein import into the mitochondrial matrix) and molecular functions of the GO analysis (the ATPase activity, NADH dehydrogenase activity, and oxidoreductase activity) and KEGG analysis (the lysosome metabolism, oxidative phosphorylation, pantothenate and CoA biosynthesis, and the fatty acid metabolism) were all involved in the resultant damage of the mitochondrial and lysosomal functions following enrofloxacin exposure.

The metabolic pathways are well recognized as important regulators of immune differentiation and activation [41–43]. Chen et al. reported that Ophiopogon japonicus increased the immune response in *E. sinensis*, inhibited the proliferation of the white spot syndrome virus (WSSV), and improved the survival of WSSV-challenged crabs [44]. Abamectin insecticides and anthelmintics inflict oxidative damage on aquatic animals and impair the immune defenses, which may further cause a sharp drop in the hemocyte counts in *E. sinensis* [29]. Therefore, immune homeostasis is crucial for *E. sinensis* survival. We determined that enrofloxacin exposure affected the biological processes, including the response to oxidative stress, regulation of autophagy, and viral release from the host cells, all of which were related to the homeostasis of the immune system [45–49]. Moreover, some DEGs related to the immune system showed significant changes in expression, including those of AKP, dual oxidase 1, and NF-κB inhibitor alpha, in the hepatopancreases of the *E. sinensis* specimens.

The widely used indicators ACP and AKP are potential indicators for evaluating the impact of pollutants on the immune defense of biological organisms [28]. Oxidative stress markers, GSH-ST, and AchE are important indicators that affect the metabolic processes of the liver and hepatopancreas in animals [50]. Overall, the ACP, AKP, GSH-ST, and AchE levels were clearly decreased following enrofloxacin exposure in the hepatopancreases of the *E. sinensis* specimens, confirming that enrofloxacin exposure affects the metabolism and immune response in the hepatopancreas of *E. sinensis*. However, the mechanisms by which enrofloxacin affects the metabolism and the immune system remain unclear and require further investigation.

5. Conclusions

In conclusion, this study not only provided novel evidence for the toxicological effects exerted on *E. sinensis* following enrofloxacin exposure but also helped elucidate the possible mechanisms underlying this toxicity and the corresponding cellular pathways that were activated, which notably involved the metabolic processes and immune responses.

Supplementary Materials: The following supporting information can be downloaded at: https://www.mdpi.com/article/10.3390/ijerph20031836/s1, Table S1: Statistics of gene annotation success rate.

Author Contributions: Conceptualization, S.L. and F.Z.; methodology, S.L., F.Z. and Q.W.; software, Q.W. and Z.X.; validation, Z.X. and Y.W.; formal analysis, Q.W., Z.X. and Y.W.; investigation, F.Z.; resources, C.H.; data curation, S.L.; writing—original draft preparation, S.L. and Q.W.; writing—review and editing, G.H., X.Z., J.L. and F.Z.; visualization, Z.X. and Y.W.; supervision, F.Z. and C.H.; project administration, S.L.; funding acquisition, S.L., C.H. and F.Z. All authors have read and agreed to the published version of the manuscript.

Funding: This research was funded by the National Key Research and Development Project (No. 2019YFC1605800), Key Subject of Ecology of the Jiangsu Province (SUJIAOYANHAN (2022) No. 2), Excellent Scientific and Technological Innovation Team of the Colleges and Universities of the Jiangsu Province (SUJIAOKE (2021) No. 1), and Key Subject of Biology of Nanjing (NINGJIAOGAOSHI (2021) No. 16).

Institutional Review Board Statement: Not applicable.

Informed Consent Statement: Not applicable.

Data Availability Statement: The data sets used and analyzed during the current study are available from the corresponding author upon reasonable request.

Conflicts of Interest: The authors declare no conflict of interest.

References

1. Song, C.; Li, L.; Zhang, C.; Qiu, L.; Fan, L.; Wu, W.; Meng, S.; Hu, G.; Chen, J.; Liu, Y.; et al. Dietary risk ranking for residual antibiotics in cultured aquatic products around Tai Lake, China. *Ecotoxicol. Environ. Saf.* **2017**, *144*, 252–257. [CrossRef] [PubMed]
2. Liangwei, X.; Zhen, L.; Keyi, M. Analysis of genetic diversity in cultured populations of the Chinese mitten crab (*Eriocheir sinensis*) by microsatellite markers. *J. Agric. Biotechnol.* **2012**, *35*, 171–176.
3. Fang, L.; Huang, Z.; Fan, L.; Hu, G.; Qiu, L.; Song, C.; Chen, J. Health risks associated with sulfonamide and quinolone residues in cultured Chinese mitten crab (*Eriocheir sinensis*) in China. *Mar. Pollut. Bull.* **2021**, *165*, 112184. [CrossRef] [PubMed]
4. Koc, F.; Uney, K.; Atamanalp, M.; Tumer, I.; Kaban, G. Pharmacokinetic disposition of enrofloxacin in brown trout (*Salmo trutta fario*) after oral and intravenous administrations. *Aquaculture* **2009**, *295*, 142–144. [CrossRef]
5. Darwish, A.M.; Farmer, B.D.; Hawke, J.P. Improved Method for Determining Antibiotic Susceptibility of *Flavobacterium columnareIsolates* by Broth Microdilution. *J. Aquat. Anim. Health* **2008**, *20*, 185–191. [CrossRef] [PubMed]
6. Martinez, M.; McDermott, P.; Walker, R. Pharmacology of the fluoroquinolones: A perspective for the use in domestic animals. *Vet. J.* **2006**, *172*, 10–28. [CrossRef]
7. Wu, G.; Meng, Y.; Zhu, X.; Huang, C. Pharmaco kinetics and tissue distribution of enrofloxacin and its metabolite ciprofloxacin in the Chinese mitten-handed crab, *Eriocheir sinensis*. *Anal. Biochem.* **2006**, *358*, 25–30. [CrossRef]

8. Liu, Y.-T.; Zhang, G.-D.; Sun, R.-Y.; Zhou, S.; Dong, J.; Yang, Y.-B.; Yang, Q.-H.; Xu, N.; Ai, X.-H. Determination of pharmacokinetic parameters and tissue distribution characters of enrofloxacin and its metabolite ciprofloxacin in *Procambarus clarkii* after two routes of administration. *Aquac. Rep.* **2022**, *22*, 100939. [CrossRef]
9. Zhang, J.; Wu, G.; Meng, Y.; Xu, X. Study on the pharmacokinetics of enrofloxacin in the Chinese mitten-handed crab, *Eriocheir sinensis*, after different administration regimes. *Aquac. Res.* **2008**, *39*, 1210–1215.
10. Tang, J.; Yang, X.-L.; Zheng, Z.-L.; Yu, W.-J.; Hu, K.; Yu, H.-J. Pharmacokinetics and the active metabolite of enrofloxacin in Chinese mitten-handed crab (*Eriocheir sinensis*). *Aquaculture* **2006**, *260*, 69–76. [CrossRef]
11. Su, H.; Wei, Y.; Sun, J.; Hu, K.; Yang, Z.; Zheng, R.; Yang, X. Effect of lactic acid on enrofloxacin pharmacokinetics in *Eriocheir sinensis*(Chinese mitten crab). *Aquac. Res.* **2019**, *50*, 1040–1046. [CrossRef]
12. Kirby, A.; Lewbart, G.A.; Hancock-Ronemus, A.; Papich, M.G. Pharmacokinetics of enrofloxacin and ciprofloxacin in Atlantic horseshoe crabs (*Limulus polyphemus*) after single injection. *J. Vet. Pharmacol. Ther.* **2018**, *41*, 349–353. [CrossRef] [PubMed]
13. Fang, W.-H.; Zhou, S.; Yu, H.-J.; Hu, L.-L.; Zhou, K.; Liang, S.-C. Phannacokinetics and tissue distribution of enrofloxacin and its metabolite ciprofloxacin in *Scylla serrata* following oral gavage at two salinities. *Aquaculture* **2007**, *272*, 180–187. [CrossRef]
14. Roca, M.; Castillo, M.; Marti, P.; Althaus, R.L.; Molina, M.P. Effect of Heating on the Stability of Quinolones in Milk. *J. Agric. Food Chem.* **2010**, *58*, 5427–5431. [CrossRef] [PubMed]
15. Liu, Y.; Ai, X.; Sun, R.; Yang, Y.; Zhou, S.; Dong, J.; Yang, Q. Residue, biotransformation, risk assessment and withdrawal time of enrofloxacin in red swamp crayfish (*Procambarus clarkii*). *Chemosphere* **2022**, *307*, 135657. [CrossRef] [PubMed]
16. Liang, J.P.; Li, J.; Li, J.T.; Liu, P.; Chang, Z.Q.; Nie, G.X. Accumulation and elimination of enrofloxacin and its metabolite ciprofloxacin in the ridgetail white prawn *Exopalaemon carinicauda* following medicated feed and bath administration. *J. Vet. Pharmacol. Ther.* **2014**, *37*, 508–514. [CrossRef]
17. Phu, P.M.; Douny, C.; Scippo, M.-L.; De Pauw, E.; Thinh, N.Q.; Huong, D.T.T.; Phuoc, V.H.; Phuong, N.T.; Dalsgaard, A. Elimination of enrofloxacin in striped catfish (*Pangasianodon hypophthalmus*) following on-farm treatment. *Aquaculture* **2015**, *438*, 1–5.
18. Su, H.; Sun, J.; Fang, S.; Wei, Y.; Zheng, R.; Jiang, Y.; Hu, K. Effects of lactic acid on drug-metabolizing enzymes in Chinese mitten crab (*Eriocheir sisnensis*) after oral enrofloxacin. *Comp. Biochem. Physiol. Part C Toxicol. Pharmacol.* **2019**, *223*, 9–14. [CrossRef]
19. Chatterjee, A.; Gupta, S. The multifaceted role of glutathione S-transferases in cancer. *Cancer Lett.* **2018**, *433*, 33–42. [CrossRef]
20. Blanchette, B.; Feng, X.; Singh, B.R. Marine Glutathione S-Transferases. *Mar. Biotechnol.* **2007**, *9*, 513–542. [CrossRef]
21. Yu, S.; Meng, S.; Xiang, M.; Ma, H. Phosphoenolpyruvate carboxykinase in cell metabolism: Roles and mechanisms beyond gluconeogenesis. *Mol. Metab.* **2021**, *53*, 101257. [CrossRef] [PubMed]
22. Nebert, D.W.; Wikvall, K.; Miller, W.L. Human cytochromes P450 in health and disease. Philosophical transactions of the Royal Society of London. *Ser. B Biol. Sci.* **2013**, *368*, 20120431. [CrossRef] [PubMed]
23. Vogt, G. Functional cytology of the hepatopancreas of decapod crustaceans. *J. Morphol.* **2019**, *280*, 1405–1444. [CrossRef] [PubMed]
24. Ried, C.; Wahl, C.; Miethke, T.; Wellnhofer, G.; Landgraf, C.; Schneider-Mergener, J.; Hoess, A. High Affinity Endotoxin-binding and Neutralizing Peptides Based on the Crystal Structure of Recombinant Limulus Anti-Lipopolysaccharide Factor. *J. Biol. Chem.* **1996**, *271*, 28120–28127. [CrossRef]
25. Li, S.; Huo, G.; Jiang, Y.; Wu, Y.; Jiang, H.; Wang, R.; Hua, C.; Zhou, F. Transcriptomics provides insights into toxicological effects and molecular mechanisms associated with the exposure of Chinese mitten crab, *Eriocheir sinensis*, to dioxin. *Dev. Comp. Immunol.* **2023**, *139*, 104540. [CrossRef]
26. Shimizu, I.; Yoshida, Y.; Suda, M.; Minamino, T. DNA Damage Response and Metabolic Disease. *Cell Metab.* **2014**, *20*, 967–977. [CrossRef]
27. Kanehisa, M.; Furumichi, M.; Tanabe, M.; Sato, Y.; Morishima, K. KEGG: New perspectives on genomes, pathways, diseases and drugs. *Nucleic Acids Res.* **2017**, *45*, D353–D361. [CrossRef]
28. Wang, T.; Yang, C.; Zhang, T.; Liang, H.; Ma, Y.; Wu, Z.; Sun, W. Immune defense, detoxification, and metabolic changes in juvenile Eriocheir sinensis exposed to acute ammonia. *Aquat. Toxicol.* **2021**, *240*, 105989. [CrossRef]
29. Hong, Y.; Yin, H.; Huang, Y.; Huang, Q.; Yang, X. Immune response to abamectin-induced oxidative stress in Chinese mitten crab, *Eriocheir sinensis*. *Ecotoxicol. Environ. Saf.* **2020**, *188*, 109889. [CrossRef]
30. Li, S.; Li, X.; Shpigelman, A.; Lorenzo, J.M.; Montesano, D.; Barba, F.J. Direct and indirect measurements of enhanced phenolic bioavailability from litchi pericarp procyanidins by Lactobacillus casei-01. *Food Funct.* **2017**, *8*, 2760–2770. [CrossRef]
31. Hong, Y.; Huang, Y.; Yan, G.; Pan, C.; Zhang, J. Antioxidative status, immunological responses, and heat shock protein expression in hepatopancreas of Chinese mitten crab, *Eriocheir sinensis* under the exposure of glyphosate. *Fish Shellfish Immunol.* **2018**, *86*, 840–845. [CrossRef] [PubMed]
32. Judge, A.; Dodd, M.S. Metabolism. *Essays Biochem.* **2020**, *64*, 607–647. [CrossRef] [PubMed]
33. Heindel, J.J.; Blumberg, B.; Cave, M.; Machtinger, R.; Mantovani, A.; Mendez, M.A.; Nadal, A.; Palanza, P.; Panzica, G.; Sargis, R.; et al. Metabolism disrupting chemicals and metabolic disorders. *Reprod. Toxicol.* **2017**, *68*, 3–33. [CrossRef]
34. Zhang, X.; Tang, X.; Tran, N.T.; Huang, Y.; Gong, Y.; Zhang, Y.; Zheng, H.; Ma, H.; Li, S. Innate immune responses and metabolic alterations of mud crab (*Scylla paramamosain*) in response to Vibrio parahaemolyticus infection. *Fish Shellfish Immunol.* **2019**, *87*, 166–177. [CrossRef]
35. Ding, Z. Lipid metabolism disorders contribute to the pathogenesis of *Hepatospora eriocheir* in the crab *Eriocheir sinensis*. *J. Fish Dis.* **2021**, *44*, 305–313. [CrossRef]
36. Thakur, C.; Chen, F. Connections between metabolism and epigenetics in cancers. *Semin. Cancer Biol.* **2019**, *57*, 52–58. [CrossRef]

37. Mulukutla, B.C.; Yongky, A.; Le, T.; Mashek, D.G.; Hu, W.-S. Regulation of Glucose Metabolism—A Perspective from Cell Bioprocessing. *Trends Biotechnol.* **2016**, *34*, 638–651. [CrossRef] [PubMed]
38. Li, X.; Kumar, A.; Carmeliet, P. Metabolic Pathways Fueling the Endothelial Cell Drive. *Annu. Rev. Physiol.* **2019**, *81*, 483–503. [CrossRef]
39. Smith, R.L.; Soeters, M.R.; Wüst, R.C.I.; Houtkooper, R.H. Metabolic Flexibility as an Adaptation to Energy Resources and Requirements in Health and Disease. *Endocr. Rev.* **2018**, *39*, 489–517. [CrossRef]
40. Ballabio, A.; Bonifacino, J.S. Lysosomes as dynamic regulators of cell and organismal homeostasis. *Nat. Rev. Mol. Cell Biol.* **2020**, *21*, 101–118. [CrossRef]
41. Huang, N.; Perl, A. Metabolism as a Target for Modulation in Autoimmune Diseases. *Trends Immunol.* **2018**, *39*, 562–576. [CrossRef] [PubMed]
42. Egan, G.; Khan, D.H.; Lee, J.B.; Mirali, S.; Zhang, L.; Schimmer, A.D. Mitochondrial and Metabolic Pathways Regulate Nuclear Gene Expression to Control Differentiation, Stem Cell Function, and Immune Response in Leukemia. *Cancer Discov.* **2021**, *11*, 1052–1066. [CrossRef] [PubMed]
43. Kumar, H. Metabolic pathways and metabolites shaping innate immunity. *Int. Rev. Immunol.* **2020**, *39*, 81–82. [CrossRef] [PubMed]
44. Chen, C.; Shen, J.-L.; Wang, T.; Yang, B.; Liang, C.-S.; Jiang, H.-F.; Wang, G.-X. Ophiopogon japonicus inhibits white spot syndrome virus proliferation in vivo and enhances immune response in Chinese mitten crab *Eriocheir sinensis*. *Fish Shellfish Immunol.* **2021**, *119*, 432–441. [CrossRef]
45. Hwang, J.; Jin, J.; Jeon, S.; Moon, S.H.; Park, M.Y.; Yum, D.-Y.; Kim, J.H.; Kang, J.-E.; Park, M.H.; Kim, E.-J.; et al. SOD1 suppresses pro-inflammatory immune responses by protecting against oxidative stress in colitis. *Redox Biol.* **2020**, *37*, 101760. [CrossRef]
46. Lauridsen, C. From oxidative stress to inflammation: Redox balance and immune system. *Poult. Sci.* **2019**, *98*, 4240–4246. [CrossRef]
47. Deretic, V. Autophagy in inflammation, infection, and immunometabolism. *Immunity* **2021**, *54*, 437–453. [CrossRef]
48. Levine, B.; Mizushima, N.; Virgin, H.W. Autophagy in immunity and inflammation. *Nature* **2011**, *469*, 323–335. [CrossRef]
49. Deretic, V.; Saitoh, T.; Akira, S. Autophagy in infection, inflammation and immunity. *Nat. Rev. Immunol.* **2013**, *13*, 722–737. [CrossRef]
50. Ataş, H.; Gönül, M.; Öztürk, Y.; Kavutçu, M. Ischemic modified albumin as a new biomarker in predicting of oxidative stress in alopecia areata. *Turk. J. Med. Sci.* **2019**, *49*, 129–138. [CrossRef]

International Journal of
Environmental Research and Public Health

MDPI

Article

Heavy Metal Pollution in Xinfengjiang River Sediment and the Response of Fish Species Abundance to Heavy Metal Concentrations

Guoxiu Shang [1], Xiaogang Wang [1,*], Long Zhu [1,*], Shan Liu [2], Hongze Li [1], Zhe Wang [1], Biao Wang [1] and Zhengxian Zhang [1]

1 Key Laboratory of Navigation Structure Construction Technology of Ministry of Transport, Nanjing Hydraulic Research Institute, Nanjing 210029, China
2 Laluo Hydro-junction and Irrigation District Administration of Tibet Autonomous Region, Lhasa 851414, China
* Correspondence: xgwang@nhri.cn (X.W.); zhulong@nhri.cn (L.Z.); Tel.: +86-025-8582-8206 (X.W.)

Abstract: Xinfengjiang River, the largest tributary of Dongjiang River, plays a key role in the water supply of Heyuan, Huizhou, Guangzhou and even the Pearl River urban agglomeration. It is crucial to determine the pollution status, potential ecological risk degree of heavy metals in Xinfengjiang river sediment and their influence on the abundance of fish species. In this paper, seven heavy metal concentrations in sediment from the Heyuan section of the Xinfengjiang river were investigated. The order of average concentration was: As > Zn > Pb > Cr > Cu > Cd > Hg. The average concentrations of Cd, Zn and Cu in the upper reaches of the Xinfengjiang Reservoir were significantly higher than those in the reservoir. The mean value order of Igeo was: Cd > Zn > Pb > As > Cu > Cr > Hg. Cd and As had the highest ecological risk index and the greatest threat to the ecological environment. Pearson correlation analysis and principal component analysis demonstrated that the pollution source of heavy metals such as Cu and Cd are much more likely to originate from the mine fields located in the northeast of the sampling sites. In addition, agriculture, electronic industry and domestic sewage also contributed to the concentration of heavy metals in different degrees. Redundancy analysis showed that the abundance of Cypriniformes was negatively correlated with Cu and Cd concentrations, suggesting that mining activities might indirectly affect the abundance of fish species.

Keywords: heavy metal; sediment; ecological risk; fish species

Citation: Shang, G.; Wang, X.; Zhu, L.; Liu, S.; Li, H.; Wang, Z.; Wang, B.; Zhang, Z. Heavy Metal Pollution in Xinfengjiang River Sediment and the Response of Fish Species Abundance to Heavy Metal Concentrations. *Int. J. Environ. Res. Public Health* **2022**, *19*, 11087. https://doi.org/10.3390/ijerph191711087

Academic Editors: Paul B. Tchounwou and Li He

Received: 4 July 2022
Accepted: 31 August 2022
Published: 4 September 2022

1. Introduction

Heavy metals have received widespread attention in the past few decades due to them being persistent hazards to aquatic ecosystems and human health, which are constantly released by natural and anthropic activities, such as mining, industrial and domestic sewage discharge, agriculture, e-waste, soil erosion as well as rock weathering [1–4]. As one of the components of an aquatic ecosystem, sediment plays an important role as both a concentrated sink of heavy metal pollution, and to some extent an inevitable source [5,6]. A significant amount of heavy metals entering water are deposited at the bottom of the water body with suspended particles, and when the physical and chemical factors (such as pH, oxidation–reduction potential and organic matter) change, they can be released again to cause secondary pollution [7–9].

The toxic effects of heavy metals could disturb the metabolism and reproduction of living organisms, cause the death of organisms and affect the aquatic biological community through bio-accumulation mechanism [10,11]. The essential micronutrients such as Cu, Zn, Cr, Mn and Fe may induce toxic effects on organisms at high concentration levels of exposure, while non-essential metals such as Cd, Pb, As, Ni and Hg enhance the overall toxic effects even at low concentration levels [12]. In addition, Pb, Cd and Hg do not perform any known functions in human biochemistry or physiology, nor do they naturally exist in organisms [13]. There is extensive literature on the accumulation of heavy metals in fish [3,10,11,14–16]. Cd, As, Cu, Zn, Pb and Cr have toxic effects to C. carpio which are sensitive to heavy metals concentration [17]; Cu, Zn, Cd and Pb in particular have the highest bioaccumulation in C. carpio [18].

Mining operations are considered the greatest threat to ecological integrity because of the lasting toxicity (hundreds of years after the cessation of mining operations) [19,20]. Furthermore, the discharge of acid mine drainage and mining tailings are mainly associated with the pollution of heavy metals in water and sediment [21]. Iron-ore extraction is typically performed through open pit mining, which offers higher productivity and lower costs and security risks compared to underground mining [22].

Xinfengjiang River is the largest tributary of Dongjiang River which serves about 28 million people by lying at the heart of drinking water supply source for Heyuan, Huizhou, part of Dongguan, Shenzhen and Hong Kong [23]. It is closely related to the survival and sustainable development of the urban agglomeration in the east of the Pearl River Delta [24]. Xinfengjiang River Basin has a long history of mining [23,25]; the river lies in a metallogenic belt rich in metals such as W, Sn, Zn and Pb. Many illegal small-scale mining sites are widely distributed in the middle reaches of the river, and there are several large-scale sources of mineral input including the Dading iron mine and the Jubankeng tungsten mine. In addition, the non-point source pollution caused by agricultural chemical fertilizers and pesticides is worthy of attention [26], and the electronic industry pollution caused by industrial parks is generally high. Jinfeng Chen et al. [23] determined the content of trace elements in the riverbed, bankside and adjacent agricultural soil of the Zhongxing River which is located in the Dongjiang River Basin. Yun-jiang Yu et al. [27] evaluated the heavy metals and pollution levels in the sediments of the Xinfengjiang Reservoir and the Dongjiang River Basin. Few researchers have addressed the problem of response of fish species to heavy metal pollution in the Xinfengjiang River. Fish are considered to be the most significant biomonitors of aquatic systems for the evaluation of heavy metal pollution levels [28]. This paper focuses on the heavy metal pollution conditions in the sediment of the Xinfengjiang River in the Heyuan city section. Geo-accumulation index, individual heavy metal ecological risk index (E_r^i) and potential ecological risk index (RI) were used to assess ecological risk levels. We then traced the main sources of heavy metals by Pearson correlation analysis and principal component analysis (PCA). Finally, redundancy analysis (RCA) was used to research the response relationship between fish species and heavy metals concentration in sediment.

2. Materials and Methods

2.1. Study Area

The Xinfengjiang River, the largest tributary of the Dongjiang River, is located in the Northeast of the Guangdong Province (Figure 1). The Xinfengjiang river is 163 km long, of which 94.16 km are in Heyuan City. The Xinfengjiang Reservoir, also known as Lvwan Lake, is the largest artificial reservoir in the Guangdong Province. The study area (23.7° N–24.4° N to 114.2° E–114.7° E) was located in Heyuan City in the middle and lower reaches of the Xinfengjiang river, accounting for 57.8% of the total length of the Xinfengjiang river, with a drainage area of 4340 km, accounting for 74.5% of the total basin area (Figure 1). Influenced by maritime climate, it belongs to a subtropical monsoon climate zone. The average temperature is 19.5~20.7 °C and the average annual precipitation is more than 1500 mm. Eleven sampling sites (S1—S11) were set up in the study area, of which four (S1–S4) were located in the upper reaches of Xinfengjiang reservoir, one (S11) in the lower reaches of the reservoir adjacent to the Dongjiang River and the other six sampling sites (S5–S10) in the reservoir (Figure 1). On the basis of consulting the research literature on heavy metals and examining the actual situation in the study area, seven target heavy metals, Cu, As, Pb, Zn, Cr, As and Hg, were selected for assessment.

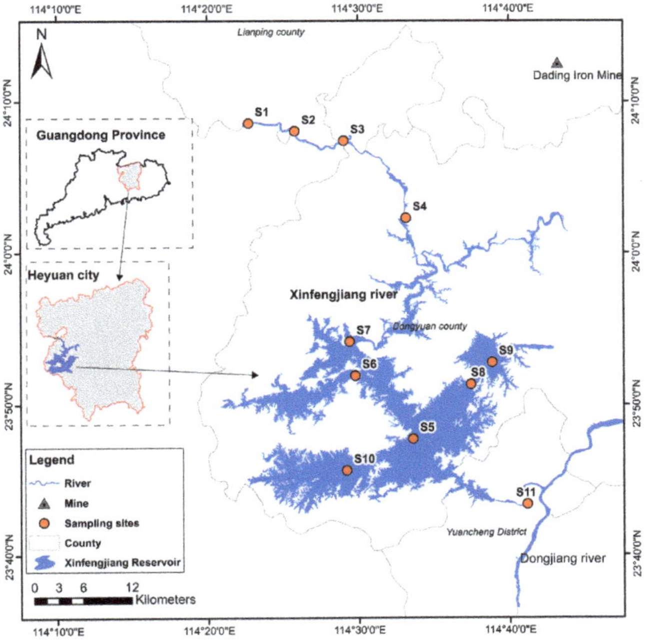

Figure 1. Study area and the location of the sampling sites.

2.2. Sample Collection and Analysis

2.2.1. Sediment Sampling

Sediment samples (0–10 cm depth) were collected at the corresponding sampling sites in August, 2021. We received a portable GPS device for recording geographic location. Three samples were collected at each site using a portable Ekman grab sampler and subsequently mixed, then placed in an acid-rinsed polyethylene plastic bag and sealed. The sediment samples were transferred to the laboratory for treatment while being kept at 4 °C.

2.2.2. Sediment Sample Analysis

A standard reference and a reagent blank were included in the heavy metal concentration test to ensure data accuracy and precision. After natural air drying in the laboratory at room temperature, gravel and plant roots in the samples were removed, ground and sifted through 100 mesh for the analysis of heavy metals, including Cu, Cd, Zn, Pb, Cr, As and Hg. The samples were digested with a microwave digestion instrument (CEM Inc., Matthews, NC, USA) for 10 h before concentration analysis. SK-2003AZ atomic fluorescence spectrophotometer(Suokun technology, Beijing, China) was used to analyze the content of As and Hg, while the other five Heavy metal concentrations were analyzed by WFX-120 atomic absorption spectrophotometer(Beifen-Ruili, Beijing, China). Each heavy metal concentration was tested three times, and the data were unacceptable until the relative standard deviation <5%.

2.2.3. Fish Sampling

Fish collection was carried out from August to September in 2021 by laying drift Gill net and ground cage, and the collection points was the same as sediment. Each drift Gill net is 50 m in length and 2 m in height, with a mesh of 3 cm. The statistical catch was recovered after placing a drift Gill net and a ground cage at each station for one night. The fish specimens collected from various survey sites were photographed and weighed at the scene, then tagged and brought back to the laboratory for identification, analysis and preservation.

2.3. *Assessment of Heavy Metal Concentration in Sediments*

2.3.1. Geo-Accumulation Index (Igeo)

The geo-accumulation index (Igeo) is one of the most widely used and simple indices to evaluate heavy metal pollution in sediment on account of its ability to reflect the enrichment situations and to provide consistent values for comparison [29]. It was calculated using the following empirical relations [30]:

$$\text{Igeo} = \log_2(C_n/1.5B_n) \quad (1)$$

where C_n is the concentration of study metal and B_n is the background concentration, as listed in Table 1 Factor 1.5 was used for the background matrix correction and reducing lithogenic effects. Table 2 showed the Igeo classes and corresponding sediment quality.

Table 1. Statistics of heavy metal concentration in sediments of XFJR.

Sample Sites	Cu (mg/kg)	Zn (mg/kg)	Pb (mg/kg)	Cd (mg/kg)	Cr (mg/kg)	Hg (mg/kg)	As (mg/kg)
S1	32.50	163.0	68.8	1.680	41.80	0.08	22.80
S2	24.20	119.0	73.8	0.640	31.20	0.09	14.30
S3	29.30	121.0	71.5	0.630	43.10	0.08	27.50
S4	145.00	186.0	143.0	2.390	59.80	0.11	327.00
S5	14.50	80.2	135.0	0.250	22.30	0.07	33.10
S6	18.50	29.7	16.3	0.080	59.30	0.07	437.00
S7	12.00	32.2	30.4	0.060	66.00	0.07	36.50
S8	1.80	12.9	25.1	0.035	6.86	0.04	10.40
S9	1.90	21.8	65.2	0.033	8.12	0.05	3.45
S10	1.78	18.7	22.3	0.038	8.26	0.06	24.90
S11	32.40	152.0	38.9	0.790	38.70	0.14	13.20
Mean	28.50	85.13	62.75	0.602	35.04	0.078	86.38
C_V(%)	141.60%	76.70%	69%	129.10%	61.90%	35.60%	172%
Background value	16.17	48.99	39.65	0.08	38.53	0.11	19.77

Table 2. Classes of contamination indices and corresponding levels.

Igeo Class [a]	Sediment Quality	E_r^i Class [b]	Potential Risk	RI Class [c]	Ecological Risk
<0	Uncontaminated	$E_r^i < 40$	Low	RI < 150	Low
0–1	Uncontaminated to moderately contaminated	$40 \le E_r^i < 80$	Moderate	$150 \le$ RI < 300	Moderate
1–2	Moderately contaminated	$80 \le E_r^i < 160$	Considerable	$300 \le$ RI < 600	Considerable
2–3 3–4 4–5	Moderately to heavily contaminated Heavily contaminated Heavily to extremely contaminated	$160 \le E_r^i < 320$ $E_r^i \ge 320$	High Very high	RI $\ge$ 600	Very high
>5	Extremely contaminated				

Igeo the geo-accumulation index, E_r^i the potential ecological risk factor of single metal, RI the potential ecological risk index, CF the contamination factor, PLI the pollution load index. [a] [30]; [b] [31]; [c] [32].

2.3.2. Potential Ecological Risk Index (RI)

The potential ecological risk index is used to assess the toxicity of metal in sediments [25,32,33]. E_r^i and RI represent individual heavy metal ecological risk and comprehensive potential ecological risk, respectively. Their calculation formulas are as follows:

$$E_r^i = T_r^i \times \left(\frac{C_i}{C_0}\right) \tag{2}$$

$$RI = \sum_{i=1}^{n} E_r^i = \sum_{i=1}^{n} T_r^i \left(\frac{C_i}{C_0}\right) \tag{3}$$

where C_i is the concentration of metal i in sediment, C_0 is the background concentration of metal, and T_r^i defines biological toxicity factor for individual metal. The E_r^i values for Cu = Pb = Ni = 5, Zn = 1, Cr = 2 and Cd = 30 [30,31], Table 2 gives the classes and levels for E_r^i and RI.

2.4. Statistical Analysis

Principal component analysis (PCA) was performed to extract significant PCs and associated loads. Pearson's correlation analysis was used to test the significant relationship between variables. Redundancy analysis (RCA) was used to analyze the response of fish abundance to heavy metal content using Canoco5.0 (Microcomputer Power, Ithaca, NY, USA). PCA and correlation analysis were implemented with R language and IBM Spss26.0.

3. Results

3.1. Heavy Metal Concentration in Sediments

The concentrations of seven heavy metals in surface sediments of Xinfengjiang River (Heyuan section) was shown in Table 1. The mean concentrations followed a decreasing ranking order of As (86.38 mg/kg) > Zn (85.13 mg/kg) > Pb (62.75 mg/kg) > Cr (35.04 mg/kg) >Cu (28.53 mg/kg) >Cd (0.602 mg/kg) >Hg (0.078 mg/kg). The variable coefficients of As, Cu and Cd concentrations with values of 172%, 141.6% and 129.1%, respectively, were higher than other four heavy metals, in other words, it is more reasonable to believe that the sources of these three heavy metals are more closely related to the effects of human activities. There was significant difference in the concentrations of most heavy metals among upstream sites (S1–S4), reservoir sites (S5–S10) and the downstream site (S11). The heavy metal contents were generally higher in the upstream than reservoir, while Hg content shows the highest values at the downstream site (S11). Particularly, S4 was the highest total heavy metals concentration.

3.2. Igeo

The distribution of Igeo value of heavy metals in the sediments is shown in Figure 2. The maximum Igeo values of Hg and Cr were below 0 and 1, respectively, indicating that the sediments were not contaminated with these two metals, or that the contamination level was low. The Igeo values of Cu (−3.76–2.58, median −0.39), Zn (−2.51–1.34, median 0.13), Pb (−1.87–1.27, median 0.13), As (−1.51–3.88, median −0.25), Cd (−1.86–4.32, median 1.06). Different colors (Figure 2b) refer to different levels of sediment pollution, with blue indicating uncontaminated, and from light yellow to orange and red indicating slightly contaminated to extremely contaminated. Cd exceeded the background value most seriously, with the highest Igeo value exceeding 4, from heavily to extremely contaminated; Cu, Zn, Pb and Cd concentrations at four sites (S1–S4) upstream of the reservoir all exceeded the regional background values. As appeared in extremely high values at points S4 and S6, exceeding other sites by a large margin. Pollution is more severe upstream (S1–S4) and downstream (S11) than at the Reservoir (S5–S10).

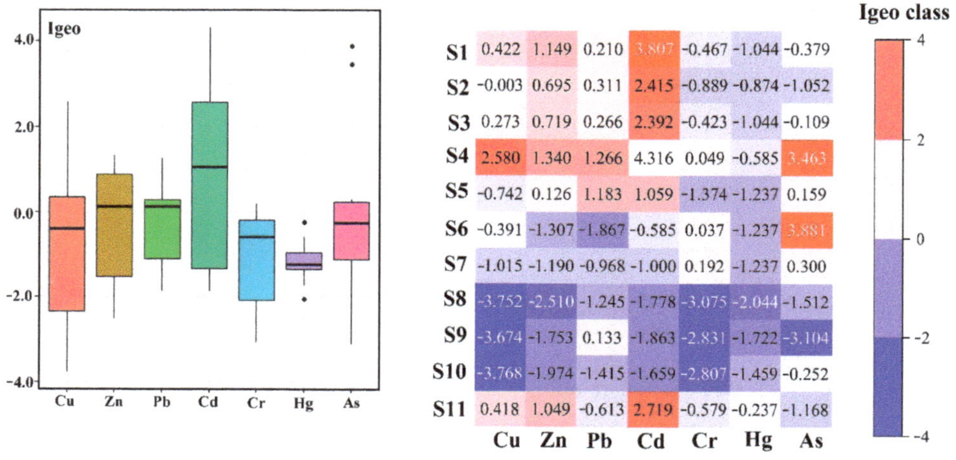

(a) (b)

Figure 2. Geo-accumulation index (Igeo) of heavy metals in sediment: (**a**) Boxplot of Igeo values; (**b**) Igeo class in different sites.

3.3. Ecological Risks and Potential Ecological Risks

The single metal ecological risk index (E_r^i) and potential ecological risk index (RI) of heavy metals in the sediments of Heyuan City section of Xinfengjiang river are shown in Table 3 and Figure 3. Zn, Pb and Cr at the low ecological risk with individual ecological risk indices lower than 40; Cu showed moderate ecological risk at S4, Hg was in the threshold of low risk (Er = 40) at S4 and moderate risk at S11; high ecological risk at S4 and S6, and the above three total metals were of low risk at the rest of the sites. Cd created the highest degree of ecological risk with six sample sites reaching moderate risk or higher, and the biggest two ecological risk values of were 630 (S1) and 896.25 (S4). The potential ecological risk indices of the sampling sites ranked in order as follow: S4 > S1 > S11 > S3 > S2 > S6 > S5 > S7 > S10 > S9 > S8, consistent with above, S4 and S1 had the highest RI, 1171.4 at S4 and 694.8 at S1, all above the high potential risk limit of 600. S7, S10, S9 and S8 were sampled in the low potential risk range. In summary, Cd and As posed the greatest threat to ecology, followed by Hg, Cu, Zn and Cr. As with the results of Igeo, the ecological risk of the four upstream sites were higher than the reservoir sites, and the downstream was in the middle. S4 was the most prominent, S5 and S6 were observed to significantly higher than other sites in reservoir.

Table 3. Individual ecological risks (E_r^i) and potential ecological risks (RI) of heavy metals.

Sites	E_r^i							RI
	Cu	Zn	Pb	Cd	Cr	Hg	As	
S1	10.05	3.33	8.68	630.00	2.17	29.09	11.53	694.8
S2	7.48	2.43	9.31	240.00	1.62	32.73	7.23	300.8
S3	9.06	2.47	9.02	236.25	2.24	29.09	13.91	302.0
S4	44.84	3.80	18.03	896.25	3.10	40.00	165.40	1171.4
S5	4.48	1.64	17.02	93.75	1.16	25.45	16.74	160.2
S6	5.72	0.61	2.06	30.00	3.08	25.45	221.04	288.0
S7	3.71	0.66	3.83	22.50	3.43	25.45	18.46	78.0
S8	0.56	0.26	3.17	13.13	0.36	14.55	5.26	37.3
S9	0.59	0.44	8.22	12.38	0.42	18.18	1.75	42.0
S10	0.55	0.38	2.81	14.25	0.43	21.82	12.59	52.8
S11	10.02	3.10	4.91	296.25	2.00	50.91	6.68	373.9

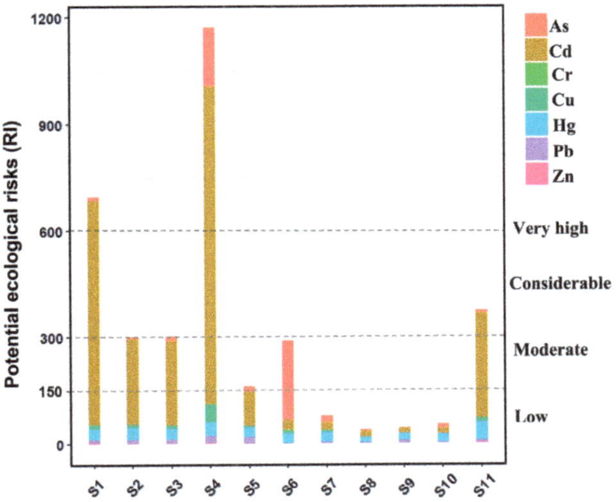

Figure 3. RI value of heavy metals in sediment (derived from the E_r^i value).

3.4. Tracing the Sources of Heavy Metal Pollution

Pearson correlation analysis and principal component analysis are widely used to explore the internal correlation of variables and help trace the pollution sources of heavy metals in sediments [34,35]. Table 4 listed the Pearson correlation matrix among sediment heavy metals. All metals showed positive correlations, with Cu-Cd (r = 0.883), Cd-Zn (r = 0.891) and Zn-Hg (0.804) showing highly significant positive correlations at $p < 0.01$, and Cu-Zn (r = 0.723), Cu-Pb (r = 0.643), Cd-Pb (r = 0.61) and Hg-Cd (r = 0.602) showing significant positive correlation at $p < 0.05$. Statistically significant heavy metals with high correlation coefficients are considered to have the same origin or similar behavior during river transport [36]. Lack of valid correlation among the other metals reveals that the contents of these metals are not controlled by a single factor.

Table 4. Pearson's correlation matrix of sediment heavy metals.

Heavy Metals	Cu	Zn	Pb	Cd	Cr	Hg	As
Cu	1						
Zn	0.723 *	1					
Pb	0.643 *	0.596	1				
Cd	0.883 **	0.891 **	0.610 *	1			
Cr	0.520	0.426	0.106	0.438	1		
Hg	0.577	0.804 **	0.281	0.602 *	0.492	1	
As	0.528	0.084	0.093	0.287	0.582	0.154	1

* Correlation is significant at the 0.05 level (2-tailed). ** Correlation is significant at the 0.01 level (2-tailed).

Principal component analysis (PCA) of heavy metals was given in Figure 4. The first principal component explained 58.8% of the variance, the second principal component explained 18.9%, and the three principal component axes together explained 89.8% of the overall variance. The loads of each heavy metal on the three components are shown in Table 5.

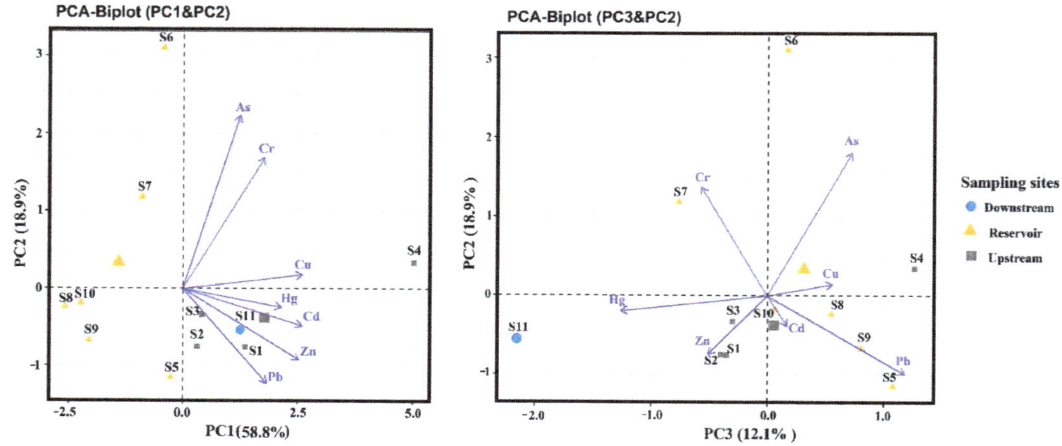

Figure 4. Principal component analysis (PCA) of heavy metals in sediment sampling sites.

Table 5. The interpretation variance of heavy metals on three principal components.

Heavy Metals	PC1	PC2	PC3
Cu	0.926	0.066	0.246
Cd	0.923	−0.169	0.076
Zn	0.898	−0.325	−0.220
Hg	0.762	−0.085	−0.549
Pb	0.647	−0.432	0.518
Cr	0.632	0.603	−0.248
As	0.442	0.792	0.319

Cu, Cd and Zn have obvious advantages on PC1, loading 0.926, 0.923 and 0.898 respectively. Different from the former, Hg, Pb and Cr also have varying degrees of loading on the other two principal components, although their loadings on the first principal component are not low (0.762, 0.647 and 0.632). Among them, Hg loads −0.549 onPC3, Pb loads −0.432 on PC2 and 0.518 on PC3, Cr loads 0.603 on PC2. It is thus clear that the sources of these three elements are significantly different from the previous three. Meanwhile, it is worth noting that As differs from the previous heavy metals in that its loading on PC2 (0.792) has a significant upper hand. In the PCA-Biplot (PC1 and PC2) (Figure 4), the reservoir sites (S5 to S10) are all distributed in the third and fourth quadrants, the upstream site S4 near the reservoir is distributed in the first quadrant, and the remaining three upstream sites (S1–S3) and one downstream site (S11) are distributed in the second quadrant. S4 and S6 are outliers, with S4 closer to PC1 and S6 closer to PC2. In the PCA-Biplot (PC3 and PC2) plot (Figure 4), S4 is closer to the positive direction of the third principal component and point S11 is closer to the negative direction of the third principal component. This indicates that the content of heavy metals in the upstream and downstream sites is higher than that in the reservoir sites, and the content of Hg at S11 is much higher than others, which is consistent with the results reflected by the ecological risk index of sampling sites.

3.5. Response of Fish Species Abundance to Heavy Metal Concentration in Sediment

A total of 23 species of fishes belonging to four orders were recorded in this survey, which were cypriniformes, perciformes, cypriniformes and siluriformes, all of them belonged to teleost. Cypriniformes are dominant at the order level, as shown in Figure 5.

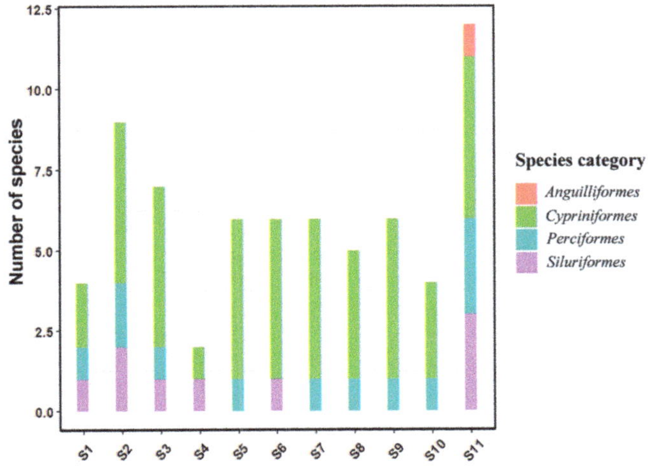

Figure 5. Distribution of fish species at sampling sites.

It has been reported that heavy metals in the environment will be enriched in fish, and there is a response relationship between the abundance of fish species and heavy metals in the environment, which affects the composition and abundance of organisms to varying degrees [37]. Redundancy analysis is a ranking method combining regression analysis with principal component analysis, which can intuitively reflect the response of species variables to environmental variables [38]. The magnitude and angle of vectors can intuitively reflect the correlation between variables. The first axis explains 54.96% of the species abundance variables, and the two axes jointly explain 86.74% of Figure 6.

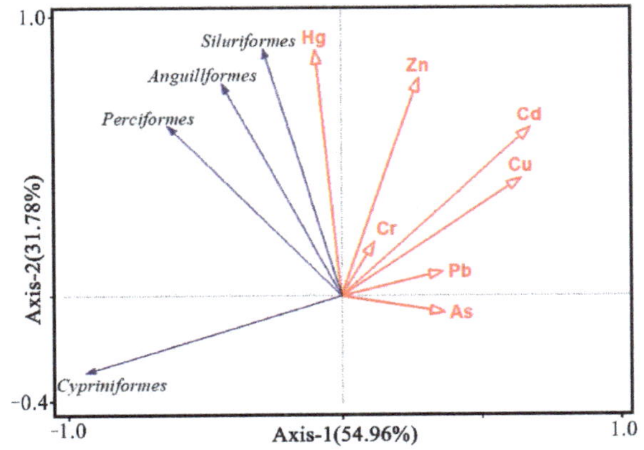

Figure 6. Redundancy analysis diagram between heavy metals and fish species.

4. Discussion

Both the geo-accumulation index and the ecological risk index demonstrate that the sediment heavy metals pollution degree in the upstream sites were generally higher than that in the reservoir area, and the downstream site was in the middle. Among them, Cd, Cu, As and Zn were the main heavy metal pollutants. They could be caused by a variety of reasons. First and foremost, the large mining sites are located in the northeast of the upstream site. Second, the upstream is closer to the community. The upstream content of heavy metal pollutants produced by agricultural chemical fertilizers and pesticides, domestic sewage discharge, breeding, etc., is also higher than that in the reservoir area. In view of the industrial characteristics of the Lianping and Dongyuan counties around the Xinfengjiang Reservoir, it was speculated that the first principal component in the principal component analysis (Figure 4) was mainly related to mineral exploitation, livestock and poultry breedin and agricultural activities. The river lies in a metallogenic belt in the northeast Guangdong province which is rich in metals such as W, Sn, Pb, Fe and Zn [39]. There are many scattered illegal small-scale mining sites as well as large and medium mining sites in the upper reaches of the reservoir, so mining activities are prevalent in this area [23]. The Lianping County is rich in mineral resources, with more than 30 kinds of proven minerals and approximately 52 mining locations. Iron ore has the greatest advantage with a reserve of 160 million tons. The Jubankeng tungsten mine, located in Lianping County, is the largest tungsten mine that has been operating intermittently since last century in the Guangdong Province. Located in the Lianping County, the Sawpan Hang tungsten mine, which mines tungsten and copper ore, is the largest tungsten mine in the Guangdong Province and has been operating intermittently since the last century, about 45 km from the upstream sampling point of the reservoir. Dading Iron Mine, located in Youxi Town, Lianping County, is the largest iron mine in the province, 25.5 km away from S4. Previous studies have shown that tailings, drainage and air suspended solids caused by mining activities will strongly affect the content of heavy metals in nearby water, sediment and soil [22]. In particular, the content of heavy metals such as Cd, Cu, Zn and Pb is generally higher than the background value, and can be up to a thousand times in Pilcomayo river [40]. Many studies confirmed the migration of trace elements from point source to downstream with sediments, and demonstrated that the contents of trace elements in downstream rivers generally showed a decreasing trend [41]. In this paper, the average concentration of the upstream points (S1–S4) of Cd, Cu and Zn is much higher than that of the reservoir points (S5–S10), which are 16 times, 6.9 times and 4.5 times higher, respectively.

According to investigation, pig farming contributes to the concentration of Cu in sediments, especially the use of copper-containing additives in pig feed, which results in high levels of copper in pig manure [42]. There are a large number of pig farms in the Xinfengjiang catchment area; on the one hand, despite sewage treatment, some of the copper and cadmium in pig manure is discharged into rivers along with the wastewater. On the other hand, pig manure, as the main source of manure, enters the soil with agricultural activities, indirectly increasing the content of heavy metals in the soil, and some of the heavy metals in the soil will be discharged into the river with the runoff. Cd and Zn have strong fluidity. Cd is a symbolic metal in pesticides and fertilizers involved in agricultural activities [43]. Besides mining, gasoline, phosphate and domestic sewage are also possible sources of Cd [44]. The second principal component may mainly represent natural erosion processes such as parent rock weathering and soil erosion. The third main component mainly associates with industrial activities, electronics, machinery manufacturing and domestic sewage discharge. There are five outfalls distributed near S11 downstream of the Xinfengjiang Reservoir Dam; meanwhile, the downstream of the reservoir converges into the Dongjiang River, in close proximity to industrial parks such as the Linjiang Industrial Zone and Zijin Industrial Park, which mainly produce electronics and electrical machinery [27]. The main pollutants are Cr, As, Cu and Pb. As is widely used in semiconductors, diodes, alloys and structural steel. Pb is the raw material of battery, and lead compounds are widely used in pigment, glass, plastic and rubber. Meanwhile, Pb is used in various metallurgical equipment and chemical manufacturing. Hg mainly comes from domestic sewage, manufacturing industry and electronic appliances.

Overall, the heavy metals in the Xinfengjiang river sediment are from a wide range of sources, including mining, agricultural activities, industrial activities, domestic sewage and rock weathering. Cu may originate from mining, farming and pesticide herbicides. Cd is correlated with mining, farming, industrial production, domestic sewage, agricultural activities, etc. Pb is mainly correlated with automobile exhaust pollution, sewage and electronic appliance pollution. As may be correlated with pesticides, Hg is mainly from domestic sewage and industrial activities.

In addition, heavy metal pollution can negatively affect fish species abundance. Cypriniformes are negatively correlated with heavy metals except Hg, and significantly negatively correlated with Cd and Cu. Perciformes are negatively correlated with As and Pb. Hg is positively correlated with Perciformes, Silphiformes and Anguilliformes, but not with Cypriniformes. Therefore, Cu and Cd have a great influence on the species of Cypriniformes, and high concentrations are detrimental to species abundance. In fact, Hg is considered as one of the most toxic elements or substances on earth. Research indicates that Hg exposure can induce various adverse effects on fish at physiological, histological, biochemical, enzymatic and genetic levels [45]. As a result of redundancy analysis, there are five sewage outlets in the section from the Xinfengjiang Reservoir Dam to the Dongjiang River, and sewage may be the biggest contributor of mercury. The abundance of fish at S11 is the highest among all points, mainly because it is close to the Dongjiang River, rather than Hg having a positive impact on the abundance of fish species. Moreover, water flow velocity is one of the important influencing factors in the living environment of fish, which in upstream points (S1–S4) is higher than that in the reservoir area (S5–S10), and the above redundancy analysis result shows Cd and Cu have a higher degree of negative influence on the Cypriniformes, indicating that the mobility of these two elements and the difference between the upstream and the reservoir area are stronger. The abundance of Cypriniformes in the upstream points was quite different, and the mean value was small. It indicates that the flow velocity has an adverse effect on the Cypriniformes, and this conclusion is consistent with the effect of heavy metals. That is to say, the flow velocity also has influence on the distribution of heavy metals such as Cd and Cu in the sediment, thereby affecting the response of fish species abundance to the concentration of heavy metals.

5. Conclusions

The pollution and toxicity of heavy metals in sediments is one of the most concerning issues in aquatic ecosystems. This study examines seven heavy metals (Cu, Zn, Pb, Cd, Cr, Hg and As) in the sediment at 11 sampling sites located in the Heyuan section of the Xinfengjiang River. Based on the results related to heavy metals and their analysis, the following conclusions could be drawn: (1) The concentrations of heavy metals and ecological risks in the upstream of the Xinfengjiang Reservoir are generally higher than those in the reservoir, and the downstream is in the middle. (2) The mean order of Igeo was: Cd>Zn>Pb>As>Cu>Cr>Hg, and the Cd at S1 and S4 reached a heavily contaminated level. The potential ecological risk index showed that Cd and As were the most serious heavy metals threatening the ecology. The order of potential risk index on sampling sites was: S4 > S1 > S11 > S3 > S2 > S6 > S5 > S7 > S10 > S9 > S8. (3) Pearson correlation analysis and principal component analysis showed that Cd, Cu and Zn had a high correlation, and Hg and Zn had a very significant correlation. The possible sources of heavy metals mainly include mining, agricultural activities, industrial activities and domestic sewage. Mine pollution should be given due attention. (4) Redundancy analysis was used to analyze the response of fish species abundance to heavy metal concentrations, and the results showed that the Cypriniformes was most affected by Cd and Cu. It is speculated that mining is the main source of heavy metals in Xinfengjiang River sediment, and indirectly affects fish species abundance. Water flow velocity is an influencing factor of fish species abundance, and may even be the main factor. In this regard, this paper is still insufficient, and it is expected that this issue can be further studied in the future.

Author Contributions: Conceptualization, X.W. and H.L.; methodology, G.S.; software, Z.W. and B.W.; formal analysis, S.L.; resources, S.L. and L.Z.; data curation, G.S. and Z.Z.; writing—original draft preparation, G.S.; writing—review and editing, X.W. and Z.W.; supervision, X.W.; funding acquisition, L.Z. All authors have read and agreed to the published version of the manuscript.

Funding: This research was funded by Research on fish passage engineering technology for the Yangtze -to-Huaihe Water Diversion, grant number YJJH-ZT-ZX-2019, and Research on key technologies of ecological fish passage in medium and low head ship locks, grant number Y120009.

Institutional Review Board Statement: Not applicable.

Informed Consent Statement: Not applicable.

Data Availability Statement: Not applicable.

Conflicts of Interest: The authors declare no conflict of interest.

References

1. Jaskuła, J.; Sojka, M.; Fiedler, M.; Wróżyński, R. Analysis of Spatial Variability of River Bottom Sediment Pollution with Heavy Metals and Assessment of Potential Ecological Hazard for the Warta River, Poland. *Minerals* **2021**, *11*, 327. [CrossRef]
2. Jumbe, A.S.; Nandini, N. Impact Assessment of Heavy Metals Pollution of Vartur Lake, Bangalore. *J. Appl. Nat. Sci.* **2009**, *1*, 53–61. [CrossRef]
3. Maurya, P.K.; Malik, D.S.; Yadav, K.K.; Kumar, A.; Kumar, S.; Kamyab, H. Bioaccumulation and Potential Sources of Heavy Metal Contamination in Fish Species in River Ganga Basin: Possible Human Health Risks Evaluation. *Toxicol. Rep.* **2019**, *6*, 472–481. [CrossRef]
4. Sandu, C.F.; Farkas, A.; Musa-Iacob, R.; Ionica, D.; Parpală, L.; Zinevici, V.; Dobre, D.; Radu, M.; Présing, M.; Casper, H.; et al. Monitoring Pollution in River Mureş, Romania, Part I. The Limitation of Traditional Methods and Community Response. *River Syst.* **2008**, *18*, 91–106. [CrossRef]
5. Huang, Z.; Zhao, W.; Xu, T.; Zheng, B.; Yin, D. Occurrence and distribution of antibiotic resistance genes in the water and sediments of Qingcaosha Reservoir, Shanghai, China. *Environ. Sci. Eur.* **2019**, *31*, 81. [CrossRef]
6. Pejman, A.; Nabi Bidhendi, G.; Ardestani, M.; Saeedi, M.; Baghvand, A. A New Index for Assessing Heavy Metals Contamination in Sediments: A Case Study. *Ecol. Indic.* **2015**, *58*, 365–373. [CrossRef]
7. Suthar, S.; Nema, A.K.; Chabukdhara, M.; Gupta, S.K. Assessment of Metals in Water and Sediments of Hindon River, India: Impact of Industrial and Urban Discharges. *J. Hazard. Mater.* **2009**, *171*, 1088–1095. [CrossRef]
8. Zhang, C.; Yu, Z.; Zeng, G.; Jiang, M.; Yang, Z.; Cui, F.; Zhu, M.; Shen, L.; Hu, L. Effects of Sediment Geochemical Properties on Heavy Metal Bioavailability. *Environ. Int.* **2014**, *73*, 270–281. [CrossRef]
9. Xu, J.; Chen, Y.; Zheng, L.; Liu, B.; Liu, J.; Wang, X. Assessment of Heavy Metal Pollution in the Sediment of the Main Tributaries of Dongting Lake, China. *Water* **2018**, *10*, 1060. [CrossRef]
10. Yi, Y.J.; Zhang, S.H. The Relationships between Fish Heavy Metal Concentrations and Fish Size in the Upper and Middle Reach of Yangtze River. *Procedia Environ. Sci.* **2012**, *13*, 1699–1707. [CrossRef]
11. Dvorak, P.; Roy, K.; Andreji, J.; Liskova, Z.D.; Mraz, J. Vulnerability Assessment of Wild Fish Population to Heavy Metals in Military Training Area: Synthesis of a Framework with Example from Czech Republic. *Ecol. Indic.* **2020**, *110*, 105920. [CrossRef]
12. Stanković, S.; Kalaba, P.; Stanković, A. Biota as Toxic Metal Indicators. *Environ. Chem. Lett.* **2013**, *12*, 63–84. [CrossRef]
13. Lenntech, R. Water Treatment and Air Purification. *Water Treatment*. 2004. Available online: http//www.excelwater.com/thp/filters/Water-Purification.htm (accessed on 10 August 2022).
14. Liu, L.; Xu, Z.; Yang, F.; Yin, X.; Wu, W.; Li, J. Comparison of Fish, Macroinvertebrates and Diatom Communities in Response to Environmental Variation in the Wei River Basin, China. *Water* **2020**, *12*, 3422. [CrossRef]
15. Sauliutė, G. Heavy Metal Interactions during Accumulation via Direct Route in Fish: A Review. *Zool. Ecol.* **2015**, *25*, 77–86.
16. Yousif, R.; Choudhary, M.I.; Ahmed, S.; Ahmed, Q. Review: Bioaccumulation of Heavy Metals in Fish and Other Aquatic Organisms from Karachi Coast, Pakistan. *Nusant. Biosci.* **2021**, *13*, 73–84. [CrossRef]
17. Gheorghe, S.; Stoica, C.; Vasile, G.G.; Nita-Lazar, M.; Stanescu, E.; Lucaciu, I.E. Metals Toxic Effects in Aquatic Ecosystems: Modulators of Water Quality. In *Water Quality*; Tutu, H., Ed.; InTech: London, UK, 2017.
18. Zafarzadeh, A.; Bay, A.; Fakhri, Y.; Keramati, H.; Hosseini Pouya, R. Heavy Metal (Pb, Cu, Zn, and Cd) Concentrations in the Water and Muscle of Common Carp (*Cyprinus Carpio*) Fish and Associated Non-Carcinogenic Risk Assessment: Alagol Wetland in the Golestan, Iran. *Toxin Rev.* **2018**, *37*, 154–160. [CrossRef]
19. Duruibe, J.; Ogwuegbu, M.O.C.; Egwurugwu, J.N. Heavy Metal Pollution and Human Biotoxic Effects. *Int. J. Phys. Sci.* **2007**, *2*, 112–118.
20. Tovar-Sánchez, E.; Hernández-Plata, I.; Martínez, M.S.; Valencia-Cuevas, L.; Galante, P.M. Heavy Metal Pollution as a Biodiversity Threat. In *Heavy Metals*; Saleh, H.E.-D.M., Aglan, R.F., Eds.; InTech: London, UK, 2018.
21. Aghili, S.; Vaezihir, A.; Hosseinzadeh, M.R. Distribution and Modeling of Heavy Metal Pollution in the Sediment and Water Mediums of Pakhir River, at the Downstream of Sungun Mine Tailing Dump, Iran. *Environ. Earth Sci.* **2018**, *77*, 128. [CrossRef]
22. Sahoo, P.K.; Powell, M.A.; Martins, G.C.; Dall'Agnol, R.; Salomão, G.N.; Mittal, S.; Pontes, P.R.M.; Guimarães, J.T.F.; de Siqueira, J.O. Occurrence, Distribution, and Environmental Risk Assessment of Heavy Metals in the Vicinity of Fe-Ore Mines: A Global Overview. *Toxin Rev.* **2022**, *41*, 675–698. [CrossRef]

23. Chen, J.; Yuan, J.; Wu, S.; Lin, B.; Yang, Z. Distribution of Trace Element Contamination in Sediments and Riverine Agricultural Soils of the Zhongxin River, South China, and Evaluation of Local Plants for Biomonitoring. *J. Environ. Monit.* **2012**, *14*, 2663. [CrossRef]
24. Wen, M.; Zhang, H.; Zhou, X.; Yang, L.; Fang, G. Water Supply Features and Security of the Xinfengjiang Reservoir:A Case Study on the Drinking Water Supply to Heyuan During 2003–2012. *Trop. Geogr.* **2014**, *34*, 518–526. (In Chinese)
25. Fu, S. Characteristic Analysis of the Trace Elements Abundance in the River-Bed Sediments near the Ferris Mine Area in the Upper Reach of Xinfengjiang. *Appl. Mech. Mater.* **2013**, *13*, 123–127. (In Chinese)
26. Zhang, B.; Zhang, Q.-Q.; Zhang, S.-X.; Xing, C.; Ying, G.-G. Emission Estimation and Fate Modelling of Three Typical Pesticides in Dongjiang River Basin, China. *Environ. Pollut.* **2020**, *258*, 113660. [CrossRef]
27. Yu, Y.; Li, L.; Li, M.; Zhang, X.; Li, Z.; Zhu, X. Heavy Metals and Arsenic in Sediments of Xinfengjiang Reservoir and East River in South China: Levels, Source and Health Risk Assessment. *Appl. Environ. Biotechnol.* **2020**, *5*, 3–13. [CrossRef]
28. Shazili, N.A.M.; Yunus, K.; Ahmad, A.S.; Abdullah, N.H.; Rashid, M.K.A. Heavy Metal Pollution Status in the Malaysian Aquatic Environment. *Aquat. Ecosyst. Health Manag.* **2006**, *9*, 137–145. [CrossRef]
29. Audry, S.; Schäfer, J.; Blanc, G.; Jouanneau, J.M. Fifty-Year Sedimentary Record of Heavy Metal Pollution (Cd, Zn, Cu, Pb) in the Lot River Reservoirs (France). *Environ. Pollut.* **2004**, *132*, 413–426. [CrossRef] [PubMed]
30. Muller, G. Index of geoaccumulation in sediments of the Rhine river. *GeoJournal* **1969**, *2*, 108–118.
31. Hakanson, L. An Ecological Risk Index for Aquatic Pollution Control. A Sedimentological Approach. *Water Res.* **1980**, *14*, 975–1001. [CrossRef]
32. Guo, W.; Liu, X.; Liu, Z.; Li, G. Pollution and Potential Ecological Risk Evaluation of Heavy Metals in the Sediments around Dongjiang Harbor, Tianjin. *Procedia Environ. Sci.* **2010**, *2*, 729–736. [CrossRef]
33. Suresh, G.; Sutharsan, P.; Ramasamy, V.; Venkatachalapathy, R. Assessment of Spatial Distribution and Potential Ecological Risk of the Heavy Metals in Relation to Granulometric Contents of Veeranam Lake Sediments, India. *Ecotoxicol. Environ. Saf.* **2012**, *84*, 117–124. [CrossRef]
34. Ekoa Bessa, A.Z.; Ngueutchoua, G.; Kwewouo Janpou, A.; El-Amier, Y.A.; Njike Njome Mbella Nguetnga, O.-A.; Kankeu Kayou, U.R.; Bisse, S.B.; Ngo Mapuna, E.C.; Armstrong-Altrin, J.S. Heavy Metal Contamination and Its Ecological Risks in the Beach Sediments along the Atlantic Ocean (Limbe Coastal Fringes, Cameroon). *Earth. Syst. Environ.* **2021**, *5*, 433–444. [CrossRef]
35. Luo, P.; Xu, C.; Kang, S.; Huo, A.; Lyu, J.; Zhou, M.; Nover, D. Heavy Metals in Water and Surface Sediments of the Fenghe River Basin, China: Assessment and Source Analysis. *Water Sci. Technol.* **2021**, *84*, 3072–3090. [CrossRef] [PubMed]
36. Suresh, G.; Ramasamy, V.; Meenakshisundaram, V.; Venkatachalapathy, R.; Ponnusamy, V. Influence of Mineralogical and Heavy Metal Composition on Natural Radionuclide Concentrations in the River Sediments. *Appl. Radiat. Isot.* **2011**, *69*, 1466–1474. [CrossRef] [PubMed]
37. Azizah, M.; Paramitha-K.W, G.A. Fish Diversity and Heavy Metal Content Mercury (Hg), Arsenic (As) on the Water and Fish in Cikaniki River, Bogor Regency. *J. Edubiotik* **2021**, *6*, 83–90. [CrossRef]
38. Huang, Z.; Liu, C.; Zhao, X.; Dong, J.; Zheng, B. Risk Assessment of Heavy Metals in the Surface Sediment at the Drinking Water Source of the Xiangjiang River in South China. *Environ. Sci. Eur.* **2020**, *32*, 23. [CrossRef]
39. Huang, S.; Tian, T.; Zou, X.; Zhou, X.; Palaninaicker, S.; Qiu, R. Bioavailability Assessment of Heavy Metals in the Vicinity of Dabaoshan Mine. *Acta Sci. Nat. Univ. Sunyatseni* **2009**, *48*, 125–129+136. (In Chinese)
40. Smolders, A.J.P.; Lock, R.A.C.; Van der Velde, G.; Medina Hoyos, R.I.; Roelofs, J.G.M. Effects of Mining Activities on Heavy Metal Concentrations in Water, Sediment, and Macroinvertebrates in Different Reaches of the Pilcomayo River, South America. *Arch. Environ. Contam. Toxicol.* **2003**, *44*, 314–323. [CrossRef]
41. Bird, G.; Brewer, P.A.; Macklin, M.G.; Şerban, M.; Bălteanu, D.; Driga, B. Heavy Metal Contamination in the Arieş River Catchment, Western Romania: Implications for Development of the Roşia Montană Gold Deposit. *J. Geochem. Explor.* **2005**, *86*, 26–48. [CrossRef]
42. Qaswar, M.; Yiren, L.; Jing, H.; Kaillou, L.; Mudasir, M.; Zhenzhen, L.; Hongqian, H.; Xianjin, L.; Jianhua, J.; Ahmed, W.; et al. Soil Nutrients and Heavy Metal Availability under Long-Term Combined Application of Swine Manure and Synthetic Fertilizers in Acidic Paddy Soil. *J. Soils Sediments* **2020**, *20*, 2093–2106. [CrossRef]
43. Irshad, M.K.; Noman, A.; Alhaithloul, H.A.S.; Adeel, M.; Rui, Y.; Shah, T.; Zhu, S.; Shang, J. Goethite-Modified Biochar Ameliorates the Growth of Rice (*Oryza sativa* L.) Plants by Suppressing Cd and As-Induced Oxidative Stress in Cd and As Co-Contaminated Paddy Soil. *Sci. Total Environ.* **2020**, *717*, 137086. [CrossRef]
44. Sekabira, K.; Origa, H.O.; Basamba, T.A.; Mutumba, G.M.; Kakudidi, E.K. Heavy Metal Assessment and Water Quality Values in Urban Stream and Rain Water. *Int. J. Environ. Sci. Technol.* **2010**, *7*, 759–770. [CrossRef]
45. Morcillo, P.; Esteban, M.Á.; Cuesta, A. Mercury and its toxic effects on fish. *AIMS Environ. Sci.* **2017**, *4*, 386–402. [CrossRef]

International Journal of
Environmental Research and Public Health

MDPI

Article

Water Quality Evaluation and Pollution Source Apportionment of Surface Water in a Major City in Southeast China Using Multi-Statistical Analyses and Machine Learning Models

Yu Zhou [1], Xinmin Wang [1], Weiying Li [2,3,*], Shuyun Zhou [4] and Laizhu Jiang [5]

1 College of Environmental Science and Engineering, Tongji University, Shanghai 200092, China
2 State Key Laboratory of Pollution Control and Resource Reuse, Tongji University, Shanghai 200092, China
3 Ministry of Education Key Laboratory of Yangtze River Water Environment, Tongji University, Shanghai 200092, China
4 Jiangsu Yinyang Stainless Steel Pipe Co., Ltd., Wuxi 214000, China
5 Fujian Qingtuo Special Steel Technology Research Co., Ltd., Fuzhou 350000, China
* Correspondence: 123lwyktz@tongji.edu.cn

Abstract: The comprehensive evaluation of water quality and identification of potential pollution sources has become a hot research topic. In this study, 14 water quality parameters at 4 water quality monitoring stations on the M River of a city in southeast China were measured monthly for 10 years (2011–2020). Multiple statistical methods, the water quality index (WQI) model, machine learning (ML), and positive matrix factorisation (PMF) models were used to assess the overall condition of the river, select crucial water quality parameters, and identify potential pollution sources. The average WQI values of the four sites ranged from 68.31 to 77.16, with a clear trend of deterioration from upstream to downstream. A random forest-based WQI model (WQI_{RF} model) was developed, and the results showed that Mn, Fe, faecal coliform, dissolved oxygen, and total nitrogen were selected as the top five important water quality parameters. Based on the results of the WQI_{RF} and PMF models, the contributions of potential pollution sources to the variation in the WQI values were quantitatively assessed and ranked. These findings prove the effectiveness of ML in evaluating water quality, and improve our understanding of surface water quality, thus providing support for the formulation of water quality management strategies.

Keywords: water quality index (WQI); machine learning; parameter selection; positive matrix factorization (PMF); source apportionment

Citation: Zhou, Y.; Wang, X.; Li, W.; Zhou, S.; Jiang, L. Water Quality Evaluation and Pollution Source Apportionment of Surface Water in a Major City in Southeast China Using Multi-Statistical Analyses and Machine Learning Models. *Int. J. Environ. Res. Public Health* **2023**, *20*, 881. https://doi.org/10.3390/ijerph20010881

Academic Editors: Paul B. Tchounwou, Xin Zhao, Xun Wang and Zhiyuan Wang

Received: 9 November 2022
Revised: 23 December 2022
Accepted: 29 December 2022
Published: 3 January 2023

1. Introduction

Surface water has historically been vital in providing water for human consumption, agriculture, and industrial requirements [1–4]. In recent decades, rapid urbanisation, industrialisation, and global population growth have led to the deterioration of surface water quality, which is a serious concern for the public and scientists [5,6]. According to a study conducted by the World Health Organization [7], at least 2 billion people worldwide use contaminated drinking water sources, 785 million people do not even have essential drinking water services, and 144 million rely on surface water.

As a water quality assessment method widely used for groundwater and surface water (especially rivers), the water quality index (WQI) method is playing an increasingly important role in water resource management [3,8–10]. Over the last several decades, various improvements have been made in the calculation of WQI values [11–13]. Compared with traditional water quality evaluation methods, the WQI method combines several environmental parameters, effectively transforming them into a single value reflecting the general water quality status, instead of comparing different evaluation results of various parameters [3].

To simplify and efficiently assess water quality, a WQI_{min} model based on a select number of representative parameters can quickly and accurately determine water quality and reduce analytical costs [14–16]. To determine the water quality parameters in the WQI_{min} model, previous studies mostly used linear regression methods based on the relationship between WQI values and various water quality parameters, and selected important indicators based on the performance of the WQI_{min} model on comprehensive evaluation values [3,10].

Machine learning (ML) models perform well in regression problems and have become very popular in recent years. In the field of environmental science, many scientists have used ML for water quality prediction. Chen et al., compared the water quality prediction performance of 10 ML models using big data from major rivers and lakes in China, identified two key water parameter sets (dissolved oxygen (DO), potassium permanganate index (COD_{Mn}), and ammonia nitrogen (NH_3-N); and COD_{Mn} and NH_3-N), and proved the superiority of random forests (RFs) [17]. Lu and Ma used two hybrid models (extreme gradient boosting and RFs) to predict six water quality indicators (water temperature, DO, pH, specific conductance, turbidity, and fluorescent dissolved organic matter) and compared the performance of each model with those of four conventional models [18]. The results showed that the RF model had a higher prediction stability. In the present study, an RF model was used for regression modelling of WQI values, and important water quality parameters were selected according to the feature importance of RFs [19–21]. Selected key water quality parameters were then applied to develop the RF-based WQI_{RFmin} model.

In addition to completing the water quality assessment and obtaining important water quality indicators, it is also necessary to explore the potential sources of water pollution. Receptor models, such as the absolute principal components score combined with multivariate linear regression (APCS–MLR) and positive matrix factorisation (PMF), have performed well in source apportionment studies [22]. The PMF approach is a multi-source analysis method for source identification and assignment that is specifically designed to process environmental data and manage the associated uncertainty and distribution [23]. The PMF method is particularly suitable for environmental data because it considers the analytical uncertainty typically associated with environmental sample measurements and renders all values and contributions in the solution to be positive, which may lead to more realistic results than other multivariate methods [24]. Previous studies [22,25] showed that PMF had a higher coefficient of determination (R^2) of prediction and a smaller proportion of unidentified sources than the APCS–MLR model, which could provide a more physically plausible source apportionment and a more realistic representation of pollution. In the last two decades, PMF has been widely used in studies related to air pollution and the atmospheric environment. In recent years, PMF has been increasingly used to apportion pollution sources in water environments [26,27]. The PMF model can describe the contributions of pollution sources to various water quality parameters; however, each water quality parameter has a different importance in different areas of research. Previous studies have rarely examined the contribution of pollution sources to WQI values, which can comprehensively assess water quality. Although some pollution sources provided a higher pollution contribution rate to water quality parameters in some studies, these sources may not be the main factor influencing water quality changes, because the concentrations of water quality parameters affected by them were too low to influence water quality changes [5].

The M River is an important river flowing through the capital city (mainly urban areas) of a province in southeast China, providing a permanent source of water for approximately 14 million people [28]. Based on the above background, WQI calculations, RF model construction, and PMF analyses were performed using a dataset of 14 water quality parameters collected on a monthly basis over 10 years (2011–2020) from four monitoring stations on the M River. The objectives of this study are to (1) analyse the spatial and temporal water quality patterns of the M River, (2) assess the comprehensive water quality condition and identify key water quality parameters of the M River, and (3) explore the potential pollution sources in the watershed and their contributions to the variation in WQI values. The results of the water quality assessment, crucial water quality parameter selection, and pollution source apportionment will be valuable for the local authorities to control and manage the water quality of the M River and to better protect it from pollution through a fixed-point traceability approach.

2. Materials and Methods

2.1. Study Area

The M River is located in the 25–29° N latitude and 116–120° E longitude region, and flows eastward through the Taiwan Strait. The river provides important assistance to people's daily lives, industry, and agriculture in the cities of southeast China [28]. As a subtropical mountain river, the M River basin has an average annual temperature of 16–20 °C, and total annual rainfall of 1500–2000 mm, which is higher than that of other plain-dominated rivers in China. In recent years, modern agriculture has developed rapidly. The overuse of chemical fertilisers and pesticides, and the reckless discharge of sewage have intensified river pollution. Meanwhile, the continuous industrialisation and urbanisation of the M River basin have led to an increase in illegal discharges of industrial wastewater and an increase in heavy metal pollution due to mining, urban construction, and the development of transportation. Inadequate management of municipal, industrial, and agricultural wastewater means that residents around the watershed are exposed to dangerous organic and inorganic contamination of their drinking water [7,10,29].

2.2. Data Preparation

The datasets were collected on a monthly basis from October 2011 to August 2020 at four monitoring stations on the M River (WWP, FWP, SWP, and CWP; Figure 1). Fourteen water quality parameters were monitored as follows: pH, water temperature (WT), DO, total nitrogen (TN), NH_3-N, nitrate-nitrogen (NO_3^--N), total phosphorus (TP), COD_{Mn}, chloride (Cl^-), sulfate ion (SO_4^{2-}), faecal coliform (*F. coli*), iron (Fe), manganese (Mn), and fluoride (F^-). The analytical methods used for each parameter are listed in Table 1.

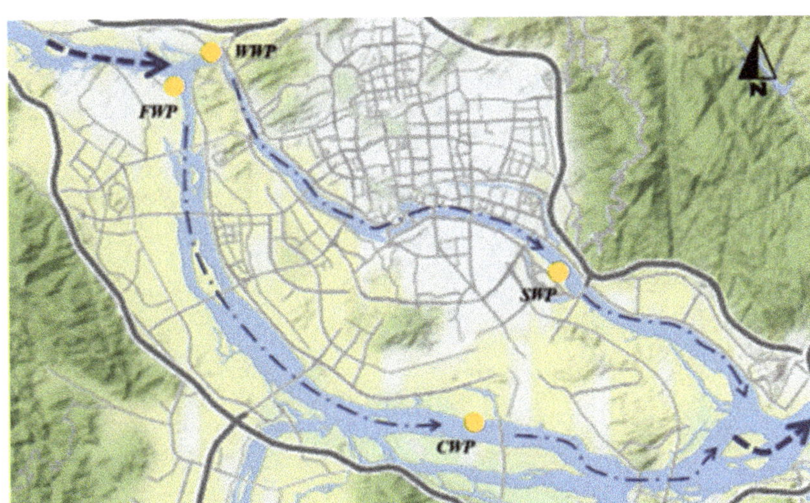

Figure 1. Locations of the water quality monitoring stations in the study area in southeast China.

Table 1. Water quality parameters measured in this study and the relevant analytical methods.

Variables	Abbreviation	Units	Testing Base
pH	pH		GB6920-1986
Water temperature	WT	°C	GB/T13195-1991
Dissolved oxygen	DO	mg/L	HJ506-2009
Total nitrogen	TN	mg/L	HJ636-2012
Ammonia	NH_3-N	mg/L	HJ665-2013
Nitrate	NO_3-N	mg/L	HJ/T84-2001
Total phosphorus	TP	mg/L	GB/T11893-1989
Permanganate index	COD_{Mn}	mg/L	GB 11892-1989
Chloride	Cl^-	mg/L	HJ/T84-2001
Sulphate	$SO_4{}^{2-}$	mg/L	HJ/T84-2001
Iron	Fe	mg/L	HJ700-2014
Manganese	Mn	mg/L	HJ700-2014
Fecal coliform	*F. coli*	colonies/L	GB/T5750.12-2006
Fluoride	F^-	mg/L	HJ/T84-2001

2.3. Water Quality Index

The calculations for the WQI in this study are based on Equation (1), which was refined and developed by Pesce and Wunderlin [16] as follows:

$$WQI = \frac{\sum_{i-1}^{n}(C_i P_i)}{\sum_{i-1}^{n} P_i} \quad (1)$$

where n is the total number of water quality parameters in the study; C_i is the normalized value of the i-th parameter; and P_i is the determined weight of the i-th parameter (the values of P_i have been verified in previous studies and are listed in Table S1).

The theory of the WQI model has been widely used and extensively discussed in previous studies [2,3,29]. The water quality status in this study was classified into five grades based on the WQI values (Table 2), which are in line with the actual water quality management standards in China [3].

Table 2. Water quality classification based on water quality index (WQI) values.

WQI value	91–100	71–90	51–70	26–50	0–25
Water quality	Excellent	Good	Moderate	Poor	Very poor

2.4. Random Forests

Random forest regressors are widely applied in ML for classification and regression, which can deal with nonlinearities and interactions, but cannot be interpreted directly [4,20,30]. It is an ensemble model based on the generation of many decision trees and their assemblage to produce the final output. Each output from the decision tree is dependent on the values of a random vector sampled independently from the same distribution of all decision trees generated in the forest. The number of predictors used to find the best split at each node is randomly chosen from a subset of all predictors [21]. The output is calculated by taking the mean and aggregation of each individual component tree [21,31]. The RF model has been found to be reliable for evaluating the ranking of the most critical predictors in trophic status prediction [32] and for predicting groundwater arsenic contamination [33].

In the construction of the decision tree, the quality of the segmentation variables and segmentation points are generally measured by the impurity of the node after segmentation.

$$G(x_i, v_{ij}) = \frac{n_{left}}{N_s} H\left(X_{left}\right) + \frac{n_{right}}{N_s} H\left(X_{right}\right) \tag{2}$$

where x_i is a segmentation variable; v_{ij} is a segmentation value of the segmentation variable; n_{left} is the number of training samples of the left child node; n_{right} is the number of training samples of the right child node; N_s is the number of training samples of the current node; X_{left} is the set of training samples of left child nodes; X_{right} is the set of training samples of the right child nodes; *H(X)* is the impurity function of the node (classification and regression generally use different impurity functions).

The mean square error (MSE) was selected by default as the impurity function of the RF regression models based on decision trees as follows:

$$G(x, v) = \frac{1}{N_s}\left[\sum_{y_i \epsilon X_{left}} \left(y_i - \overline{y}_{left}\right)^2 + \sum_{y_j \epsilon X_{right}} \left(y_i - \overline{y}_{right}\right)^2\right] \tag{3}$$

The importance of a node is given by:

$$n_k = w_k \times G_k - w_{left} \times G_{left} - w_{right} \times G_{right} \tag{4}$$

where w_k is the ratio of the number of training samples to the total number of training samples in node *k*; w_{left} is the ratio of the number of training samples in the left child node of node *k* to the total number of training samples in node *k*; w_{right} is the ratio of the number of training samples in the right child node of node *k* to the total number of training samples in node *k*; G_k is the impurity of node *k*; G_{left} is the impurity of the left child node of node *k*; and G_{right} is the impurity of the right child node of node *k*.

After calculating the importance of each node, the importance of a certain feature can be obtained as follows:

$$f_i = \frac{\sum_{j \epsilon nodes\ split\ on\ feature\ i} n_j}{\sum_{k \epsilon all\ nodes} n_k} \tag{5}$$

To ensure that the importance of all features will add up to one, the importance of each feature must be normalised:

$$f_{ni} = \frac{f_i}{\sum_{j \epsilon all\ features} f_j} \tag{6}$$

In this study, the WQI_{RFmin} model based on the key parameters selected by the RF regression model was also developed. The RF in this study consisted of 500 trees and was applied to train the WQI_{RF} model with the values of water quality indicators as the feature input model and the corresponding WQI as the label (predicted value), which were built using the Scikit-learn v.0.23.1 package in Python 3.8.3. Metrics including R^2, MSE, MAE, and MAPE were adopted to evaluate the performance of the regressor on the testing dataset.

2.5. Positive Matrix Factorisation

The PMF method is a multivariate statistical analysis tool [23], which is usually used to decompose the sample data matrix into two matrices: factor contributions and factor profiles, with the following formula:

$$X_{nm} = E_{nm} + \sum_{j=1}^{p} G_{np} \times F_{pm} \tag{7}$$

where X_{nm} is the original matrix ($n \times m$), representing n samples and m monitoring variables, which can be decomposed into two matrices G_{np} $(n \times p)$ and F_{pm} $(p \times m)$; p is the number of calculated sources (extraction factor); G is the source contribution matrix; F is the source component spectral matrix (factor load); E_{nm} $(n \times m)$ is the residual matrix representing the difference between the analytical result and the measured value.

The results are constrained by a penalty function such that no sample can have a negative source contribution, and no species can have a negative concentration in any source profile. A detailed description of the PMF model is provided in Paatero and Tapper [23]. The researchers have explained the PMF model in detail, thus no more detailed description here. This study used the PMF 5.0 software recommended by the US EPA for data analysis.

2.6. Contribution of Potential Pollution Sources to the Variation in WQI Values

According to the principle of RFs described in the previous section, the WQI_{RF} model based on water quality parameters was developed to quantitatively calculate the feature importance of each water quality parameter. The PMF model can quantitatively evaluate the contribution of each source to water quality; however, the WQI_{RF} model has calculation errors; therefore, $(1 - MAPE_{WQI_{RF}})$ should be added as the error correction factor for the contribution of potential pollution sources to the variation in WQI values, as follows:

$$p_j = (1 - MAPE_{WQI_{RF}}) \times \sum (f_{ni} \times c_{ji}) \quad (8)$$

where p_j is the contribution of pollution source j to the comprehensive water quality evaluation based on WQI values; $MAPE_{WQI_{RF}}$ is the mean absolute percentage error of the WQI_{RF} model developed by RFs; and c_{ji} is the contribution of pollution source j to water quality parameter i.

3. Results

3.1. Analysis of Water Quality Characteristics Based on Individual Parameters

The descriptive statistics of the original data for the selected 14 water quality parameters are listed in Table S2. For water quality comparison, the surface water quality standards of GB3838-2002 (State Environment Protection Bureau of China 2002a) are also included in Table S2. The statistical analysis results of each water quality parameter from 2011 to 2020 showed that, excluding TN, Fe, Mn, and *F. coli*, most of the water quality parameters were better than the Class III water quality standards over the long term.

Water pH indicates an acidic or basic nature and is an important parameter for assessing the quality of drinking water and irrigation water. It has profound effects on water quality, affecting the solubility of metals, alkalinity, and water hardness. From the analysis results, the incoming water from the four monitoring stations in River M over the past 10 years was relatively weakly acidic. The pH values ranged from 6.47 to 7.6, with 64% of the samples having a pH less than 7. Although it is in line with the surface water environmental quality standard GB3838-2002 (6–9 pH), but as a drinking water intake point, it is not enough to meet the surface water standard, but also needs to meet the drinking water hygiene standard GB5749-2022 (6.5–8.5 pH), which could only be said to just satisfy. As we all know, long-term consumption of acidic or weakly acidic water not only leads to the potential risk of erosive tooth wear, but also leads to gradually acidic body fluids, increased blood viscosity and imbalance of the acid–base balance of the human body. Many studies have shown that a low pH of the water supply system has a strong corrosive effect on metal pipes, which can easily lead to 'yellow water' and pipe bursts.

The values of TP, SO_4^{2-}, NO_3^-, F^-, COD_{Mn}, Cl^-, NH_3-N, and DO were lower than the respective Class III standards. For TN, 75% of the samples exceeded the Class III standards. The highest TN concentration (4.76 mg/L) was 4-, 2-, and 1.5-times higher than the standards of classes III, IV, and V, respectively. We observed that the multi-year average concentration of TN was 1.54 mg/L, with 48% and 23% of all observed samples exceeding the Class IV and V surface water standards, respectively (Figure 2). When TN and TP in surface water exceed their respective standards, microorganisms proliferate, plankton grow vigorously, and waterbodies are prone to eutrophication. Considering that the TN concentration did not increase significantly from upstream to downstream, the background value of the upstream water was the main factor. The causes of pollution may have been due to agricultural fertiliser (NO_3^--N fertiliser) pollution, residential sewage, and farming wastewater pollution.

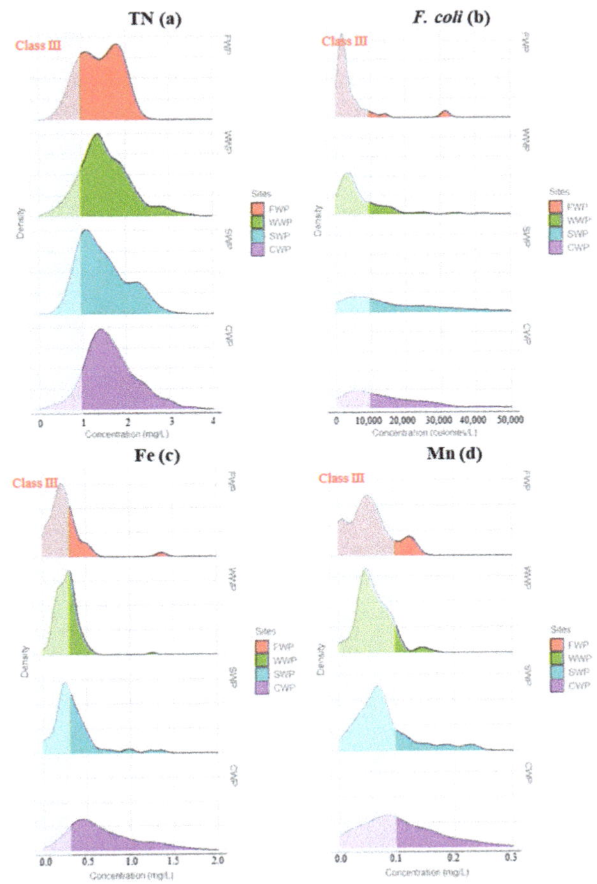

Figure 2. Density distributions of (**a**) TN, (**b**) *F. coli*, (**c**) Fe, and (**d**) Mn concentrations.

In addition, Mn, Fe, and *F. coli* exceeded the Class III standards to different degrees. The *F. coli* concentrations in the downstream region were significantly higher than those in the upstream regions, implying that the urban section of the city is a source of faecal coliform pollution to the river, although the background value of upstream water cannot be ignored.

Trace metals may be present in natural surface water and groundwater, and can be sourced from either natural processes or human activities. Multiple metal ion analyses were performed, but only Fe and Mn concentrations were found to be above the analytical detection limits. The Fe and Mn concentrations of water samples ranged from 1.26 mg/L to 3.2 mg/L and 0.16 mg/L to 1.52 mg/L, respectively. The exceedance rates of the Fe and Mn concentrations at the WWP and FWP monitoring sites in the upper reach were significantly lower than those at the CWP and SWP monitoring sites in the lower reach. The Mn and Fe concentrations at the WWP and FWP sites were likely related to the interaction between water and ophiolitic rocks in the basin, whereby relatively high levels of Mn and Fe in the surrounding ore-bearing landmass could provide a source of these elements to the rivers flowing over this terrain. The relatively high Mn and Fe concentrations at the downstream sites of CWP and SWP were probably mainly influenced by anthropogenic contaminants.

The coefficient of variation (CV) is the most discriminating factor in the variability description; it can eliminate the influence caused by the difference of units and the mean value between two or more datasets. As shown in Table S2, all parameters showed a CV value of between 3.5% and >100%, indicating great variability. Among them, Cl^- and *F. coli* had the largest variabilities, indicating that these water quality parameters were extremely unevenly distributed throughout the basin and were affected by external sources of pollution. In addition, most analysed parameters in water samples presented spatiotemporal variabilities, whereby the concentrations of Mn, Fe, and *F. coli* in the lower reach were significantly higher than those in the upper reach (Figure 2).

3.2. Water Quality Assessment Based on the WQI

To calculate the WQI values at each sampling point, the weight values were determined for each water quality parameter according to their relative importance in terms of the overall drinking water quality (Table S3). A weight of 3 was assigned to the trace metals, which can have major effects on water quality, especially for drinking purposes [15]. The accumulation of trace metals in water indicates both natural or anthropogenic sources, and may affect human health at high levels. The parameters of COD_{Mn}, NH_4-N, and *F. coli* were also each assigned a weight of 3 by taking into consideration their importance in water quality [10,14]. The exceedance of these indicators could lead to the presence of excessive organic pollutants in surface water [15], causing lasting toxic effects on aquatic organisms, and compromising drinking water safety for humans. The lower weights of 1 and 2 were assigned to WT, pH, TN, NO_3-N, TP, Cl^-, SO_4^{2-}, and F^- because of their low importance in water quality [3,10]. Then, the relative weights (P_i) were computed for each parameter. The WQI values were calculated using Equation (1), and the water quality types were determined for each sampling point (Table S3).

The WQI results showed the spatial profiles and annual patterns of the variations in surface water quality (Figure 3). A violin plot is a collection of box-line and density plots, which can be used to show the percentile points of the data by thinking in terms of box lines, and a density plot to show the 'contour' effect of the data distribution, where the larger the 'contour' is, the more concentrated the data is. Based on the WQI scores, 58.2% of water samples were rated as 'good', with an average WQI value of 72.1, while the remaining water samples were rated as 'moderate'.

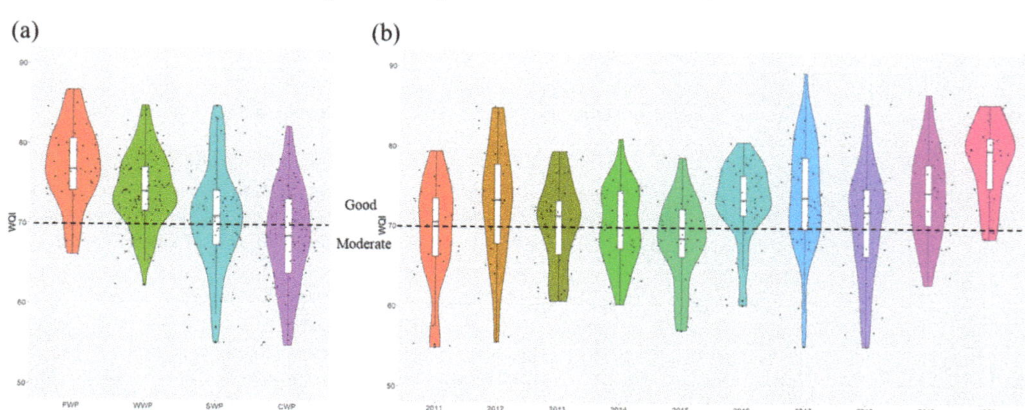

Figure 3. Spatial (**a**) and annual (**b**) variations of the WQI during 2011–2020.

Regarding the spatial variation in the calculated WQI values, the water quality exhibited a clear trend of deterioration from upstream to downstream. The mean WQI values at the FWP (upstream), WWP (upstream), SWP (downstream), and CWP (downstream) sites were 77.2, 74.1, 71.2, and 68.3, respectively. Overall, 86.4%, 76.5%, 51.2%, and 34.5% of water samples from the FWP, WWP, SWP, and CWP sites were rated as 'good', respectively. From the above analysis, Fe, Mn, and *F. coli* increased from upstream to downstream. As these water quality parameters accounted for high weightings in the calculation of the WQI, they were largely responsible for the decline in the WQI.

The annual changes in WQI values suggested that the median and interquartile range of WQI values shifted upward during the study period, and the wide part of the distribution density also shifted upward, indicating that the water quality was continuously improved with time. During 2011–2015, 54.2% of water samples were rated as 'moderate'. In 2015, only 27.8% of water samples were rated as 'good'. However, 70% of WQI values exceeded 70 (i.e., 'good') after 2016. The water quality in 2020 was the best, and the average WQI was 78.5, with 87.5% of water samples being rated as 'good'.

3.3. Selection of Key Water Quality Parameters

The WQI_{RF} model was developed using RFs with all 14 water quality parameters (training data:testing data = 9:1), and the results showed that Mn made the most significant contribution to the WQI values (Figure 4). The parameters of Fe, *F. coli*, and DO were selected sequentially, and the R^2 values of the models were considerably increased. Additionally, TN slightly enhanced the performance of the model. Hence, Mn, Fe, *F. coli*, DO, and TN were established as essential and critical parameters in the training of the WQI_{RFmin} model.

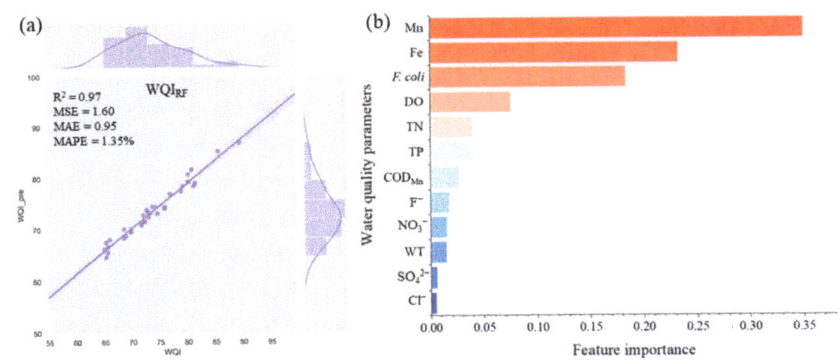

Figure 4. WQI_{RF} model results. (**a**) The predicted results on testing data and (**b**) the feature importance of key water quality parameters.

According to the constructed judgement of RFs on the importance of water quality parameters, two, three, four, and five parameters were selected to develop WQI_{RFmin} models using RFs. The performance of each WQI_{RFmin} model was based on a comprehensive evaluation of the R^2, MSE, MAE, and MAPE values (Table 3, Figure 5), indicating that increases in the parameters could better explain the variation in the WQI. Among the WQI_{RFmin} models, the WQI_{RFmin} model comprising Mn, Fe, *F. coli*, DO, and TN had the best R^2 (0.96), MSE (1.77), MAE (1.06), and MAPE (1.47%) values, indicating that it was the best WQI_{RFmin} model for the study area.

Based on the results of measured water parameters, water quality can be accurately assessed by some procedures; however, it is costly and time-consuming to measure all water parameters in all types of surface water because of the various analytical requirements. Therefore, it is more practical to measure key parameters indicative of water quality rather than completely following the guidelines of GB3838-2002 to understand water quality. Moreover, it is of great significance to predict water quality based on the selection of indicative fundamental water parameters. The five water quality parameters extracted by RFs in this study could determine the WQI with a very high accuracy.

Table 3. Parameter selection results of the WQI_{RF} models based on the training dataset.

Parameters	Feature Importance	R^2	MSE	MAE	MAPE (%)
Mn	0.35	—	—	—	—
Mn + Fe	0.58	0.73	20.01	3.66	5.09
Mn + Fe + *F. coli*	0.76	0.84	11.26	2.76	3.88
Mn + Fe + *F. coli* + DO	0.84	0.93	2.99	1.41	1.98
Mn + Fe + *F. coli* + DO + TN	0.88	0.96	1.77	1.06	1.47
All water quality parameters	1	0.97	1.60	.0.95	1.35

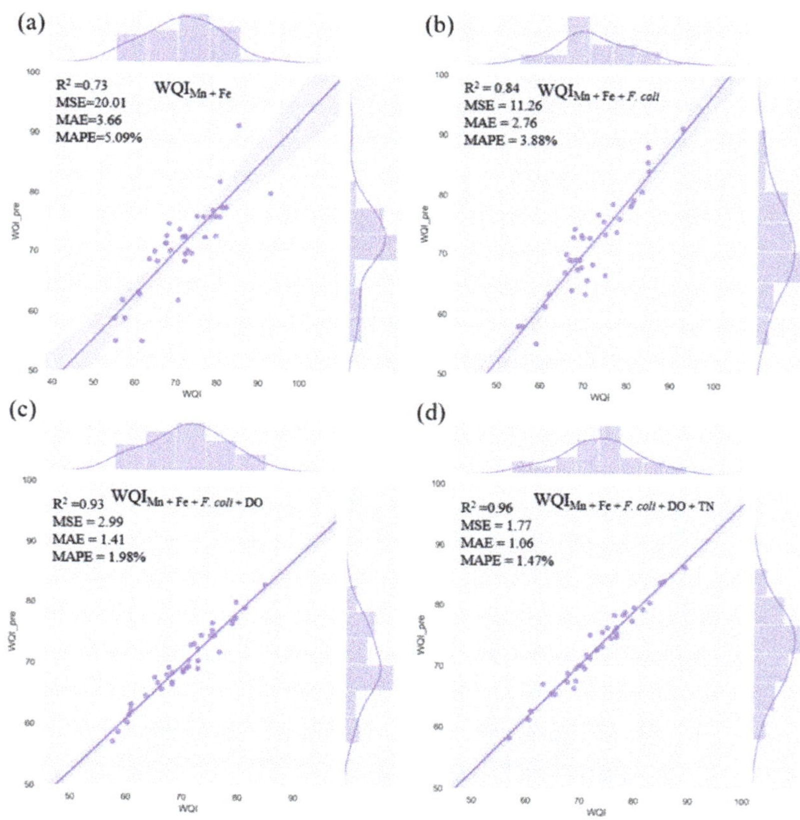

Figure 5. Comparison of the WQI_{RFmin} values based on different groups of parameters. (**a**) Mn + Fe, (**b**) Mn + Fe + *F. coli*, (**c**) Mn + Fe + *F. coli* + DO, (**d**) Mn + Fe + *F. coli* + DO + TN.

3.4. Pollution Source Apportionment Using the PMF Model

According to a quantitative analysis of pollution sources based on PMF, five factors were determined for the surface water of the study area (Figure 6). F1 was characterised as microbial contamination because of the high percentage contribution of *F. coli* (87.4%), which could be attributed to sewage discharge, potentially from a leak due to a sewer system malfunction [5]. F2 was characterised by high weightings of TN (67.2%), F^- (61.3%),

SO_4^{2-} (81.6%), Cl^- (80.6%), and NO_3^- (69.0%). A large amount of rural land is distributed in the upstream region of the M River. Considering that fertilisers might be transported with surface runoff and discharged into the river, frequent agricultural activities might have been the main cause of the high levels of nitrogen [25], and F2 could be attributed to non-point source agricultural pollution [26]. F3 was the main contributor of WT (53.6%), DO (58.5%), and COD_{Mn} (56.4%), as well as TP and TN; therefore, F3 may correspond to unexplainable variability, which may be the result of a combination of natural factors and urban domestic sewage [22]. F4 was characterised by a significant contribution of TP (73.3%), which is an important indicator of eutrophication; hence, F4 may represent nutrient pollution, which could include runoff pollution from urban areas [34]. The contribution rates of F5 were concentrated on Fe (79.3%) and Mn (93.7%), representing the impact of heavy metal pollution. The Fe and Mn concentrations in the M River increased significantly from upstream to downstream, indicating the external input of heavy metals in the study area, for example, from the local mining industry.

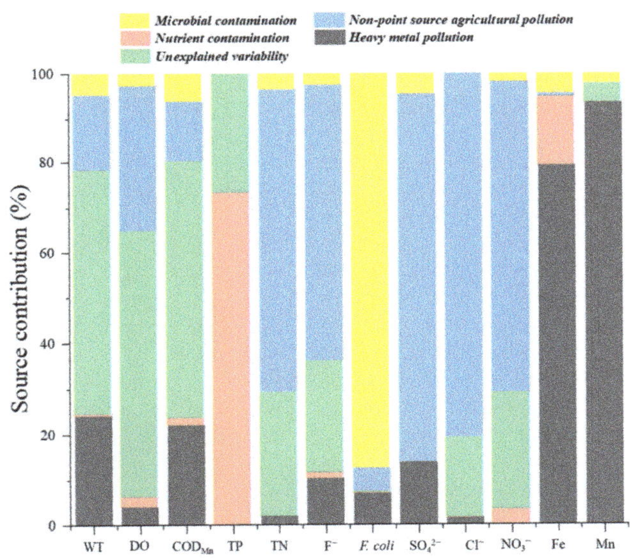

Figure 6. Contributions of pollution sources to the selected water quality variables.

3.5. Contribution of Pollution Sources to Variation of WQI Value

The contributions of each potential pollution source to the variation in the WQI values were calculated (Table 4). Heavy metal pollution had the greatest impact on the WQI values, with a contribution of 53.18%, and the Fe and Mn concentrations increased significantly from the upper reach to the lower reach, which had a significant impact on the overall water quality. Therefore, close attention should be given to heavy metal pollution of the M River. The second largest contributor was microbial contamination (*F. coli*, 18.15%), which fluctuated widely in the M River and played a critical role in the WQI value. Non-point source agricultural pollution contributed significantly to many water quality parameters, but its contribution to the variation in the WQI values was only 9.64%. The concentrations of F^-, SO_4^{2-}, Cl^-, and NO_3^- were generally stable. The TN concentration was relatively high for a long time and severely exceeded the Class III standard; however, its impact on the water quality evaluation was not significant. The contribution of nutrient contamination was 6.73%, which was primarily due to TP; however, TP was of a relatively good status for a long time and did not play a key role in the comprehensive evaluation of water quality. Unexplained variability contributed 10.95% to the variation in the WQI values, in which DO was a crucial water quality parameter for the WQI.

Table 4. Contribution of pollution sources to the variation in WQI values.

Pollution Sources	Microbial Contamination	Non-Point Source Agricultural Pollution	Unexplained Variability	Nutrient Contamination	Heavy Metal Pollution	Model Error
Contribution (%)	18.15	9.64	10.95	6.73	53.18	1.35

4. Discussion

4.1. Quantitative Assessment of the Impact of Pollution Sources on Water Quality

The WQI can comprehensively evaluate the status of water quality. For the trained WQI_{RF} model based on RFs, according to the analysis of the model's feature importance, the proposed WQI_{RFmin} model in this study consisted of five key water quality parameters, that is, Mn, Fe, *F. coli*, DO, and TN, and exhibited a very good performance for water quality evaluations. The selected parameters of the WQI_{RFmin} model should be able to comprehensively explain the overall variations and characteristics of water quality and should be conducive for efficiently evaluating water quality with relatively lower measurement costs [3]. Five potential pollution sources were obtained using the PMF method. Because the RF model could assess the importance of each parameter in the model, the feature importance of each water quality parameter in the WQI_{RF} could be calculated. The contribution of each potential pollution source to the variation in the WQI values was quantitatively assessed by multiplying the feature importance of each water quality indicator by the contribution of the source to each water quality indicator in the PMF model and then accumulating them.

Previous studies have used the WQI to assess surface water quality in many areas [2,3,8,9,35], and many studies have also analysed potential pollution sources of surface water [36–38]. However, the determination of most pollution sources and their effects are usually based on the personal experience of the researcher and the qualitative judgement of the local survey information [26].

Few studies have quantitatively analysed the impact of pollution sources on the water quality assessment. Although some pollution sources provided a higher pollution contribution rate to water quality parameters in this study, the contribution of the pollution source to the WQI values was not enough to change the WQI values; this, the actual impact of these sources on the water quality assessment was not significant. Through the quantitative analysis of the relationship between pollution sources and the WQI values, it is possible to (i) obtain the pollution sources that have a substantial impact on water quality evaluation, (ii) clarify the focus of water pollution management, and (iii) provide relevant departments with a reasonable water resource protection strategy.

From the perspective of water quality evaluation, this study systematically analysed the water quality of the M River basin and obtained five important water quality indicators through the ML method. From the perspective of pollution source analysis, this study identified potential pollution sources and quantitatively analysed the impact of pollution sources on water quality evaluation.

The method used in this study identified the most important potential sources of pollution in terms of their effect on the WQI score. Nevertheless, the disadvantage of using the receptor PMF model to determine the potential sources of pollution in surface water is that the source of pollution to a waterbody cannot be clearly identified. If the potential sources of pollution can be identified by this method for targeted pollution control, and subsequent water samples can be collected and compared for water quality analysis, the results of present studies could be verified. Moreover, the important water quality indicators and water quality characteristics could also be analysed before and after pollution control.

4.2. Advantages and Innovation of RFs in the Construction of the WQI_{min} Model

In previous studies, scholars generally used the stepwise multiple linear regression method to develop the WQI_{min} models [3,10], which were evaluated based on R^2, MSE, and percentage error (PE) values to select important water quality indicators. Compared with previous studies, the data distribution of WQI values in the present study was wide and the model was relatively difficult to construct. The WQI_{min} obtained with the above method did not perform well on the testing set, in which PE > 10% [10].

In recent years, ML has shown excellent performance in regression models, and has attracted increasing attention for use in academia and industry. The RF-based WQI_{RFmin} model in this study exhibited a better performance and yielded more stable results compared with the traditional stepwise multiple linear regression method (Figure S1). In recent years, some research has focused on combining ML with individual water quality indicators. Chen et al. used ML methods to classify surface water quality with only a few water quality parameters [17]. However, the national standards for surface water quality evaluation in China still use a single-indicator evaluation method. There are relatively few studies on the combination of ML and comprehensive water quality assessment. The use of RFs combined with the WQI method in this study is a novel attempt to use ML for water quality assessment. Given the rapid development of artificial intelligence and big data, ML and deep learning can be combined with water quality assessment, water quality warning systems, and other related water quality research in the future.

5. Conclusions

The main conclusions are as follows: (1) The main water quality parameters of the M River that exceeded the Class III standards were TN, *F. coli*, Fe, and Mn. The WQI results indicated that the water quality of the M River was 'good' overall, with an overall average WQI value of 72.11. The average WQI values of the four monitoring stations ranged from 68.31 to 77.16, and there was a clear trend of deterioration from upstream to downstream. (2) The feature importance of each water quality parameter in the WQI_{RF} model was quantitatively assessed, and five parameters (Mn, Fe, *F. coli*, DO, and TN) were selected as key water quality parameters for establishing the WQI_{RFmin} model, which had good accuracy ($R^2 = 0.96$). (3) The PMF method was applied to identify five pollution sources and to apportion their contributions to each water quality parameter. (4) Quantitative assessment of the impact of pollution sources on water quality showed that pollutions sources were ranked as: heavy metal pollution (53.18%) > microbial contamination (18.15%) > non-point source agricultural (9.64%) > nutrient contamination (6.73%), while the unexplained variability accounted for 10.95% of the total.

The methods used in this study to analyse the water quality of the M River could reduce the measurement cost of water quality assessment and effectively improve the measurement efficiency. In addition, the findings provide support for formulating water quality management strategies. The methods of selecting key water quality parameters and of assessing the quantitative contributions of pollution sources to the variation in the WQI values could be practically applied to other surface waters to greatly improve our understanding of the overall water quality condition. Additional studies will be required to assess precisely the unidentified sources of pollution and variation of further water quality parameters that were not analyzed in this study.

However, water pollution is a complex process, and more factors will affect the migration and transformation of pollutants. Therefore, we should continue to improve the research methods and technical means, and explore the methods and theories of traceability of exceeded pollutants at both qualitative and quantitative levels. It is necessary to verify and analyze the existing results, optimize the sampling scheme, and establish a model of the relationship between environmental variables and water pollutants. This will be a major direction for future development.

Supplementary Materials: The following supporting information can be downloaded at: https://www.mdpi.com/article/10.3390/ijerph20010881/s1, Figure S1: Comparison of the WQI and WQI_{LRmin} values from the stepwise multiple linear regression based on the testing dataset; Table S1: Water quality characteristic; Table S2. Weights and normalization factors of the parameters used in the calculation of the water; Table S3. The parameter selection results of the WQI_{LRmin} models from the stepwise multiple linear regression.

Author Contributions: All authors contributed to the study conception and design. Conceptualization, methodology, software, formal analysis, and writing: Y.Z.; validation and writing: X.W.; methodology and software: S.Z.; validation: L.J.; supervision, project administration, funding acquisition: W.L. All authors have read and agreed to the published version of the manuscript.

Funding: This research was funded by the National Natural Science Foundation of China (No. 51979194) and Cross-regional Joint Pre-vention and Control Mechanism and Strategic Scientific Research Program for Water Quality Bi-osafety Risks in Upper Yangtze River (No. 2021-YB-CQ-3). We also thank the research on water quality stability characteristics and countermeasures of the Fuzhou Water Supply System (Project NO. 20203000) from Fuzhou Water Group Co. Ltd., China and the Comparative study on corrosion characteristics of pipes and microbial safety of water quality in water supply system (Project NO. kh0040020191200).

Institutional Review Board Statement: Not applicable.

Informed Consent Statement: Not applicable.

Data Availability Statement: Not applicable.

Conflicts of Interest: The authors declare no conflict of interest.

Nomenclature

WQI: Water quality index; PMF: Positive matrix factorization; ML: Machine learning; RF: Random forests; APCS-MLR: Absolute principal components score combined with multivariate linear regression; WWP: West District Water Plant; FWP: Fei Fengshan Water Plant; SWP: Southeast District Water Plant; CWP: Chengmen Water Plant; MSE: Mean square error; MAE: Mean absolute error; MAPE: Mean absolute percentage error; CV: Coefficient of variation; WT: Water temperature; TN: Total nitrogen; TP: Total phosphorus; DO: Dissolved oxygen.

References

1. Niu, A.P.; Song, L.Y.; Xiong, Y.H.; Lu, C.J.; Junaid, M.; Pei, D.S. Impact of water quality on the microbial diversity in the surface water along the Three Gorge Reservoir (TGR), China. *Ecotoxicol. Environ. Saf.* **2019**, *181*, 412–418. [CrossRef] [PubMed]
2. Qu, X.; Chen, Y.S.; Liu, H.; Xia, W.T.; Lu, Y.; Gang, D.D.; Lin, L.S. A holistic assessment of water quality condition and spatiotemporal patterns in impounded lakes along the eastern route of China's South-to-North water diversion project. *Water Res.* **2020**, *185*, 116275. [CrossRef] [PubMed]
3. Nong, X.Z.; Shao, D.G.; Zhong, H.; Liang, J.K. Evaluation of water quality in the South-to-North Water Diversion Project of China using the water quality index (WQI) method. *Water Res.* **2020**, *178*, 115781. [CrossRef] [PubMed]
4. Shukla, K.; Dadheech, N.; Kumar, P.; Khare, M. Regression-based flexible models for photochemical air pollutants in the national capital territory of megacity Delhi. *Chemosphere* **2021**, *272*, 129611. [CrossRef] [PubMed]
5. Liu, L.L.; Dong, Y.C.; Kong, M.; Zhou, J.; Zhao, H.B.; Tang, Z.; Zhang, M.; Wang, Z.P. Insights into the long-term pollution trends and sources contributions in Lake Taihu, China using multi-statistic analyses models. *Chemosphere* **2020**, *242*, 125272. [CrossRef] [PubMed]
6. Rodell, M.; Famiglietti, J.S.; Wiese, D.N.; Reager, J.T.; Beaudoing, H.K.; Landerer, F.W.; Lo, M.H. Emerging trends in global freshwater availability. *Nature* **2018**, *557*, 650. [CrossRef]
7. McMillen, C. Water and the death of ambition in global health, c.1970–1990. *Hist. Cienc. Saude-Manguinhos* **2020**, *27*, 211–230. [CrossRef]
8. Gao, Q.; Li, Y.; Cheng, Q.Y.; Yu, M.X.; Hu, B.; Wang, Z.G.; Yu, Z.Q. Analysis and assessment of the nutrients, biochemical indexes and heavy metals in the Three Gorges Reservoir, China, from 2008 to 2013. *Water Res.* **2016**, *92*, 262–274. [CrossRef]
9. Hurley, T.; Sadiq, R.; Mazumder, A. Adaptation and evaluation of the Canadian Council of Ministers of the Environment Water Quality Index (CCME WQI) for use as an effective tool to characterize drinking source water quality. *Water Res.* **2012**, *46*, 3544–3552. [CrossRef]

10. Wu, Z.S.; Wang, X.L.; Chen, Y.W.; Cai, Y.J.; Deng, J.C. Assessing river water quality using water quality index in Lake Taihu Basin, China. *Sci. Total Environ.* **2018**, *612*, 914–922. [CrossRef]
11. Noori, R.; Berndtsson, R.; Hosseinzadeh, M.; Adamowski, J.F.; Abyaneh, M.R. A critical review on the application of the National Sanitation Foundation Water Quality Index. *Environ. Pollut.* **2019**, *244*, 575–587. [CrossRef] [PubMed]
12. Nong, X.Z.; Shao, D.G.; Xiao, Y.; Zhong, H. Spatio-Temporal Characterization Analysis and Water Quality Assessment of the South-to-North Water Diversion Project of China. *Int. J. Environ. Res. Public Health* **2019**, *16*, 2227. [CrossRef]
13. Uddin, M.G.; Nash, S.; Olbert, A.I. A review of water quality index models and their use for assessing surface water quality. *Ecol. Indic.* **2021**, *122*, 107218. [CrossRef]
14. Kocer, M.; Sevgili, H. Parameters selection for water quality index in the assessment of the environmental impacts of land-based trout farms. *Ecol. Indic.* **2014**, *36*, 672–681. [CrossRef]
15. Pak, H.Y.; Chuah, C.J.; Tan, M.L.; Yong, E.L.; Snyder, S.A. A framework for assessing the adequacy of Water Quality Index—Quantifying parameter sensitivity and uncertainties in missing values distribution. *Sci. Total Environ.* **2021**, *751*, 141982. [CrossRef] [PubMed]
16. Pesce, S.F.; Wunderlin, D.A. Use of water quality indices to verify the impact of Cordoba City (Argentina) on Suquia River. *Water Res.* **2000**, *34*, 2915–2926. [CrossRef]
17. Chen, K.Y.; Chen, H.X.; Zhou, C.L.; Huang, Y.C.; Qi, X.Y.; Shen, R.Q.; Liu, F.R.; Zuo, M.; Zou, X.Y.; Wang, J.F.; et al. Comparative analysis of surface water quality prediction performance and identification of key water parameters using different machine learning models based on big data. *Water Res.* **2020**, *171*, 115454. [CrossRef] [PubMed]
18. Lu, H.F.; Ma, X. Hybrid decision tree-based machine learning models for short-term water quality prediction. *Chemosphere* **2020**, *249*, 126169. [CrossRef]
19. Roguet, A.; Eren, A.M.; Newton, R.J.; McLellan, S.L. Fecal source identification using random forest. *Microbiome* **2018**, *6*, 185. [CrossRef]
20. Tan, Q.W.; Li, W.Y.; Chen, X. Identification the source of fecal contamination for geographically unassociated samples with a statistical classification model based on support vector machine. *J. Hazard. Mater.* **2021**, *407*, 124821. [CrossRef]
21. Breiman, L. Random forests. *Mach. Learn.* **2001**, *45*, 5–32. [CrossRef]
22. Zhang, H.; Cheng, S.Q.; Li, H.F.; Fu, K.; Xu, Y. Groundwater pollution source identification and apportionment using PMF and PCA-APCA-MLR receptor models in a typical mixed land-use area in Southwestern China. *Sci. Total Environ.* **2020**, *741*, 140383. [CrossRef] [PubMed]
23. Paatero, P.; Tapper, U. Positive Matrix Factorization—A Nonnegative Factor Model with Optimal Utilization of Error-Estimates of Data Values. *Environmetrics* **1994**, *5*, 111–126. [CrossRef]
24. Yang, L.P.; Mei, K.; Liu, X.M.; Wu, L.S.; Zhang, M.H.; Xu, J.M.; Wang, F. Spatial distribution and source apportionment of water pollution in different administrative zones of Wen-Rui-Tang (WRT) river watershed, China. *Environ. Sci. Pollut. Res.* **2013**, *20*, 5341–5352. [CrossRef] [PubMed]
25. Salim, I.; Sajjad, R.U.; Paule-Mercado, M.C.; Memon, S.A.; Lee, B.Y.; Sukhbaatar, C.; Lee, C.H. Comparison of two receptor models PCA-MLR and PMF for source identification and apportionment of pollution carried by runoff from catchment and sub-watershed areas with mixed land cover in South Korea. *Sci. Total Environ.* **2019**, *663*, 764–775. [CrossRef] [PubMed]
26. Gholizadeh, M.H.; Melesse, A.; Reddi, L. Water quality assessment and apportionment of pollution sources using APCS-MLR and PMF receptor modeling techniques in three major rivers of South Florida. *Sci. Total Environ.* **2016**, *566*, 1552–1567. [CrossRef] [PubMed]
27. Xia, F.; Zhang, C.; Qu, L.Y.; Song, Q.J.; Ji, X.L.; Mei, K.; Dahlgren, R.A.; Zhang, M.H. A comprehensive analysis and source apportionment of metals in riverine sediments of a rural-urban watershed. *J. Hazard. Mater.* **2020**, *381*, 121230. [CrossRef]
28. Jian, X.; Zhang, W.; Yang, S.Y.; Kao, S.J. Climate-Dependent Sediment Composition and Transport of Mountainous Rivers in Tectonically Stable, Subtropical East Asia. *Geophys. Res. Lett.* **2020**, *47*, e2019GL086150. [CrossRef]
29. Wu, Z.S.; Lai, X.J.; Li, K.Y. Water quality assessment of rivers in Lake Chaohu Basin (China) using water quality index. *Ecol. Indic.* **2021**, *121*, 107021. [CrossRef]
30. Gislason, P.O.; Benediktsson, J.A.; Sveinsson, J.R. Random Forests for land cover classification. *Pattern. Recogn. Lett.* **2006**, *27*, 294–300. [CrossRef]
31. Strobl, C.; Boulesteix, A.L.; Kneib, T.; Augustin, T.; Zeileis, A. Conditional variable importance for random forests. *BMC Bioinform.* **2008**, *9*, 307. [CrossRef] [PubMed]
32. Parkhurst, D.F.; Brenner, K.P.; Dufour, A.P.; Wymer, L.J. Indicator bacteria at five swimming beaches—Analysis using random forests. *Water Res.* **2005**, *39*, 1354–1360. [CrossRef] [PubMed]
33. Bindal, S.; Singh, C.K. Predicting groundwater arsenic contamination: Regions at risk in highest populated state of India. *Water Res.* **2019**, *159*, 65–76. [CrossRef]
34. Zhu, G.W.; Wang, F.; Gao, G.; Zhang, Y.L. Variability of Phosphorus Concentration in Large, Shallow and Eutrophic Lake Taihu, China. *Water Environ. Res.* **2008**, *80*, 832–839. [CrossRef] [PubMed]
35. Milojkovic, J.V.; Popovic-Djordjevic, J.B.; Pezo, L.L.; Brceski, I.D.; Kostic, A.Z.; Milosevic, V.D.; Stojanovic, M.D. Applying multi-criteria analysis for preliminary assessment of the properties of alginate immobilized Myriophyllum spicatum in lake water samples. *Water Res.* **2018**, *141*, 163–171. [CrossRef]

36. Zanotti, C.; Rotiroti, M.; Fumagalli, L.; Stefania, G.A.; Canonaco, F.; Stefenelli, G.; Prevot, A.; Leoni, B.; Bonomi, T. Groundwater and surface water quality characterization through positive matrix factorization combined with GIS approach. *Water Res.* **2019**, *159*, 122–134. [CrossRef]
37. Jafarabadi, A.R.; Raudonyte-Svirbutaviciene, E.; Toosi, A.S.; Bakhtiari, A.R. Positive matrix factorization receptor model and dynamics in fingerprinting of potentially toxic metals in coastal ecosystem sediments at a large scale (Persian Gulf, Iran). *Water Res.* **2021**, 188. [CrossRef]
38. Zhang, Y.; Guo, C.S.; Xu, J.; Tian, Y.Z.; Shi, G.L.; Feng, Y.C. Potential source contributions and risk assessment of PAHs in sediments from Taihu Lake, China: Comparison of three receptor models. *Water Res.* **2012**, *46*, 3065–3073. [CrossRef]

International Journal of
Environmental Research and Public Health

MDPI

Article

An Assessment of Temporal and Spatial Dynamics of Regional Water Resources Security in the DPSIR Framework in Jiangxi Province, China

Mengtian Lu [1], Siyu Wang [2], Xiaoying Wang [3], Weihong Liao [4,*], Chao Wang [4], Xiaohui Lei [4] and Hao Wang [1,4]

1 Institute of Municipal Engineering, College of Civil Engineering and Architecture, Zhejiang University, Hangzhou 310030, China; 11512052@zju.edu.cn (M.L.); wanghao@iwhr.com (H.W.)
2 Department of Psychology, University of Arizona, Tucson, AZ 85721, USA; sywangr@email.arizona.edu
3 Anhui Water Conservancy Technical College, Hefei 231603, China; wxy769662707@163.com
4 Key Laboratory of Simulation and Regulation of Water Cycle in River Basin, China Institute of Water Resources and Hydropower Research, Beijing 100038, China; wangchao@iwhr.com (C.W.); lxh@iwhr.com (X.L.)
* Correspondence: liaowh@iwhr.com

Abstract: Water resources are critical for the survival and prosperity of both natural and socioeconomic systems. A good and informational water resources evaluation system is substantial in monitoring and maintaining sustainable use of water. The Driver-Pressure-State-Impact-Response (DPSIR) framework is a widely used general framework that enabled the measurement of water resources security in five different environmental and socioeconomic subsystems: driver, pressure, state, impact, and response. Methodologically, outcomes of water resources evaluation based on such framework and using fuzzy set pair analysis method and confidence interval rating method depend critically on a confidence threshold parameter which was often subjectively chosen in previous studies. In this work, we demonstrated that the subjectivity in the choice of this critical parameter can lead to contradicting conclusions about water resources security, and we addressed this caveat of subjectivity by proposing a simple modification in which we sample a range of thresholds and pool them to make more objective evaluations. We applied our modified method and used DPSIR framework to evaluate the regional water resource security in Jiangxi Province, China. The spatial-temporal analysis of water resources security level was carried out in the study area, despite the improvement in Pressure, Impact, and Response factors, the Driver factor is found to become less safe over the years. Significant variation of water security across cities are found notably in Pressure and Response factors. Furthermore, we assessed both cross-sectionally and longitudinally the inter-correlations among the DPSIR nodes in the DPSIR framework. The region-specific associations among the DPSIR nodes showed important deviances from the general DPSIR framework, and our analysis showed that in our study region, although Responses of regional government work effectively in improving Pressure and State security, more attention should be paid to improving Driver security in future regional water resources planning and management in Jiangxi Province, China.

Keywords: water resources security; DPSIR; confidence threshold method

Citation: Lu, M.; Wang, S.; Wang, X.; Liao, W.; Wang, C.; Lei, X.; Wang, H. An Assessment of Temporal and Spatial Dynamics of Regional Water Resources Security in the DPSIR Framework in Jiangxi Province, China. *Int. J. Environ. Res. Public Health* **2022**, *19*, 3650. https://doi.org/10.3390/ijerph19063650

Academic Editor: Dong-Chan Koh

Received: 19 January 2022
Accepted: 16 March 2022
Published: 19 March 2022

1. Introduction

Water resources are the basis of human survival and development and are irreplaceable natural resources for sustainable economic and social development [1]. Since the 1970s, the rapid growth of world population and the rapid development of the global economy have led to the rapid growth of global water consumption and water pollution [2,3]. In recent years, under the influence of global climate change and high-intensity human activities, the water cycle and the spatial and temporal distribution of water resources have undergone complicated changes. The complexity of hydrological characteristics and the insecurity of water resources increased substantially [4,5]. Therefore, water resources security evaluation and the selection of appropriate evaluation methods is of critical importance in monitoring the sustainable use of water resources and guiding countries and regions to maintain socially sustainable development [6,7].

To evaluate water resources security, scholars have come up with various indicators to measure the degree of regional water resources security, such as per capita water resources and water resources vulnerability index [8,9]. Measuring water resources security via per capita water resources is proposed by Falkenmark et al. [10]. The vulnerability index of water resources refers to the percentage of annual freshwater resources taken up in the total amount of available or renewable freshwater resources. Raskin et al. [11] used a vulnerability index and classified water resource pressure as low, medium low, medium high, and high based on the degree of water resources usage. Other commonly used water resources security evaluation indicators include the water resources development and utilization index [12], the water allocation and priority strategy index [13], and water poverty index [14].

Recently, more and more studies evaluated water resources security from a multi-dimensional perspective that utilizes a system of indicators from different domains [15–20]. Various multi-dimensional water resources evaluation frameworks and methods have been developed, including methods based on catastrophe theory [15], system dynamics model (SDM) [16,17], process analysis method (PAM) [21], WaterGAP3 modeling framework [18], projection pursuit model [19], and multistage integrated method [20]. For a comprehensive review of water resources evaluation tools, see [22]. Among them, one of the most commonly used frameworks is the Driver-Pressure-State-Impact-Response (DPSIR) framework. Compared to other frameworks like SDM and PAM, the DPSIR framework includes more measures and is more flexible [23]. The DPSIR framework has been widely applied in water resources and ecological security assessment studies [19,24–44]. Some primitive versions and new variants of the DPSIR framework were also used in the literature, e.g., PSR model [45,46], DPSI model [33], PSIR model [47,48], DPSR model [49], and DPSIRM model [36,50].

The DPSIR model was proposed by the European Environmental Agency in 1995 [51,52] and has been widely used in policymaking and research. The DPSIR model has the advantage of linking among several components in the water resources security assessment system, and it allows for analyzing the coupling relationship between natural environment resources and human activities. The DPSIR model aims to establish a causal chain of Driver-Pressure-State-Impact-Response, and these five different sub-systems have different implications [24,27,30,33,53]. "Driver" refers to the socio-economic or socio-cultural factors that promote the increase or decrease of water system pressure. The Driver sub-system includes factors like population growth, prosperity level, social or technological change, etc. "Pressures" is mainly reflected by the direct pressure of human behavior on natural resources and the environment. The Pressures sub-system includes factors like water usage and wastewater discharge. "State" is the condition of the environment under various pressure factors. The State sub-system includes factors like water resource quantity and quality. "Impact" refers to the consequences of environmental conditions, which represents the observable positive or negative results, such as human health impact or vegetation damage. "Response" indicates the countermeasures taken by mankind in the process of promoting sustainable development, such as improving resource utilization efficiency, reducing pollution, increasing investment, etc. In summary, the DPSIR model evaluates threats from social, economic, and human activities to regional water resources security and the human responses to these threats.

China is a country with serious uneven spatial and temporal distribution of water resources, and water resources problems are very prominent. Water resources shortage, drought, and flood disasters and water ecological environment problems have become important factors restricting China's economic development. Water resources security evaluation has received significant interests among scholars [7,16,20,27,30,36,45,54–60]. Regional water resources security evaluation has been done at both river basins [37,43,44,61–63] and urban areas which are analyzed both at the level of individual cities [15,41,45,46,58,64,65] and at the level of provinces [16,20,27,47,54,56,65–68]. Compared to other provinces in China, Jiangxi is a province relatively rich in water resources [65]. However, the temporal and spatial distribution of water resources in Jiangxi is uneven, and seasonal water resources are scarce. With the rapid development of the economy and the acceleration of urbanization, the contradiction between supply and demand of water resources is becoming increasingly prominent, the quality of water environment is declining year by year, and water pollution emergencies occur from time to time [66]. How to reasonably develop water resources and achieve sustainable utilization is an important and arduous task.

In this paper, we evaluated the regional water resources security in Jiangxi Province, China, using the DPSIR framework. The study period was chosen based on the availability of data in Jiangxi Province, China. The entropy weight method [69] was used to calculate the indices weight, and the fuzzy set pair analysis method [70] was used to evaluate the water resources security. Instead of using a subjective confidence threshold parameter to draw boundaries between Safe vs. Unsafe, which has been used by many studies in the literature [38–41], we demonstrated that the limitation of such method is that subjectivity in the choice of the confidence threshold could lead to contradictory conclusions. Furthermore, we addressed this issue of subjectivity by proposing a modified method that samples a range of thresholds to obtain a more objective measure of water resources security. We evaluated the temporal and spatial dynamics of water resources security in the Driver force, Pressure, State, Impact, and Responses domains in 11 cities in Jiangxi Province over the period of 2010–2018 using our modified method. In addition, we empirically assessed the inter-correlations among the DPSIR nodes over time and space using repeated measures correlation. Our analysis revealed a more complicated and region-specific flows of interactions in the DPSIR framework. Our approach estimates region-specific sensitivities and associations among the DPSIR sub-systems and can use this information to better guide local policy makers on improving the carrying capacity of water resources and strengthening the sustainable development of economy, society, and water resources in the Jiangxi Province.

2. Study Area and Datasets

2.1. Study Area

The Jiangxi Province (113°34′36″–118°28′58″ E, 24°29′14″–30°04′41″ N) lies in the southern bank of the middle and lower reaches of the Yangtze River and is one of the most important inland provinces in China. It belongs to East China, with a land area of 166,900 square kilometers, accounting for 1.8% of the national land area, with a total population of more than 46.47 million and jurisdiction over 11 prefecture level cities, namely, Nanchang, Jingdezhen, Pingxiang, Jiujiang, Xinyu, Yingtan, Ganzhou, Ji'an, Yichun, Fuzhou, and Shangrao. The geographical locations of the cities are shown in Figure 1. In this paper, we will evaluate the water resources safety of these 11 cities in Jiangxi Province using the DPSIR framework.

2.2. Datasets

The dataset used in this paper to conduct the water resources security evaluation comes from the annual statistical data and official documents approved by regional governmental departments, including China Statistical Yearbook, Jiangxi statistical yearbook, and Jiangxi water resources bulletin. It could be accessed through China's economic and social big data research platform (https://data.cnki.net/NewHome/index (accessed on 20 November 2020)) and National Bureau of Statistics website (http://data.stats.gov.cn/ (accessed on 20 November 2020)). Given the integrity and availability of data, the study period is chosen to be from 2010 to 2018. Among them, most evaluation index data of driver (D), pressure (P), state (S), influence (I), response (R) can be obtained directly, and some indexes can be obtained by index calculation method.

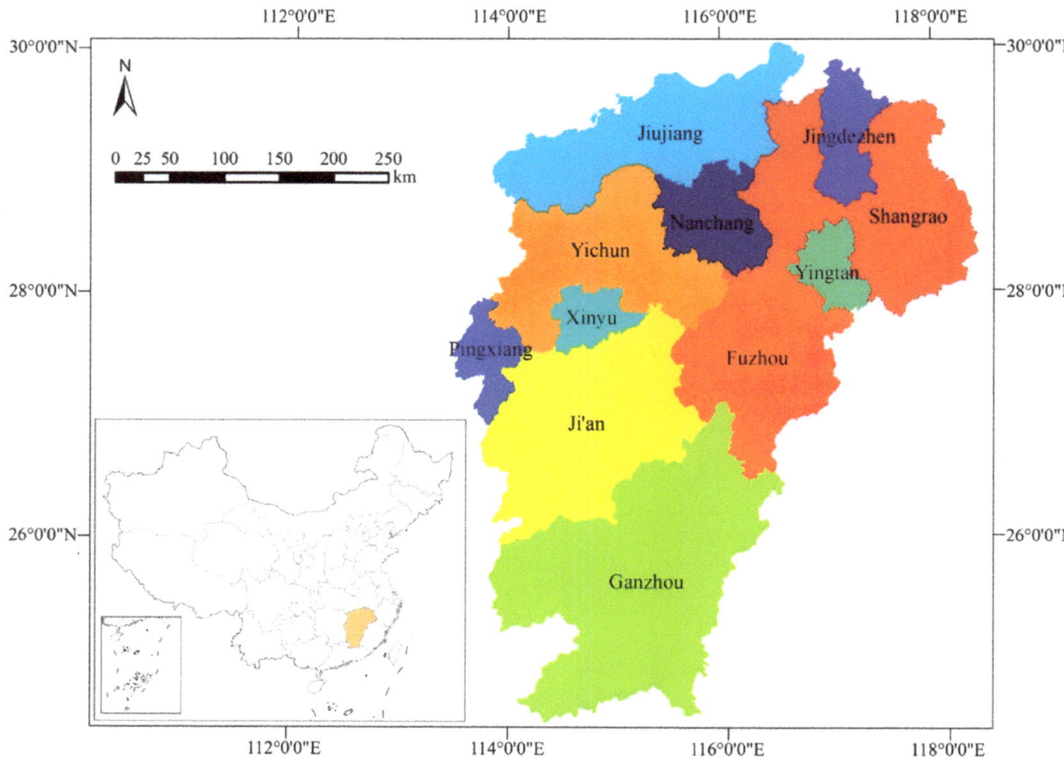

Figure 1. Study area: Jiangxi Province, China.

The whole year of Jiangxi has mild climate, with sufficient sunshine, plentiful rainfall, and long frost-free period, which belongs to the subtropical warm and humid monsoon climate. The average annual temperature is about 16.3–19.5 °C. Jiangxi Province is one of the rainy provinces in China, with annual precipitation of 1341–1943 mm. Moreover, 97.7% of the area of Jiangxi belongs to the Yangtze River Basin, with rich water resources and dense river network. There are more than 2400 large and small rivers in Jiangxi Province, with a total length of about 18,400 km. The average annual water resources amount to 156.5 billion cubic meters. The total runoff is the seventh in China, the per capita is the fifth, and the cultivated land is the sixth. The annual average value of groundwater natural resources is more than 21.2 billion cubic meters, accounting for 2.3% of the national natural groundwater resources.

3. Methodology

3.1. Water Resources Security Evaluation Index

In this paper, water resources security is evaluated using the DPSIR framework. One domain-specific index is calculated respectively for each of the five subdomains, namely, Driver, Pressure, State, Impact, and Response. These indices are computed for each of the 11 cities and for each year separately. Fifteen social economic factors are used in the calculation of the indices of water resources security in Jiangxi Province. These factors are classified based on the nature of the factor into one of the five subsystems of DPSIR (see Table 1). All the factors used in this paper were well-accepted in the field and were used by multiple studies in the previous literature; the relevant references to each of the 15 factors are listed in the last column of Table 1.

Table 1. Social economic factors used in the DPSIR system. Here, index type "+" indicates that higher values are more desirable for the particular factor whereas "-" indicates that lower values are more desirable.

Subsystem	Factors	Unit	Calculation	Meaning of Index	Index Type	References
Driver	Per capita GDP (D1)	yuan/person	GDP/population	It indicates the driver of economic development on water resources security	+	[47,48,71,72]
	Population density (D2)	person/km^2	Total population/ land area	It indicates the driver of population density on water security	-	[47,60,72,73]
	Urbanization rate (D3)	%	Non-agricultural population/total population	It indicates the driver of regional development on water resources security	+	[15,47,48,60]
	Annual GDP growth rate (D4)	%	Statistical data	It indicates the driver of economic development intensity to water resources security	+	[15,47,60]
Pressure	Water use for each 10,000 yuan of GDP (P1)	m^3/10,000 yuan	Total amount of water use/ total GDP	It indicates the pressure of economic development intensity on the quantity of water resources	-	[47,60,71,72]
	Wastewater discharge for each 10,000 yuan of GDP (P2)	m^3/10,000 yuan	Wastewater discharge/ total GDP	It indicates the pressure of industrial development on the quality of water resources	-	[47,60]
	Water use for each 10,000 yuan of agricultural output (P3)	m^3/10,000 yuan	Total amount of irrigated water use/total output value of farming	It indicates the pressure of agricultural development on the quality of water resources	-	[47,60,71]
	Per capita daily consumption of tap water for residential use (P4)	L/day	Statistical data	It indicates the pressure of human life on the quantity of water resources	-	[47,60,73,74]
State	Per capita water resource quantity (S1)	m^3/person	Total amount of water resources/total population	It indicates the per capita state of water resources	+	[15,23,47,60,71, 74–77]
	Per unit area water resource quantity (S2)	10,000 m^3/km^2	Total amount of water resources/ land area	It indicates the per unit area water resource state.	+	[23,72]
Impact	Energy consumption for each 10,000 yuan of GDP (I1)	Tons of SCE /10,000 yuan	Total Energy Composition/ total GDP	It indicates the Potential impact of resource utilization on water resources	-	[74]
	Rate of green coverage area to developed area (I2)	%	Statistical data	It indicates the response of surface water storage to water resources	+	[60,71,74]
	Proportion of tertiary industry in GDP (I3)	%	Statistical data	It indicates the impact of water resources system on industrial structure	+	[47,60]
Response	Utilization rate of water resources (R1)	%	Total amount of water use/total amount of water resources	It indicates the response of water resources quantity security	-	[15,47,72]
	Urban sewage treatment rate (R2)	%	Statistical data	It indicates the response of standard discharge of sewage	+	[60,71,74–76]

GDP: gross domestic product; SCE: standard coal equivalent.

3.2. Individual Factor Grade Intervals

Based on the current domestic research results in China and the regional situation of the Jiangxi Province [57], as the first step of the pipeline, for each of the 15 factors (See Table 2), we assign the following 5 security levels: 1—safe, 2—generally safe, 3—barely safe, 4—unsafe, and 5—very unsafe. The specific boundaries used here are chosen based on both standards that have been used in the literature and standards and planning objectives issued by the local government and protection requirements of river water system [58,65,66]. For details, see Table 2, for specific water resources safety classification boundaries.

Table 2. Grades of water resource security evaluation.

Factor Level	Index Level	Index Type	1—Safe	2—Generally Safe	3—Barely Safe	4—Unsafe	5—Very Unsafe
Driver (D)	D1 (yuan)	+	>75,000	55,000–75,000	35,000–55,000	15,000–35,000	<15,000
	D2 (person/km^2)	-	<250	250–2000	2000–3750	3750–5500	>5500
	D3 (%)	+	>70	50–70	30–50	10–30	<10
	D4 (%)	+	>10	8–10	5–8	3–5	<3
Pressure (P)	P1 (m^3)	-	<300	300–600	600–1000	1000–1500	>1500
	P2 (m^3)	-	<20	20–30	30–40	40–60	>60
	P3 (m^3)	-	<500	500–1000	1000–1500	1500–2000	>2000
	P4 (L/day)	-	<70	70–120	120–170	170–220	>220
State (S)	S1 (m^3)	+	>3000	1700–3000	1000–1700	500–1000	<500
	S2 (10,000 m^3/km^2)	+	>200	200–150	150–100	100–50	<50
Impact (I)	I1 (Tons of SCE)	-	<0.5	0.5–1	1–2	2–5	>5
	I2 (%)	+	>40	30–40	20–30	10–20	<10
	I3 (%)	+	>60	40–60	20–40	5–20	<5
Response(R)	R1 (%)	-	<5	5–15	15–30	30–45	>45
	R2 (%)	+	>90	80–90	70–80	60–70	<60

3.3. Fuzzy Security Level Using Fuzzy Set Pair Analysis

To better quantify the dynamics of the DPSIR factors, instead of directly using the classification threshold in Table 2, the water security levels of each of the 5 DPSIR categories are calculated using Fuzzy Set Pair Analysis (FSPA). FSPA is a special case of Set Pair Analysis (SPA) which is a systematic analysis method established by Zhao 1989 [78]. SPA has been applied to various canonical set pairs in the field of hydrology and water resources, e.g., flood vs. drought, qualified vs. unqualified, safety vs. danger [17,70,79,80]. FSPA applies fuzzy logic theory to SPA and accounts for system uncertainty in addition to identity, by considering difference and opposition of two sets of each set pair. In dealing with problems with uncertainty such as evaluating the water security level, instead of using a hard coded boundary to categorize factors into Safe vs. Unsafe, FSPA represents the security level by quantifying the similarity of the current value to that of each security category. FSPA is proved to be effective and has been successfully applied to water resources evaluation in many studies [54,56,81–86].

3.3.1. Calculation of Index Connection Degree

Assume that the set of water resources security evaluation index system composed of n index values $X = \{x_1, x_2, \cdots, x_n\}$, $x_i \in \{1, 2, 3, \ldots, K\}$. K is the number of the evaluation grade standard, in our case $K = 5$ since we have 5 different water security levels. The evaluation grade standard is B_k $(k = 1, 2, \cdots, K)$, and here, the set of the kth level of the evaluation grade standard is $B_k = \{k, k, \ldots, k\}$. The similarity between the two sets X and B_k represents how close X is to a security level of k.

Here, we define the index connection degree between X and B_k, for each k. First, we compute the difference between each element of X and B_k, $|x_i - k|$. Then, we count the number of occurrences that $|x_i - k| = i$, and denote it as F_i. In particular F_0 is the number of times that $x_i = k$, and F_{k-1} is the number of times that $\{x_i, k\} = \{1, k\}$. Then the K-element connection degree of (X, B_k) is

$$\mu_{X \sim B_k} = b_0 I_0 + b_1 I_1 + b_2 I_2 + \cdots + b_{K-2} I_{K-2} + b_{K-1} I_{K-1} \tag{1}$$

where $b_i = \frac{F_i}{n}$ and $b_0 + b_1 + b_2 + \cdots + b_{K-2} + b_{K-1} = 1$

$I_1, I_2, \ldots, I_{k-1}$ are the coefficients of difference degree component, which can be determined by uniform value method e.g., $I_i = 1 - \frac{2i}{K-1}$; I_{K-1} is the coefficient of contrary degree which is usually taken as -1.

In this paper, B_k is specified as the set B_1 composed of the first level evaluation criteria of a certain index. In case of (X, B_1), $b_0 = \frac{F_0}{n}$ is the identical degree of (X, B_k) which represents the possibility of index X belonging to the 1st level of the standard; $b_1 = \frac{F_1}{n}$, $b_2 = \frac{F_2}{n}$, $b_{K-2} = \frac{F_{K-2}}{n}$ are the difference degree which represents the possibility of index X belonging to the 2nd, 3rd, and the $(K-1)$ level of the standard; $b_{K-1} = \frac{F_{K-1}}{n}$ is the contrary degree which represents the possibility of index x_t belonging to the Kth level.

Because the boundary of grade standard is fuzzy, the degree of connection $\mu_{X_t \sim B_1}$ can be calculated using Equations (2) and (3). X_t is the factor of interest (D1–D4, P1–P4, S1–S2, I1–I3, R1–R2) at year t. $s_1, s_2, \ldots, s_{K-1}$ are the grade boundaries for factor X_t as indicated in Table 2.

Generally, the indicators of water resources security evaluation can be divided into cost indicators (negative indicators) and benefit indicators (positive indicators). For the cost indicators (the smaller the better), when $K > 2$, the K-element connection degree of (X_t, B_1) is

$$\mu_{X_t \sim B_1} = \begin{cases} 1I_0 + 0I_1 + 0I_2 + \cdots + 0I_{K-2} + 0I_{K-1}, & x_t \le s_1; \\ \frac{s_1+s_2-2x_t}{s_2-s_1} I_0 + \frac{2x_t-2s_1}{s_2-s_1} I_1 + 0I_2 + \cdots + 0I_{K-2} + 0I_{K-1}, & s_1 < x_t \le \frac{s_1+s_2}{2}; \\ 0I_0 + \frac{s_2+s_3-2x_t}{s_3-s_1} I_1 + \frac{2x_t-s_1-s_2}{s_3-s_1} I_2 + \cdots + 0I_{K-2} + 0I_{K-1}, & \frac{s_1+s_2}{2} < x_t \le \frac{s_2+s_3}{2}; \\ \cdots & \\ 0I_0 + 0I_1 + \cdots + \frac{2s_{K-1}-2x_t}{s_{K-1}-s_{K-2}} I_{K-2} + \frac{2x_t-s_{K-1}-s_{K-2}}{s_{K-1}-s_{K-2}} I_{K-1}, & \frac{s_{K-2}+s_{K-1}}{2} < x_t \le s_{K-1}; \\ 0I_0 + 0I_1 + 0I_2 + \cdots + 0I_{K-2} + 1I_{K-1}, & x_t > s_{K-1} \end{cases} \quad (2)$$

where $s_1 \le s_2 \le \cdots \le s_{K-1}$.

For the benefit indicators (the bigger the better), when $K > 2$, the K-element connection degree of (X_t, B_1) is

$$\mu_{X_t \sim B_1} = \begin{cases} 1I_0 + 0I_1 + 0I_2 + \cdots + 0I_{K-2} + 0I_{K-1}, & x_t \ge s_1; \\ \frac{2x_t-s_1-s_2}{s_1-s_2} I_0 + \frac{2s_1-2x_t}{s_1-s_2} I_1 + 0I_2 + \cdots + 0I_{K-2} + 0I_{K-1}, & \frac{s_1+s_2}{2} \le x_t < s_1; \\ 0I_0 + \frac{2x_t-s_2-s_3}{s_1-s_3} I_1 + \frac{s_1+s_2-2x_t}{s_1-s_3} I_2 + \cdots + 0I_{K-2} + 0I_{K-1}, & \frac{s_2+s_3}{2} \le x_t < \frac{s_1+s_2}{2}; \\ \cdots & \\ 0I_0 + 0I_1 + \cdots + \frac{2x_t-2s_{K-1}}{s_{K-2}-s_{K-1}} I_{K-2} + \frac{s_{K-1}+s_{K-2}-2x_t}{s_{K-2}-s_{K-1}} I_{K-1}, & s_{K-1} \le x_t < \frac{s_{K-2}+s_{K-1}}{2}; \\ 0I_0 + 0I_1 + 0I_2 + \cdots + 0I_{K-2} + 1I_{K-1}, & x_t < s_{K-1} \end{cases} \quad (3)$$

where $s_1 \ge s_2 \ge \cdots \ge s_{K-1}$.

3.3.2. Determination of Index Weights

After the establishment of the evaluation index system, it is necessary to select an appropriate method to determine the evaluation index weight. At present, the widely used methods include analytic hierarchy process (AHP) [36], expert investigation method, BP neural network technology [87], and entropy weight method [69]. Compared with commonly used alternative methods such as AHP, which involves the construction of a subjective evaluation matrix, the entropy weight method is more objective and can instead objectively weigh the different indices based on implicit information in the index data. The entropy method has been adopted by many previous studies in water resources evaluation [42,44,54,59,60,69]. Thus, in this study, we used the entropy weight method.

Entropy weight method is often used in information theory to calculate index weight. In information theory, information is a measure of the order degree of the system, and entropy represents the disorder degree of the system. When the entropy of an index is smaller, the difference among the index data is more systematic, which indicates a greater influence of the index on the evaluated object, and hence, a greater weight should be assigned. On the contrary, the greater the entropy, the more random the index data, indicating a smaller influence of the index on the evaluated object, and hence, a smaller weight of the index should be assigned. The steps are as follows:

(1) Standardize the evaluation index:

Assume the water resources security evaluation system includes n evaluation objects (for each DPSIR domain, n is the number of years of evaluation) and m evaluation indexes (for each DPSIR domain, we have m factors). If the jth evaluation index of the ith evaluation object is x_{ij} $(i = 1, 2, \cdots, n; j = 1, 2, \cdots, m)$, then the index matrix X composed of x_{ij} is as follows:

$$X = (x_{ij})_{n \times m} = \begin{bmatrix} x_{11} & \cdots & x_{1m} \\ \vdots & \ddots & \vdots \\ x_{n1} & \cdots & x_{nm} \end{bmatrix} \quad (4)$$

We compute for the positive index a normalized value:

$$v_{ij} = \frac{x_{i,j} - x_{\min j}}{x_{\max j} - x_{\min j}} \quad (5)$$

We compute for the negative index a normalized value:

$$v_{ij} = \frac{x_{\max j} - x_{ij}}{x_{\max j} - x_{\min j}} \quad (6)$$

Here, $x_{\max}$ is the maximum value of the index, and $x_{\min}$ is the minimum value of the index for column j.

After normalization, the $n \times m$ value matrix V is obtained:

$$V = (v_{ij})_{n \times m} = \begin{bmatrix} v_{11} & \cdots & v_{1m} \\ \vdots & \ddots & \vdots \\ v_{n1} & \cdots & v_{nm} \end{bmatrix} \quad (7)$$

(2) Determine the value of evaluation index entropy:

We compute the relative ratio of v_{ij} for each evaluation object i as p_{ij}, and the calculation formula is as follows:

$$p_{ij} = \frac{v_{ij}}{\sum_{i=1}^{n} v_{ij}} \quad (8)$$

Then, the entropy E_j of the jth evaluation index is defined as:

$$E_j = -\frac{1}{\ln n} \sum_{i=1}^{n} p_{ij} \cdot \ln p_{ij} \quad (9)$$

A special case is that if $p_{ij} = 0$, then $p_{ij} \cdot \ln p_{ij} = 0$.

(3) Determine the weight of index:

The entropy weight ω_j of each evaluation index can be expressed as:

$$\omega_j = \frac{1 - E_j}{\sum_{j=1}^{m} (1 - E_j)} \quad (10)$$

3.3.3. Calculation of Connection Degree

If the evaluation sample is set X, then the K-element connection degree of (X, B_1) can be defined as [44]:

$$\begin{aligned}\mu_{X\sim B_1} = & \sum_{j=1}^{n} \omega_j \mu_{X_t\sim B_1} \\ & = \sum_{j=1}^{n} \omega_j b_{j,0} I_0 + \sum_{j=1}^{n} \omega_j b_{j,1} I_1 + \sum_{j=1}^{n} \omega_j b_{j,2} I_2 + \cdots + \sum_{j=1}^{n} \omega_j b_{j,K-2} I_{K-2} + \sum_{j=1}^{n} \omega_j b_{j,\,K-1} I_{K-1}\end{aligned} \tag{11}$$

If we let $f_1 = \sum_{j=1}^{n} \omega_j b_{j,0}$, $f_2 = \sum_{j=1}^{n} \omega_j b_{j,1}$, $\cdots, f_{K-1} = \sum_{j=1}^{n} \omega_j b_{j,K-2}$, $f_K = \sum_{j=1}^{n} \omega_j b_{j,K-1}$, then Equation (11) can be transformed into:

$$\mu_{X\sim B_1} = f_1 I_0 + f_2 I_1 + f_3 I_2 + \cdots + f_{K-1} I_{K-2} + f_K I_{K-1} \tag{12}$$

where f_K represents the possibility that the evaluation sample belongs to the Kth level of the standard. We computed a continuous measure of security rating μ, the connection degree, by setting $I_i = 1 - \frac{2i}{K-1}$. As a result, we can rewrite Formula (12) as

$$\mu_{X\sim B_1} = \sum_{i=1}^{K} f_i\left(1 - \frac{2(i-1)}{K-1}\right) = f_1 + f_2\left(1 - \frac{2}{K-1}\right) + f_3\left(1 - \frac{4}{K-1}\right) + \cdots + f_{K-1}\left(-1 + \frac{2}{K-1}\right) - f_K \tag{13}$$

The connection degree is computed for each DPSIR component, for each year at each city. The connection degree is linearly transformed from [0, 1] to [1, 5] to be compared with other measures in Section 3.3.4.

3.3.4. Measures of Water Resources Security

In this paper, we computed a discrete measure of overall water security rating (1 to 5) for each city and each year using the confidence level grading method [88]:

$$\min_k h_k = f_1 + f_2 + \cdots + f_k > \lambda\,, \qquad \text{where} \quad k = 1, 2, \cdots, K \tag{14}$$

Here, h_k is the property measure, and λ is the confidence threshold. λ should not be too large, otherwise the evaluation results tend to be conservative and stable; λ should not be too small, otherwise the reliability of the results becomes poor, and the evaluation results tend to be over-positive. It is generally recommended that λ be taken in [0.5, 0.7] [58,59,89]. This threshold parameter λ is often subjectively chosen in this range of [0.5, 0.7] in the previous literature [38–41]. For a selected λ, the evaluation security rating is selected as the minimal k such that h_k is greater than λ. For example, if $h_1 = f_1 < \lambda$ and $h_2 = f_1 + f_2 > \lambda$, then the evaluation security rating is $k = 2$.

In order to avoid the subjectivity of the choice of λ, we proposed a simple modification of the confidence level grading method. We sampled 100 λs uniformly from the empirical range of [0.5, 0.7], and we repeated the calculations of the minimal k such that $h_k > \lambda$ for 100 times for each of the 100 different λs. Then, we averaged the results from these 100 λs to determine the grade of each index. This gave us a continuous measure of water resources security. We showed in Figure 2 that this measure correlates strongly with the connection degree.

3.4. Statistical Analysis

3.4.1. Repeated Measures Correlation

Repeated measures correlation (rmcorr) was first introduced by Bland and Altman in biostatistics to analyze the correlations between paired repeated measures, which are two corresponding measures assessed for each individual/site on multiple occasions [90,91]. For repeated measures, the independence assumption adopted by simple correlation/regression is often violated. Repeated measure correlation addresses this non-independence among observations by using analysis of covariance (ANCOVA) to statistically adjust for inter-site variability. Conceptually, rmcorr is similar to a regression model that assumes a common slope and varying intercept for each group (in our case, city or year). By removing measured variance between sites, rmcorr provides the best-fit parallel regression lines with varying intercepts and the same slope. The rmcorr coefficient ranges from −1 to 1, similar to Pearson correlation [92].

In our study, we used rmcorr to analyze the correlations among pairs of DPSIR water security ratings (for example, between D and P) that were computed for each year and each city. We assessed the repeated correlations both longitudinally and cross-sectionally. In the longitudinal assessment, we used rmcorr to adjust for the variances across cities. In other words, we analyzed how a pair of factors (e.g., D and P) covaried over the years. In the cross-sectional assessment, we used rmcorr to adjust for the variances across years instead. In other words, we analyzed how a pair of factors (e.g., D and P) covaried across cities. Repeated measures correlation analysis was implemented in R programming language using the "rmcorr" package [92].

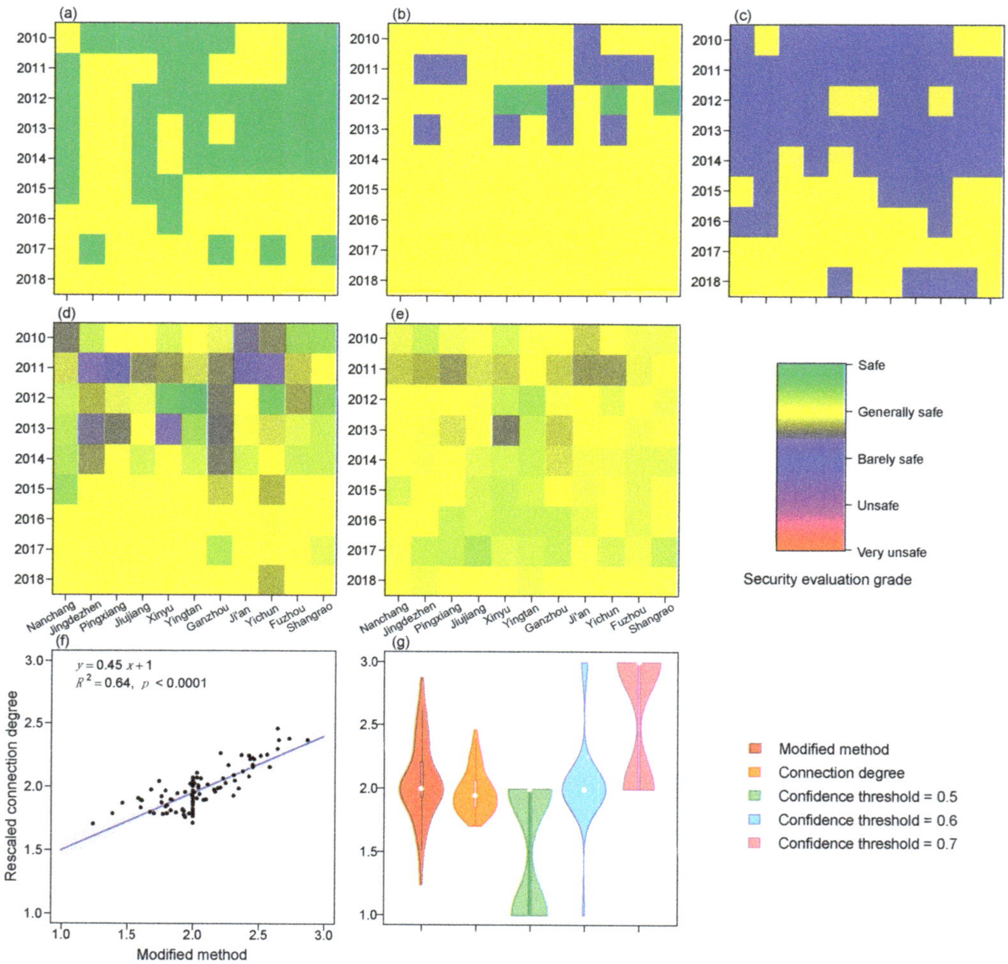

Figure 2. (**a–c**) Water resource security grade with λ = 0.5, 0.6, 0.7; (**d**) water resources security using the graded confidence threshold; (**e**) connection degree rescaled from 0–1 to 1–5; (**f**) correlation between the modified method and the connection degree; (**g**) distributions of water security scores in all methods; here, 1 is Safe, and 3 is Barely safe.

3.4.2. Linear Mixed Models

In this study, Linear Mixed Models (LMMs) [93] were used to assess the temporal trend of the DPSIR factors over the years. In our case

$$f = \beta_{year} Y + \tau_{city} + \epsilon$$

Here, f stands for D, P, S, I, or R, and the LMMs analysis was applied for each factor separately. β_{year} is the slope parameter that quantifies the temporal trend of each factor, Y is the year, τ_{city} is the city ID, and ϵ is the random error. We assume a fixed effect of Year and a random effect of City. The analysis was implemented in R programming language using the "lmer" function in package "lme4" [94].

4. Results

4.1. Water Resource Security Evaluation in Jiangxi Province Using the Modified Confidence Threshold Method

Using Fuzzy set pair analysis and the entropy weight method, we calculated the water resource security for each city and each year. In our dataset, the calculated weights for each DPSIR factor are shown in the table below (Table 3).

Table 3. Weights of each DPSIR factor.

D1	D2	D3	D4	P1	P2	P3	P4	S1	S2	I1	I2	I3	R1	R2
0.060	0.067	0.065	0.150	0.052	0.051	0.049	0.040	0.075	0.070	0.045	0.062	0.079	0.050	0.086

The water resource security evaluation method described in this paper depends critically on a confidence level parameter λ. It has been suggested in the literature that λ should be chosen from [0.5, 0.7]. However, the subjective selection of λ can have consequential impacts on the estimated water security. To illustrate the caveat of this subjectivity in the selection of λ, we compared five different measures of water resource security, three measures using subjective thresholds λ = 0.5, 0.6, 0.7, the measure using our proposed modified method, and finally the connection degree measure. The estimated water resource security with $\lambda = 0.5, 0.6, 0.7$ is shown in Figure 2a–c, respectively. A lower λ tends to give optimistic estimates that the water security is classified as Safe or Generally Safe for all cities and all years, whereas a higher λ tends to give pessimistic estimates that the water security is classified as Barely Safe for significant proportions of cities and years. Furthermore, this subjective selection of λ can lead to contradictory conclusions. When comparing the water resource security in 2010–2014 and 2015–2018, with $\lambda = 0.5$, most cities change from Safe to Generally Safe, showing a declining temporal trend in safety (Figure 2a). However, with $\lambda = 0.7$, most cities change from Barely Safe to Generally Safe, showing an increasing temporal trend in safety (Figure 2c). To avoid this subjectivity in parameter selection of λ, we used a modified method that utilizes a range of uniformly sampled λs (Figure 2d). We further compared this modified method and the connection degree. The connection degree is rescaled from 0–1 to 1–5 for equal comparison (Figure 2e). The modified method is strongly correlated with the connection degree (Figure 2f, R^2 = 0.64, p < 0.001). The distributions of water security scores across cities and years in all methods (the modified method, the connection degree, and old methods with $\lambda = 0.5, 0.6, 0.7$ are shown in Figure 2g. The score from the modified method has a shape that is closest to a normal distribution and lends itself to more convenient statistical tools.

Overall, we find that the water resource security of 11 cities in Jiangxi province is generally safe, ranging from Barely Safe to Safe (Figure 2d–e). We did not find Unsafe or Very Unsafe grades in any of the 11 cities from 2010–2018. In addition, we found that the water resource security changed from a more inconsistent and volatile grade (Barely Safe to Safe) before the year of 2014, to a more consistent and stable grade (Generally Safe) since the year of 2014. This change reflected the regional water resource managers' efforts in water resources management. The overall degree of environmental remediation is strengthening over the years.

4.2. Temporal and Spatial Dynamics of Water Resource Security in Jiangxi Province

We further investigated the DPSIR components of the water resources security and how they varied over the years and across different cities. Temporally, we showed the DPSIR scores of each city as a function of year (Figure 3, upper panel) to investigate the temporal trend. Spatially, we showed the variation of the DPSIR scores across years as a function of city (Figure 3, lower panel) to investigate the spatial differences in water resources management. Each color denotes a city, and the green line in the upper row showed the yearly averages of water security score across the eleven cities.

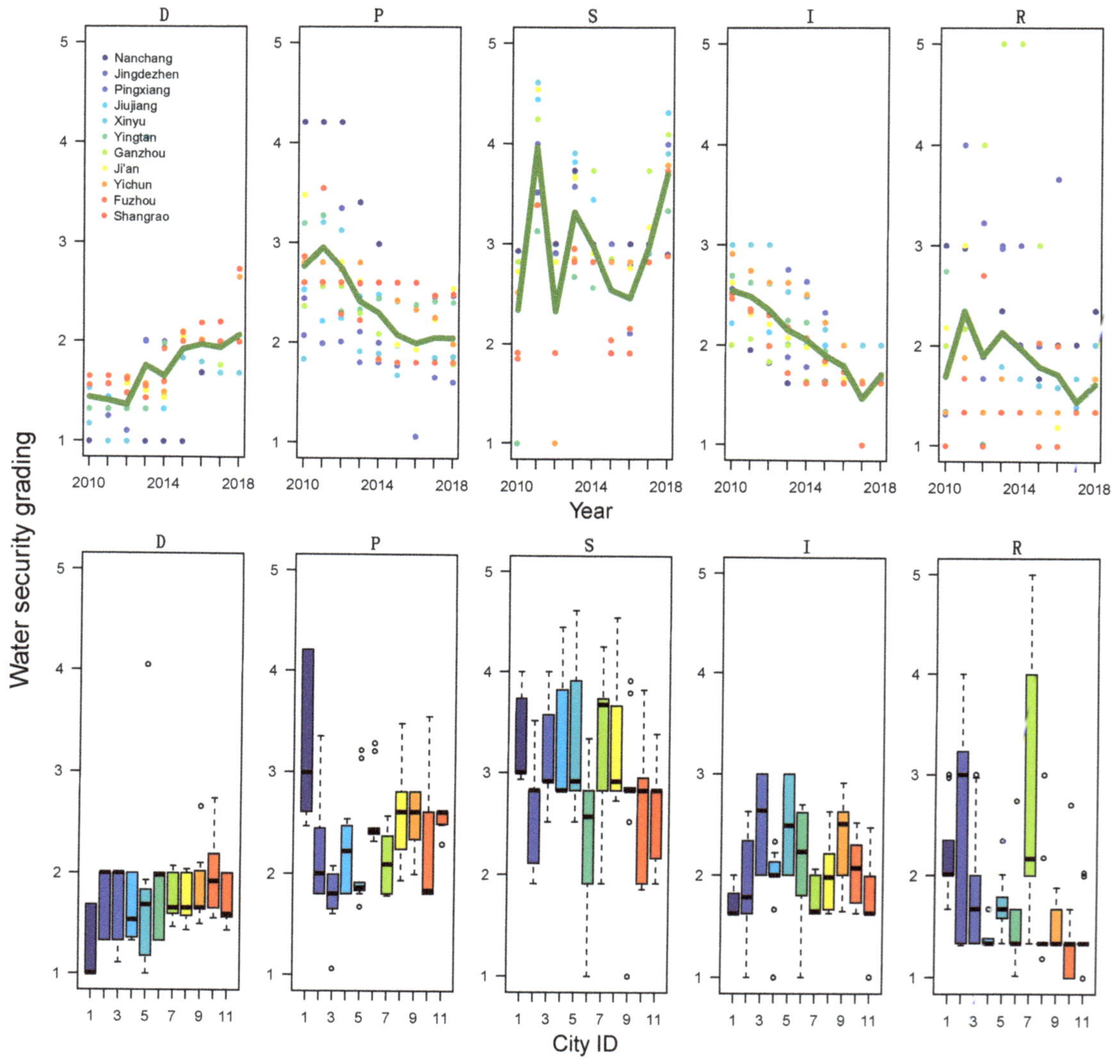

Figure 3. Time series and boxplot of water security grading of each city during the years 2010–2018.

Linear Mixed Models (LMMs) were used in analysis to account for repeated measures and assess the statistically significance of the temporal trends of the DPSIR factors. Specially, we showed a significant descending trend (from 5-Unsafe to 1-Safe) in the Pressures, Impact, and Response factors. In addition, we showed a significantly ascending trend (from 1-Safe to 5-Unsafe) in the Driver factor. The State factor remains unchanged during the time period of the study (Table 4). The Pressures factor significantly improved over the years ($p < 0.001$). Considering the four water resources pressure indices used in this study, the decrease in the security level shows that the utilization rate of water resources in various industries is getting higher, and the pressure on water resources security is gradually weakened. The Impact factor significantly improved over the years ($p < 0.001$). The increasing in index I indicates that the urbanization level increases, and the greening rate and the water supply assurance rate of these cities are getting higher over the study period. The response factor also significantly improved over the years ($p = 0.017$), showing an increase in the laws and rules in favor of sustainable water usage. The safety of the Driver factors significantly declined over the years ($p < 0.001$). This indicates that the economic and industrial structure of the cities in Jiangxi Province rapidly developed but exposed more risks to water security over the study period. The State factor remained unchanged during the period of study ($p > 0.05$), indicating a sustainable water quality.

Table 4. LMM results of temporal trend analysis of the DPISR water resources security.

	D		P		S		I		R	
Predictors	Estimates	*p*	Estimates	*p*	Estimates	*p*	Estimates	*p*	Estimates	*p*
(Intercept)	1.27 (0.87–1.67)	**<0.001**	2.99 (2.58–3.40)	**<0.001**	2.79 (2.39–3.20)	**<0.001**	2.70 (2.30–3.10)	**<0.001**	2.15 (1.74–2.57)	**<0.001**
years	0.09 (0.04–0.14)	**<0.001**	−0.12 (−0.17–−0.07)	**<0.001**	0.03 (−0.02–0.08)	0.212	−0.13 (−0.18–−0.08)	**<0.001**	−0.06 (−0.11–−0.01)	**0.017**
Random Effects										
σ^2	0.11		0.11		0.45		0.04		0.43	
ICC	0.14		0.51		0.16		0.62		0.32	
N	11citys		11citys		11citys		11citys		11citys	
Observations	99		99		99		99		99	
Marginal R^2 / Conditional R^2	0.299/0.397		0.304/0.660		0.013/0.172		0.518/0.815		0.040/0.346	

ICC: Intraclass correlation coefficient. The significant *p* values are bolded.

The variations of DPSIR factors over the years are computed for each city. We showed a significant difference among the cities. As the capital city of the province, the capital city Nanchang (City #1 in dark blue, Figure 3) has the most industry and economic growth among the 11 cities (the safest/lowest D factor); the heaviest pollution problem, land-use charges, and population growth (the least safe/highest P factor); and a relatively low water quality (a relatively unsafe/high S factor). On the other hand, Fuzhou (City #10 in red, Figure 3) is a city under-development with a high Driver factor rating, a low Pressures factor rating, and a low State factor rating, the opposite of Nanchang. This anti-correlations among factors are further explored in the next section. Moreover, we noticed that some cities went through more changes over the years compared to other cities. In particular, the cites Jingdezhen (City #2) and Ganzhou (City #7) have a significant higher variance in the Response factor compared to other cities, showing a significant impact of taxes and environmental laws on water management in these cities.

4.3. The Cross-Sectional and Longitudinal Inter-Correlations among the DPSIR Subsystems

In this paper, we assessed empirically the inter-connections among the 5 subsystems in the DPSIR framework, namely, Driver, Pressure, State, Impact, and Response. Temporal trends were removed for each factor, and the inter-correlations were computed based on residuals. In particular, we assessed the inter-correlations among the DPSIR components both cross-sectionally and longitudinally, by controlling the temporal (year) and spatial (city) factor, respectively, using repeated measures correlation. The diagonal panels in Figure 4 showed the histograms of the DPSIR security ratings across years and cities. The upper triangular of Figure 4 showed the repeated measures correlations among DPSIR factors over the years (city is treated as a random effect). Data points in these plots are grouped by cities (the color scheme is the same as in Figure 3); different colors here are different cities, and a separate regression line is drawn for each city. The lower triangular of Figure 4 showed the repeated measures correlations among DPSIR factors across cities (year is treated as a random effect). Data points in these plots are grouped by years, and the different intensity of the grey indicates different years (more recent years are represented by darker colors). A separate regression line was drawn for each year. The regression's coefficient and associated p-values are shown in the title of each subplot (Figure 4).

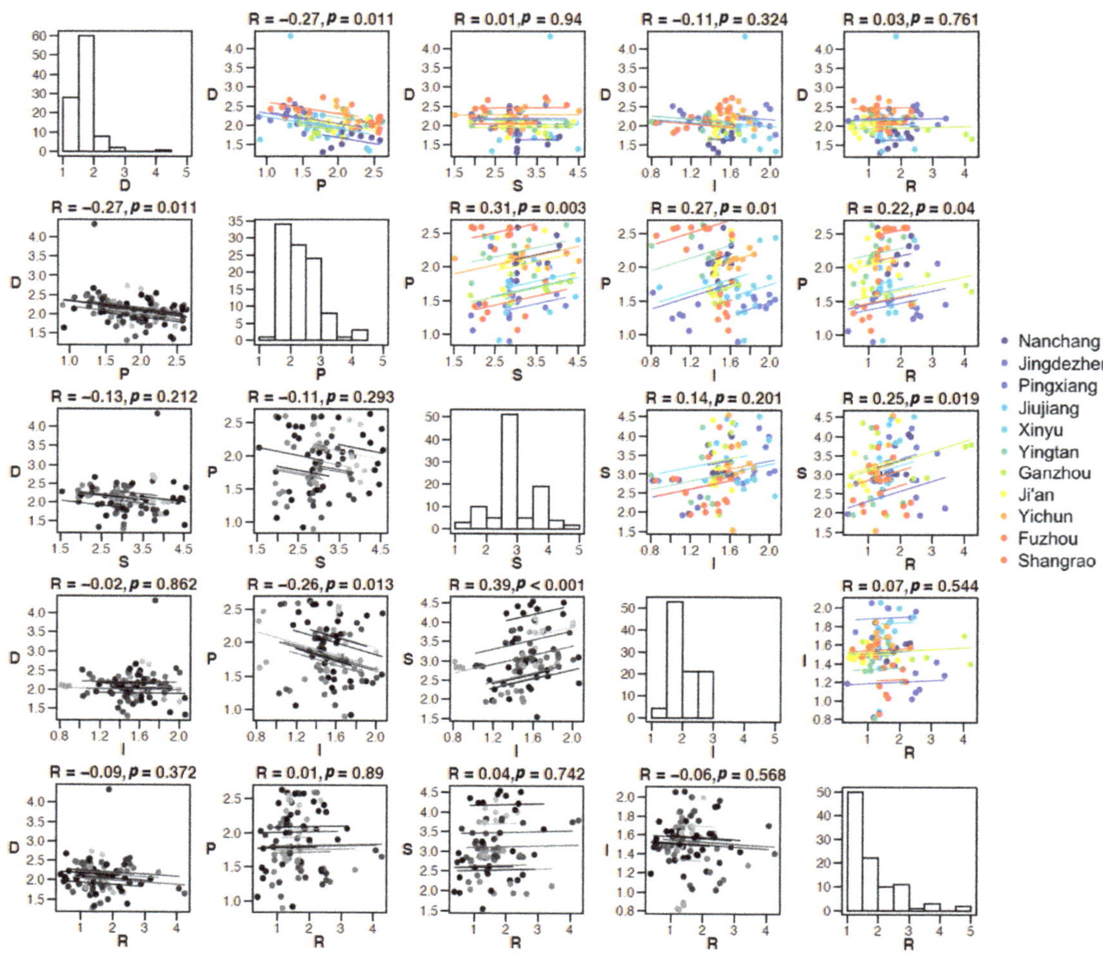

Figure 4. The repeated measures correlations among the DPSIR water security estimates of the 11 cities in Jiangxi Province during 2010–2018; data in the upper triangular are grouped by cities, and different colors indicate different cities; data in the lower triangular are grouped by years, and different intensity of the grey indicates different years.

Driver and Pressures factors are negatively correlated, both over the years (R = −0.27, $p = 0.011$) and across cities (R = −0.27, $p = 0.011$). This is in line with the DPSIR framework that an increase in Driver should lead to a decrease in the Pressures score (e.g., high industry development leads to more pollution). Increasing Drivers such as population and economic and social development promotes the development of the city but at the same time imposes pressures on water supply demand and sewage treatment, which would decrease the safety level in Pressure factors. Pressure and States correlate over the years (R = 0.31, $p = 0.003$).

Higher Pressure would lead to Poorer State (e.g., lower water quantity and quality). State factor does not correlate significantly with Impact factor over the years ($p > 0.05$) but correlates significantly with Impact factor across cities ($R = 0.39$, $p < 0.001$). Pressure and Impact positively correlate over the years ($R = 0.27$, $p = 0.01$) and negatively correlate across cities ($R = -0.26$, $p = 0.013$). Over the years, increasing water demand due to agricultural and residential water use leads to a decrease in the safety rating of the Impact of water security (e.g., decreasing greening coverage). Across the cities, however, cities with high Pressure safety have low Impact safety. Finally, the Response factor correlated significantly with both Pressure ($R = 0.22$, $p = 0.04$) and State ($R = 0.25$, $p = 0.019$) over the years. Good responses to water threats can indeed decrease water pressure and lead to better water/soil states. Impact factor did not significantly react to responses in the study region ($p > 0.05$).

5. Discussion

In this paper, we used the DPSIR framework to evaluate regional water resources security in Jiangxi Province in China. Our study constructed a water resources security evaluation index system and evaluation grade standards for 11 cities in Jiangxi Province. Fuzzy set pair analysis method and entropy weight method were used to evaluate water security through a modified confidence-threshold method. We analyzed the temporal trend and spatial variations of water resources security, separately for the Driver, Pressure, State, Impact, and Response subsystems. In addition, we empirically assessed the inter-correlations among the five subsystems of the DPSIR system, both longitudinally and cross-sectionally by controlling for city and year respectively. We revealed a more complicated and region-specific DPSIR flow and suggested that such region-specific networks are informational in guiding local policy making.

The evaluation of water resources security in the DPSIR framework depends critically on a confidence threshold parameter λ. This threshold λ was usually subjectively chosen in the literature. The subjective choice of λ varied from study to study, some of the choices are 0.55 [39], 0.6 [40], and 0.7 [41]. In our paper, we showed that this subjectivity would lead to inconsistent and even opposite conclusions in water resources evaluation and is prone to errors (Figure 2). To address this caveat, we proposed a simple modification that averages water security estimates over a range of λs. In this method, researchers no longer had to specify a single subjective λ value. We compared our modified method with connection degree and found that these two measures correlate strongly with each other. We recommend future researchers to avoid using subjective λs and use alternative continuous measures such as our modified method or the connection degree.

We comprehensively evaluated Jiangxi Province's water resources security of all cities from 2010 to 2018. We can see that since 2014, Jiangxi Province's water resources become more stable and consistent across cities and mostly fall in the category of "Generally Safe". Through trend analysis, we were able to identify differential temporal trends among the DPSIR factors. The water resources security in Jiangxi Province has been improving over the years in many aspects. The water security levels of the indices pressure (P), impact (I), and Response (R) have decreasing trends over time and are becoming safer over the years. This is related to the continued construction of water-saving reconstruction projects in Jiang provinces in recent years. Moreover, the development and utilization rate of water resources in Jiangxi Province is increasing, which also makes the water resources security safer over the years. The index driver (D) has an increasing trend and is becoming less safe, and more attention is needed in regulating the Driver. This manifests in a number of factors including the increase of population density and the decrease of GDP growth rate. Jiangxi Province has attracted a large number of migrant populations in recent years, resulting in population aggregation, which in turn results in a relative increase of domestic water consumption of urban residents and a relative reduction of per capita water and economic resources. The state (S) remains unchanged.

Despite the general temporal trend, we identified significant differences among cities. The water resources security in some cities are relatively poorer than others and require more attention. Cities like Nanchang, Ganzhou, and Jiujiang are developing rapidly socio-economically; the GDP growth rates are large; these cities not only attract a large number of foreign populations, but also have local population aggregation, resulting in the relative increase of domestic water consumption of urban residents. The per capita water resources are relatively reduced, and the contradiction between supply and demand of water resources become prominent. However, cities like Yingtan, Shangrao and Fuzhou are rich in annual rainfall and per capita water resources. The domestic water consumption of urban residents and the water consumption of 10,000-yuan industrial added value are less, and the reuse rate of industrial water and sewage treatment rate are also higher than the average value in the province.

Furthermore, we empirically assessed the inter-correlations among the DPSIR subsystems. We argue that despite the general causal flow in the DPSIR framework, the sensitivities of changes between different subsystems should vary from region to region. It is important to assess the region-specific intercorrelations among the DPSIR subsystems empirically. In this paper, we tested such correlations both longitudinally and cross-sectionally using repeated measures correlations. Longitudinal correlations reflect more the intrinsic associations between the rise and fall of different pairs of factors, whereas cross-sectional correlations reflect more the regional heterogeneity in water resources management. The results are summarized in Figure 5. The general DPSIR framework has a forward loop in the order of D-P-S-I-R, and feedback edges (R, D), (R, P) and (R, S) represent the Responses in response to Drivers, Pressures, and States. In our study region, there are informational deviations from this standard framework in inter-correlations among the DPSIR subsystems. For longitudinal correlations, Impact seems to be detached from the rest of the 4 subsystems. In particular, Responses do not seem to arise from Impacts but rather from Pressures and State directly. In addition, the fact that Responses correlate with Pressure and State but not Driver, indicate that the local policies are less targeted at regulating Driver compared to managing Pressures and State. Our analysis suggested that future policies should consider regulating Driver. We suggest that relevant departments can increase investment in economic development, accelerate the development of advanced water resources technology, tap the exploitable potential of water resources, and resolve the problem of increasing demand for water resources from the source. Moreover, it is beneficial to formulate effective water-saving strategic plans, promote the recycling of water resources, and improve the optimal allocation of water resources. For cross-sectional correlations, we observed significant correlations between (S, I) and (P, I). This suggests the regional heterogeneity in the causal influences on the Impact factor. We think that these region-specific interconnections and their deviances from the standard DPSIR framework carry useful information about the local water resources system and can assist local governors in making better local water resources policies. Our approach of empirical assessment of the region-specific connectivity in various subsystems over space and time can serve as a reference for future regional water resources evaluation studies.

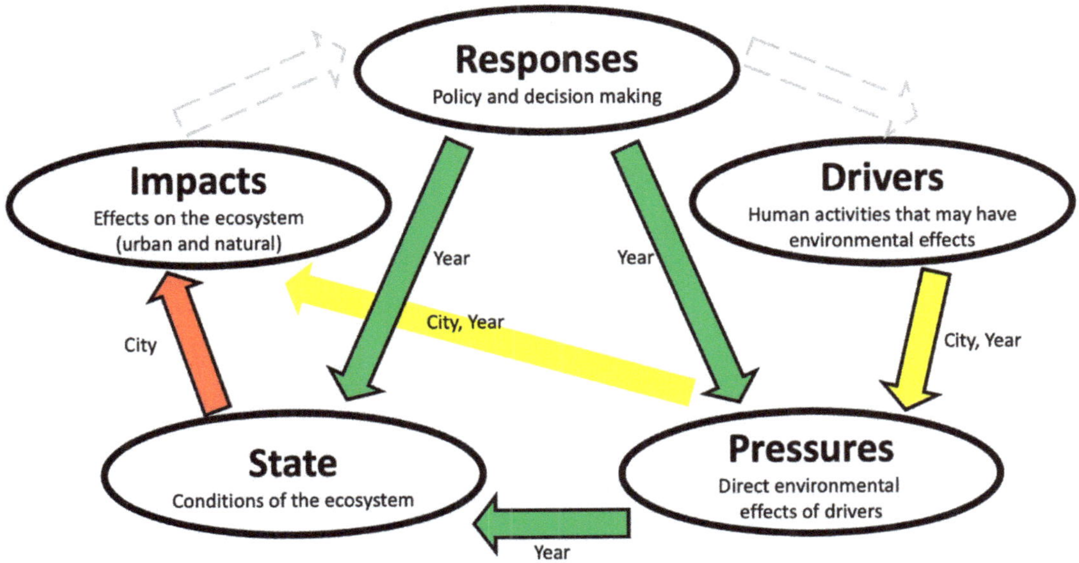

Figure 5. A causal network of DPSIR model used for water resources security gradings in this study. The green arrows represent correlations over years, the red arrows represent correlations across cities, and the yellow arrows represent correlations both over years and across cities. The dashed arrows are links in the DPSIR framework that did not show empirical correlations in our study region.

6. Conclusions

This study used the DPSIR framework and assessed the temporal and spatial dynamics of water resources security in Jiangxi Province, China. In this study, we demonstrated that methods using the subjective confidence threshold in assessing water resources security can lead to contradictory conclusions depending on the choice of the confidence threshold parameter. To address this limitation, we proposed a modified method that samples a range of threshold values and pools them to obtain more objective measures of water resources security.

From the analysis of the temporal trend and spatial variation of DPSIR-based water security level in Jiangxi Province, we found that while Pressures, Impacts, and Responses are becoming safer over the years, the Driver factor is becoming less safe and requires more attention. The increase in population density and the decrease in per capita water resources have made the contradiction between supply and demand of water resources more prominent in Jiangxi Province.

Through longitudinal and cross-sectional correlational analysis among the DPSIR subsystems in Jiangxi Province, we found regional deviances of inter-correlations among the DPSIR nodes from the general DPSIR causal chain framework. In particular, the Responses in this area correlate with Pressure and State but not with Driver. This suggests that local government should direct future policies towards improving the security level of Driver, and relevant departments should increase the development of water resources potential, formulate water-saving strategic measures, and optimize the allocation of water resources.

Author Contributions: Conceptualization, M.L., X.W. and W.L.; methodology, M.L. and S.W.; formal analysis, M.L. and C.W.; writing—original draft preparation, M.L. and S.W.; writing—review and editing, M.L., S.W., X.W. and W.L.; visualization, M.L. and S.W.; supervision, W.L., X.L. and H.W. All authors have read and agreed to the published version of the manuscript.

Funding: This paper was jointly supported by the National Key R&D Program of China (2017YFC040-6004) and the National Natural Science Fund (52179027).

Institutional Review Board Statement: Not applicable.

Informed Consent Statement: Not applicable.

Data Availability Statement: Publicly available datasets were analyzed in this study. This data can be found here: https://data.cnki.net/NewHome/index, accessed on 20 November 2020 and http://data.stats.gov.cn/, accessed on 20 November 2020.

Conflicts of Interest: The authors declare no conflict of interest.

References

1. Qin, D.; Lu, C.; Liu, J.; Wang, H.; Wang, J.; Li, H.; Chu, J.; Chen, G. Theoretical framework of dualistic nature–social water cycle. *Chin. Sci. Bull.* **2014**, *59*, 810–820. [CrossRef]
2. John, C.K.; Pu, J.H.; Moruzzi, R.; Pandey, M. Health-risk assessment for roof-harvested rainwater via QMRA in Ikorodu area, Lagos, Nigeria. *J. Water Clim. Chang.* **2021**, *12*, 2479–2494. [CrossRef]
3. John, C.K.; Pu, J.H.; Pandey, M.; Moruzzi, R. Impacts of sedimentation on rainwater quality: Case study at Ikorodu of Lagos, Nigeria. *Water Supply* **2021**, *21*, 3356–3369. [CrossRef]
4. Yang, D.; Yang, Y.; Xia, J. Hydrological cycle and water resources in a changing world: A review. *Geogr. Sustain.* **2021**, *2*, 115–122. [CrossRef]
5. Chaturvedi, A.; Pandey, B.; Yadav, A.K.; Saroj, S. Chapter 5—An overview of the potential impacts of global climate change on water resources. In *Water Conservation in the Era of Global Climate Change*; Elsevier: Amsterdam, The Netherlands, 2021; pp. 99–120. [CrossRef]
6. Di Baldassarre, G.; Sivapalan, M.; Rusca, M.; Cudennec, C.; Garcia, M.; Kreibich, H.; Konar, M.; Mondino, E.; Mård, J.; Pande, S.; et al. Sociohydrology: Scientific Challenges in Addressing the Sustainable Development Goals. *Water Resour. Res.* **2019**, *55*, 6327–6355. [CrossRef] [PubMed]
7. Li, P.; Qian, H. Water resources research to support a sustainable China. *Int. J. Water Resour. Dev.* **2018**, *34*, 327–336. [CrossRef]
8. Anandhi, A.; Kannan, N. Vulnerability assessment of water resources—Translating a theoretical concept to an operational framework using systems thinking approach in a changing climate: Case study in Ogallala Aquifer. *J. Hydrol.* **2018**, *557*, 460–474. [CrossRef]
9. Xinchun, C.; Mengyang, W.; Xiangping, G.; Yalian, Z.; Yan, G.; Nan, W.; Weiguang, W. Assessing water scarcity in agricultural production system based on the generalized water resources and water footprint framework. *Sci. Total Environ.* **2017**, *609*, 587–597. [CrossRef]
10. Falkenmark, M.; Widstrand, C. Population and water resources: A delicate balance. *Popul. Bull.* **1992**, *47*, 1–36.
11. Raskin, P.; Gleick, P.; Kirshen, P.; Pontius, G.; Strzepek, K. *Comprehensive Assessment of the Freshwater Resources of the World*; Stockholm Environment Institute: Stockholm, Sweden, 1997.
12. Alcamo, J.; Döll, P.; Henrichs, T.; Kaspar, F.; Lehner, B.; Rösch, T.; Siebert, S. Development and testing of the WaterGAP 2 global model of water use and availability. *Hydrol. Sci. J.* **2003**, *48*, 317–337. [CrossRef]
13. Ashton, P.J. Avoiding conflicts over Africa's water resources. *Ambio* **2002**, *31*, 236–242. [CrossRef] [PubMed]
14. Sullivan, C.A. Calculating a Water Poverty Index. *World Dev.* **2002**, *30*, 1195–1210. [CrossRef]
15. Wang, X.-j.; Zhang, J.-y.; Shamsuddin, S.; Xia, X.-h.; He, R.-m.; Shang, M.-t. Catastrophe theory to assess water security and adaptation strategy in the context of environmental change. *Mitig. Adapt. Strateg. Glob. Chang.* **2014**, *19*, 463–477.
16. Yin, S.; Dongjie, G.; Weici, S.; Weijun, G. Integrated assessment and scenarios simulation of urban water security system in the southwest of China with system dynamics analysis. *Water Sci. Technol.* **2017**, *76*, 2255–2267. [CrossRef]
17. Su, F.; Li, P.; He, X.; Elumalai, V. Set Pair Analysis in Earth and Environmental Sciences: Development, Challenges, and Future Prospects. *Expo. Health* **2020**, *12*, 343–354. [CrossRef]
18. Flörke, M.; Schneider, C.; McDonald, R.I. Water competition between cities and agriculture driven by climate change and urban growth. *Nat. Sustain.* **2018**, *1*, 51–58. [CrossRef]
19. Carr, E.R.; Wingard, P.M.; Yorty, S.C.; Thompson, M.C.; Jensen, N.K.; Roberson, J. Applying DPSIR to sustainable development. *Int. J. Sustain. Dev. World Ecol.* **2007**, *14*, 543–555. [CrossRef]
20. Li, X.; Su, X.; Wei, Y. Multistage integrated water security assessment in a typical region of Northwestern China. *J. Clean. Prod.* **2019**, *220*, 732–744. [CrossRef]
21. Jaramillo, P.; Nazemi, A. Assessing urban water security under changing climate: Challenges and ways forward. *Sustain. Cities Soc.* **2018**, *41*, 907–918. [CrossRef]
22. Octavianti, T.; Staddon, C. A review of 80 assessment tools measuring water security. *WIREs Water* **2021**, *8*, e1516. [CrossRef]
23. Jensen, O.; Wu, H. Urban water security indicators: Development and pilot. *Environ. Sci. Policy* **2018**, *83*, 33–45. [CrossRef]
24. Cao, Y.; Bian, Y. Improving the ecological environmental performance to achieve carbon neutrality: The application of DPSIR-Improved matter-element extension cloud model. *J. Environ. Manag.* **2021**, *293*, 112887. [CrossRef] [PubMed]
25. Haque, M.N.; Mamun, A.; Saroar, M.; Roy, T.S. Application of "DPSIR" framework to assess the status and role of blue ecosystem services (BES) in Khulna city. *J. Eng. Sci.* **2019**, *10*, 49–60.

26. Kagalou, I.; Leonardos, I.; Anastasiadou, C.; Neofytou, C. The DPSIR approach for an integrated river management framework. A preliminary application on a Mediterranean site (Kalamas River-NW Greece). *Water Res. Manag.* **2012**, *26*, 1677–1692. [CrossRef]
27. Liu, S.; Ding, P.; Xue, B.; Zhu, H.; Gao, J. Urban Sustainability Evaluation Based on the DPSIR Dynamic Model: A Case Study in Shaanxi Province, China. *Sustainability* **2020**, *12*, 7460. [CrossRef]
28. Mimidis, K.; Andrikakou, P.; Kallioras, A.; Pliakas, F. The DPSIR approach to groundwater management for sustainable development in coastal areas: The case of Nea Peramos aquifer system, Kavala, Greece. *Water Util. J.* **2017**, *16*, 67–80.
29. Mosaffaie, J.; Jam, A.S.; Tabatabaei, M.R.; Kousari, M.R. Trend assessment of the watershed health based on DPSIR framework. *Land Use Policy* **2020**, *100*, 104911. [CrossRef]
30. Qinlin, X.; Chao, T.; Yanjun, W.; Xiuqing, L.; Liming, X. Measurement and Comparison of Urban Haze Governance Level and Efficiency Based on the DPSIR Model: A Case Study of 31 Cities in North China. *J. Resour. Ecol.* **2020**, *11*, 549–561. [CrossRef]
31. Nezami, S.R.; Nazariha, M.; Moridi, A.; Baghvand, A. Environmentally Sound Water Resources Management in Catchment Level using DPSIR Model and Scenario Analysis. *Int. J. Environ. Res.* **2013**, *7*, 569–580. [CrossRef]
32. Shi, S.; Tao, X.; Chen, X.; Chen, H.; Fitri, A.; Yang, X. Evaluation of urban water security based on DPSIR model. In *IOP Conference Series: Earth and Environmental Science*; IOP Publishing: Bristol, UK, 2021.
33. Sun, S.; Wang, Y.; Liu, J.; Cai, H.; Wu, P.; Geng, Q.; Xu, L. Sustainability assessment of regional water resources under the DPSIR framework. *J. Hydrol.* **2016**, *532*, 140–148. [CrossRef]
34. Tscherning, K.; Helming, K.; Krippner, B.; Sieber, S.; Gomez y Paloma, S. Does research applying the DPSIR framework support decision making? *Land Use Policy* **2012**, *29*, 102–110. [CrossRef]
35. Vannevel, R. Using DPSIR and Balances to Support Water Governance. *Water* **2018**, *10*, 118. [CrossRef]
36. Zhang, F.; Zhang, J.; Wu, R.; Ma, Q.; Yang, J. Ecosystem health assessment based on DPSIRM framework and health distance model in Nansi Lake, China. *Stoch. Hydrol. Hydraul.* **2015**, *30*, 1235–1247. [CrossRef]
37. Apostolaki, S.; Koundouri, P.; Pittis, N. Using a systemic approach to address the requirement for Integrated Water Resource Management within the Water Framework Directive. *Sci. Total Environ.* **2019**, *679*, 70–79. [CrossRef]
38. Labianca, C.; De Gisi, S.; Todaro, F.; Wang, L.; Tsang, D.C.; Notarnicola, M. A holistic DPSIR-based approach to the remediation of heavily contaminated coastal areas. *Environ. Pollut.* **2021**, *284*, 117129. [CrossRef]
39. Xiang, X.; Li, Q. Water Resources Vulnerability Assessment and Adaptive Management Based on Projection Pursuit Model. *J. Coast. Res.* **2020**, *103*, 431–435. [CrossRef]
40. Yu, H.; Yang, Z.; Li, B. Sustainability Assessment of Water Resources in Beijing. *Water* **2020**, *12*, 1999. [CrossRef]
41. Zhao, R.; Fang, C.; Liu, H.; Liu, X. Evaluating urban ecosystem resilience using the DPSIR framework and the ENA model: A case study of 35 cities in China. *Sustain. Cities Soc.* **2021**, *72*, 102997. [CrossRef]
42. Jia, Y.; Shen, J.; Wang, H.; Dong, G.; Sun, F. Evaluation of the Spatiotemporal Variation of Sustainable Utilization of Water Resources: Case Study from Henan Province (China). *Water* **2018**, *10*, 554. [CrossRef]
43. Wang, S.; Wei, Y. Water resource system risk and adaptive management of the Chinese Heihe River Basin in Asian arid areas. *Mitig. Adapt. Strat. Glob. Chang.* **2019**, *24*, 1271–1292. [CrossRef]
44. Ruan, J.; He, G. Comprehensive evaluation of water resources security of the Huaihe Eco-economic Belt. *Water Supply* **2021**, *22*, 1047–1061. [CrossRef]
45. Chen, L.; Shi, J. Analysis and predication of urban water security: A case study of Chengdu City, China. *IOP Conf. Series Earth Environ. Sci.* **2016**, *39*, 012027. [CrossRef]
46. Zi-Ying, Z.; An-Yan, H. Construction method and application of water environment security assessment system in Taizhou bay of Zhejiang province. *J. Environ. Prot. Ecol.* **2018**, *19*, 515–526.
47. Zhou, F.; Su, W.; Zhang, F. Influencing Indicators and Quantitative Assessment of Water Resources Security in Karst Region Based on PSER Model—The Case of Guizhou. *Sustainability* **2019**, *11*, 5671. [CrossRef]
48. Van Ginkel, K.C.H.; Hoekstra, A.Y.; Buurman, J.; Hogeboom, R.J. Urban Water Security Dashboard: Systems Approach to Characterizing the Water Security of Cities. *J. Water Resour. Plan. Manag.* **2018**, *144*, 04018075. [CrossRef]
49. Zhao, Y.; Wang, Y.; Wang, Y. Comprehensive evaluation and influencing factors of urban agglomeration water resources carrying capacity. *J. Clean. Prod.* **2021**, *288*, 125097. [CrossRef]
50. Sun, D.; Wu, J.; Zhang, F.; Su, W.; Hui, H. Evaluating Water Resource Security in Karst Areas Using DPSIRM Modeling, Gray Correlation, and Matter–Element Analysis. *Sustainability* **2018**, *10*, 3934. [CrossRef]
51. Stanners, D.; Bourdeau, P. *Europe's Environment: The Dobris Assessment*; EEA: Copenhagen, Denmark, 1995.
52. Holten-Andersen, J.; Paaby, H. *Recommendations on Strategies for Integrated Assessment of Broad Environmental Problems*; Report submitted to the European Environment Agency (EEA); National Environmental Research Institute (NERI): Aarhus, Denmark, 1995.
53. Gebremedhin, S.; Getahun, A.; Anteneh, W.; Bruneel, S.; Goethals, P. A Drivers-Pressure-State-Impact-Responses Framework to Support the Sustainability of Fish and Fisheries in Lake Tana, Ethiopia. *Sustainability* **2018**, *10*, 2957. [CrossRef]
54. Cai, J.; He, Y.; Xie, R.; Liu, Y. A footprint-based water security assessment: An analysis of Hunan province in China. *J. Clean. Prod.* **2019**, *245*, 118485. [CrossRef]
55. Ding, T.; Qian, W. Geographical patterns and temporal variations of regional dry and wet heatwave events in China during 1960–2008. *Adv. Atmos. Sci.* **2011**, *28*, 322–337. [CrossRef]
56. Guo, E.; Zhang, J.; Ren, X.; Zhang, Q.; Sun, Z. Integrated risk assessment of flood disaster based on improved set pair analysis and the variable fuzzy set theory in central Liaoning Province, China. *Nat. Hazards* **2014**, *74*, 947–965. [CrossRef]

57. Liang, Y.; Lv, A. Risk assessment of water resource security in China. *Resour. Sci.* **2019**, *4*, 775–789.
58. Yang, Z.-h.; Zhou, Q.-w.; Guo, Y.; Su, W.-c.; Zhang, F.-t. Dynamic assessment of water resources security in karst area based on SPA-MC model-a case study of Guiyang city. *China Environ. Sci.* **2017**, *37*, 1589–1600.
59. Zhang, F.; Wang, L.; Su, W. Evaluation on the safety of epikarst dualistic water resources by coupling model of entropy weight set pair. *China Environ. Sci.* **2015**, *35*, 3511–3520. (In Chinese)
60. Zhang, Z.; He, W.; An, M.; Degefu, D.M.; Yuan, L.; Shen, J.; Liao, Z.; Wu, X. Water Security Assessment of China's One Belt and One Road Region. *Water* **2019**, *11*, 607. [CrossRef]
61. Zhong, S.; Elzarka, H. Retrospective Evaluation of the Vulnerability of Watershed Sustainable Water Development Using a Time-Series-Based Space Geometry Model: Xiang Jiang Watershed, China. *J. Hydrol. Eng.* **2021**, *26*, 05021023. [CrossRef]
62. Pagan, J.; Pryor, M.; Deepa, R.; Grace, J.M., III; Mbuya, O.; Taylor, R.; Dickson, J.O.; Ibeanusi, V.; Chauhan, A.; Chen, G.; et al. Sustainable Development Tool Using Meta-Analysis and DPSIR Framework—Application to Savannah River Basin, U.S. *JAWRA J. Am. Water Resour. Assoc.* **2020**, *56*, 1059–1082. [CrossRef]
63. Sarindizaj, E.E.; Karamouz, M. Dynamic Water Balance Accounting-Based Vulnerability Evaluation Considering Social Aspects. *Water Resour. Manag.* **2022**, *36*, 659–681. [CrossRef]
64. Pan, B.; Han, M.; Li, Y.; Wang, M.; Du, H. An Analysis on the Trend of Sustainable Utilization of Water Resources in Dongying City, China. *Water Resour.* **2021**, *48*, 158–166. [CrossRef]
65. Xiao, J. Safety Evaluation of Water Resources in Jiangxi Province. Master's Thesis, Changsha University of Science & Technology, Changsha, China, 2008; p. 108. (In Chinese).
66. Zhiping, Y.; Lihong, M. Evaluation of Environmental Carrying Capacity of Water Resources in Jiangxi Province. *Agric. Sci. Technol.* **2015**, *16*, 2849–2853.
67. Sun, M.; Kato, T. *City Water Resources Vulnerability: The Case of Jinan and Qingdao in Shandong Province, China*; Springer: Singapore, 2020.
68. Tan, C.; Peng, Q.; Ding, T.; Zhou, Z. Regional Assessment of Land and Water Carrying Capacity and Utilization Efficiency in China. *Sustainability* **2021**, *13*, 9183. [CrossRef]
69. Lei, X.; Qiu, G. Empirical study about the carrying capacity evaluation of regional resources and environment based on entropy-weight TOPSIS model. *Huanjing Kexue Xuebao/Acta Sci. Circumstantiae* **2016**, *36*, 314–323.
70. Wensheng, W.; Yueqing, L.; Juliang, J.; Jing, D. *Set-Pair Analysis of Hydrology and Water Resources*; Science Publishing House: Beijing, China, 2010.
71. Babel, M.S.; Shinde, V.R.; Sharma, D.; Dang, N.M. Measuring water security: A vital step for climate change adaptation. *Environ. Res.* **2020**, *185*, 109400. [CrossRef] [PubMed]
72. Wang, X.; Chen, Y.; Li, Z.; Fang, G.; Wang, Y. Development and utilization of water resources and assessment of water security in Central Asia. *Agric. Water Manag.* **2020**, *240*, 106297. [CrossRef]
73. Arreguin-Cortes, F.I.; Saavedra-Horita, J.R.; Rodriguez-Varela, J.M.; Tzatchkov, V.G.; Cortez-Mejia, P.E.; Llaguno-Guilberto, O.J.; Sainos-Candelario, A.; Sandoval-Yoval, L.; Ortega-Gaucin, D.; Mendoza-Cazares, E.Y.; et al. Municipal level water security indices in Mexico. *SN Appl. Sci.* **2019**, *1*, 1194. [CrossRef]
74. Aboelnga, H.T.; El-Naser, H.; Ribbe, L.; Frechen, F.-B. Assessing Water Security in Water-Scarce Cities: Applying the Integrated Urban Water Security Index (IUWSI) in Madaba, Jordan. *Water* **2020**, *12*, 1299. [CrossRef]
75. Khan, S.; Guan, Y.; Khan, F.; Khan, Z. A Comprehensive Index for Measuring Water Security in an Urbanizing World: The Case of Pakistan's Capital. *Water* **2020**, *12*, 166. [CrossRef]
76. Nie, R.-X.; Tian, Z.-P.; Wang, J.-Q.; Zhang, H.-Y.; Wang, T.-L. Water security sustainability evaluation: Applying a multistage decision support framework in industrial region. *J. Clean. Prod.* **2018**, *196*, 1681–1704. [CrossRef]
77. Assefa, Y.T.; Babel, M.S.; Sušnik, J.; Shinde, V.R. Development of a Generic Domestic Water Security Index, and Its Application in Addis Ababa, Ethiopia. *Water* **2018**, *11*, 37. [CrossRef]
78. Zhao, K.Q. Set pair and set pair analysis—A new concept and systematic analysis method. In Proceedings of the National Conference on System Theory and Regional Planning, Baotou, China, January 1989.
79. Zhang, X.Q.; Li, N.N. Research on water resources carrying capacity based on set pair analysis evaluation model. In Proceedings of the 3rd International Conference on Information Management, Innovation Management and Industrial Engineering, Kunming, China, 26–28 November 2010.
80. Yang, F.; Shao, D.; Xiao, C.; Tan, X.; Gu, W.; Wang, K. Improvement of set pair analysis evaluation method and its application on urban water security. In Proceedings of the 2011 International Symposium on Water Resource and Environmental Protection, Xi'an, China, 20–22 May 2011.
81. Yue, W.; Cai, Y.; Rong, Q.; Li, C.; Ren, L. A hybrid life-cycle and fuzzy-set-pair analyses approach for comprehensively evaluating impacts of industrial wastewater under uncertainty. *J. Clean. Prod.* **2014**, *80*, 57–68. [CrossRef]
82. Zhang, X.; Du, X.; Li, Y. Comprehensive evaluation of water resources carrying capacity in ecological irrigation districts based on fuzzy set pair analysis. *Desalin. Water Treat.* **2020**, *187*, 63–69. [CrossRef]
83. Zhou, Y.; Gao, M. An improved fuzzy-set pair analysis in water quality evaluation. *Chin. J. Environ. Eng.* **2015**, *9*, 749–755.
84. Hongwei, W.; Xin, Z.; Junnan, Q. Application of fuzzy set pair analysis method in safety evaluation of water resources. *J. Northwest Agric. For. Univ.* **2011**, *39*, 210–214.

85. Bo, Y.; Junjie, Y.; Ye, M.; Tao, W. A comprehensive evaluation and research of water resources allocation plans based on fuzzy set pair analysis. In Proceedings of the International Conference on Logistics Engineering, Management and Computer Science, Shenyang, China, 24–26 May 2014.
86. Xie, Y.-X.; Wang, W.-S. Fractal fuzzy set pair analysis assessment model of urban flood-waterlogged vulnerability. *J. Shenzhen Univ. Sci. Eng.* **2012**, *29*, 12–17. [CrossRef]
87. Guo, Q.; Wang, J.; Zhang, B. Comprehensive evaluation of the water resources carrying capacity based on DPSIMR. *J. Nat. Res.* **2017**, *32*, 484–493.
88. Wang, J.; Li, S.-C.; Li, L.-P.; Lin, P.; Xu, Z.-H.; Gao, C.-L. Attribute recognition model for risk assessment of water inrush. *Bull. Eng. Geol. Environ.* **2017**, *78*, 1057–1071. [CrossRef]
89. Liu, J.; Chen, J.; Yuan, Z.; Xu, J.; Wang, Y.; Lin, Y. Water resource security evaluation of the Yangtze River Economic Belt. *Water Supply* **2020**, *20*, 1554–1566. [CrossRef]
90. Bland, J.M.; Altman, D.G. Calculating correlation coefficients with repeated observations: Part 2—Correlation between subjects. *BMJ* **1995**, *310*, 633. [CrossRef]
91. Bland, J.M.; Altman, D.G. Statistics notes: Calculating correlation coefficients with repeated observations: Part 1—Correlation within subjects. *BMJ* **1995**, *310*, 446. [CrossRef]
92. Bakdash, J.Z.; Marusich, L. Repeated Measures Correlation. *Front. Psychol.* **2017**, *8*, 456. [CrossRef]
93. Pinheiro, J.; Bates, D. *Mixed-Effects Models in S and S-PLUS*; Springer Science & Business Media: Cham, Switzerland, 2006.
94. Bates, D. Fitting linear mixed models in R. *R News* **2005**, *5*, 27–30.

International Journal of
Environmental Research and Public Health

MDPI

Article

The Preparation of N-Doped Titanium Dioxide Films and Their Degradation of Organic Pollutants

Yanyan Dou [1], Yixuan Chang [1], Xuejun Duan [1,*], Leilei Fan [2], Bo Yang [3,*] and Jingjing Lv [1]

1 School of Energy and Environment, Zhongyuan University of Technology, Zhengzhou 450007, China
2 Department of Resources and Environment, Zunyi Normal College, Zunyi 563006, China
3 People's Government of Donganggezhuang Town, Luanzhou City, Tangshan 063000, China
* Correspondence: 5529@zut.edu.cn (X.D.); 2021108299@zut.edu.cn (B.Y.)

Abstract: N-doped TiO_2 films supported by glass slides showed superior photocatalytic efficiency compared with naked TiO_2 powder due to them being easier to separate and especially being responsive to visible light. The films in this study were prepared via the sol–gel method using TBOT hydrolyzed in an ethanol solution and the nitrogen was provided by cabamide. The N-doped TiO_2 coatings were prepared via a dip-coating method on glass substrates (30 × 30 × 2 mm) and then annealed in air at 490 °C for 3 h. The samples were characterized using X-ray diffraction (XRD), scanning electron microscopy (SEM) and UV-vis. The doping rate of N ranged from 0.1 to 0.9 (molar ratio), which caused redshifts to a longer wavelength as seen in the UV-vis analysis. The photocatalytic activity was investigated in terms of the degradation of phenol under both UV light and visible light over 4 h. Under UV light, the degradation rate of phenol ranged from 86% to 94% for all the samples because of the sufficient photon energy from the UV light. Meanwhile, under visible light, a peak appeared at the N-doping rate of 0.5, which had a degrading efficiency that reached 79.2%, and the lowest degradation rate was 32.9%. The SEM, XRD and UV-vis experimental results were consistent with each other.

Keywords: photocatalysis; N-doped films; visible light; titanium dioxide

Citation: Dou, Y.; Chang, Y.; Duan, X.; Fan, L.; Yang, B.; Lv, J. The Preparation of N-Doped Titanium Dioxide Films and Their Degradation of Organic Pollutants. *Int. J. Environ. Res. Public Health* **2022**, *19*, 15721. https://doi.org/10.3390/ijerph192315721

Academic Editors: Xun Wang, Zhiyuan Wang and Xin Zhao

Received: 20 October 2022
Accepted: 23 November 2022
Published: 25 November 2022

1. Introduction

In recent years, substances that cause water pollution are mainly refractory pollutants in industrial and agricultural wastewater and domestic sewage, such as volatile halogenated hydrocarbons, phenols, nitrobenzene and polycyclic aromatic hydrocarbons. Much effort has been devoted to the photocatalysis field in water treatment to efficiently remove nonbiodegradable compounds and avoid secondary pollution. The photocatalytic reaction transfers an electron from a valance band electron to an empty conduction band by absorbing photon energy equal to or more than the semiconductor band gap. The resulting electron–hole pairs contribute to the degradation reaction of the pollutant. Phenolic compounds are highly toxic and carcinogenic; they are present in wastewater, mainly in effluents from the production of pharmaceuticals, plastics, pesticides, oil and petrochemicals. Phenol was used as a model form of pollution since its molecular structure contains a benzene ring, making it quite stable and difficult to be biodegraded.

Common semiconductor photocatalysts include TiO_2, WO_3, ZnO and NiO. In order to improve the photocatalytic degradation activity, many technologies were proposed, such as metal or nonmetal doping, the development of microspheres and coupling semiconductors together. A pure TiO_2 photocatalyst often has some shortcomings, such as easy recombination of photogenerated electrons and holes, low catalytic efficiency and only responds to UV light, which limits its practical application. To solve these problems, current research focuses on metal/nonmetallic element doping, precious metal deposition and the construction of composite photocatalysts. TiO_2 of ultrafine nanoparticle size is considered to be the most useful photocatalyst for its excellent properties, such as high-light-conversion efficiency, chemical stability, nontoxic nature and low cost [1–13].

The band gap energy of TiO_2 is 3.2 eV (for anatase) and 3.0 eV (for rutile), and the maximum absorption wavelength of TiO_2 is 387.5 nm (anatase); that is to say, TiO_2 can only assimilate UV light rather than generate electrons (e^-) and holes (h^+), which can subsequently induce redox reactions for the degradation of nonbiodegradable organics in water [14]. Under the irradiation of UV light, electrons are promoted from the valence band to the conduction band of the semiconductor, creating electron–hole pairs, which can cause highly oxidizing hydroxyl and highly reduced superoxide radicals [15]. While the energy of UV light only takes 4–5% of the solar energy, how to enlarge the maximum absorption wavelength of TiO_2 for visible light and cause TiO_2 to absorb more visible light has become a research hotspot in recent years [16]. Ion doping, semiconductor complexes and surface photosensitized methods were employed to cause TiO_2 to be responsive to visible light. Nonmetallic ion doping is one of the widely studied ways of inducing new electronic bands and optical transitions, which involves the inclusion or substitution of a foreign atom, such as nitrogen, sulfur, fluorine or sulfur, that replaces the oxygen atom in the TiO_2 crystal lattice [17–19]. Doping TiO_2 with nitrogen can create a redshift in the absorption wavelength from UV to the visible range because of the formation of new states inside the TiO_2 bandgap. This shift could enable photocatalytic reactions to produce a high degradation rate under sunlight illumination [20–23].

Different nano-TiO_2 photocatalytic systems in suspension were studied, such as nanoparticles, nanobelts and nanotubes, where TiO_2 has a larger specific surface area and high absorption of light, and all show increased photocatalytic activity when excited under visible light toward the degradation of different chemical species [24–26]. However, difficulties in separation and recycling lead to the smaller possibility of industrial application. Furthermore, the dosage of TiO_2 is difficult to control, where too little will lead to low photocatalytic efficiency and too much will cause light scattering that influences the absorption of light. Thus, much effort was expended to immobilize photocatalysts in the form of thin films on a stable support to avoid the problems associated with disposing photocatalyst suspensions [27]. Different contents of doped nitrogen have different impacts on the photocatalytic efficiency, for disparities in the replaced oxygen atom can lead to variations in the photocatalyst activity. Frequently used nitrogen sources include urea, triethylamine, ammonia and ethylmethylamine [25].

N-doped TiO_2 thin films can effectively solve the previous problems by enlarging the maximum absorption wavelength and realizing immobilization together. Nitrogen is doped into TiO_2 using the sol–gel method, which is easy to operate and the reaction condition is mild. The formation of catalyst needs to go through the procedure of dipping, pulling out, drying, annealing, and chilling [28].

The purpose of this research was to prepare nitrogen-doped TiO_2 thin films on sheet substrates of different nitrogenous amounts using the sol–gel method. The advantages of this method are that TiO_2 can be stimulated with visible light, it solves the traditional issue of photocatalytic technology being difficult to control, problems such as high cost and effective components being easily lost are solved, and it greatly improves the possibility of a large-scale practical application in wastewater treatment engineering. We compared the degradation properties of these samples, investigated degradation results and their characterization consequences, analyzed the influence of the amount of nitrogen, and addressed the relationship between the concentration of zymolyte and the reaction time.

2. Materials and Methods

2.1. Materials and Reagent

Tetrabutyltitanate (CP, 98%), acetylacetone (AR, 98.0%), polyethylene glycol, acetone (AR, 99.5%) and hydrofluoric acid were provided by GuangFu Fine Chemical Research Institution in TianJin. Anhydrous ethanol (AR, 99.7%) and carbamide (AR, 99.5%) were made by ShuangShuang Chemical Co., Ltd., in YanTai, China. Deionized water was created in our lab. The glass substrates used were common commercial glass sheets cut to the needed size.

2.2. Preparation of N-Doped TiO_2 Films

The dip-coating method was utilized to immobilize the films. Precursor solutions for N-doped TiO_2 coatings were prepared with tetrabutyltitanate, anhydrous ethanol, acetylacetone, carbamide, deionized water and polyethylene glycol (1 wt%). Tetrabutyltitanate (20 mL), anhydrous ethanol (80 mL) and acetylacetone (3 mL) were mixed using magnetic stirring for 20 min, followed by a certain quality of carbamide added into the solution as the donor of nitrogen (the molar ratios of N/Ti used were 0.1, 0.3, 0.5, 0.7 and 0.9). Deionized water (15 mL) was dropwise added at the speed of one drop per second and left to stir for one hour; then, the procedure was finished after the addition of polyethylene glycol (1 mL) to avoid fracturing of the films [28]. Undoped TiO_2 sol with the same reagents and procedures was also prepared to compare the photocatalytic activity with N-doped coatings. The sol was aged for 24 h at 35 °C and then used to make the coatings.

Commercial glass substrates (30 mm × 30 mm × 2 mm) cleaned with acetone (10 wt%) and eroded with hydrofluoric acid (10 wt%) for 3 h were used as the support of the N-doped TiO_2 films. The substrates were maintained in the aged sol for 10 min and then pulled out at the rate of 2 cm/s using a dip coater (CZ-4200, Qingdao Zhongrui Intelligent Instrument Co., LTD, Qingdao, China). After the films dried out, the previous process was repeated another three times; the product was then calcined for 3 h at 490 °C. The films were finally even and tight on the glass sheet. The coatings obtained were named TN0, TN1, TN3, TN5, TN7 and TN9 according to the amount of doped nitrogen.

2.3. Characterization

After the deposition of the TiO_2 films onto glass substrates, all of the remaining sol was dried at 80 °C for 20 h in order to obtain dried gels, which were then calcined at 490 °C for 3 h to prepare the powders with the same crystal phase as the coatings. These powders were synthesized to analyze the phase compositions of titania by means of X-ray diffraction (XRD) using a D/Max-2200 Powder X-ray Diffractometer (XRD, D/Max-2200, Nippon Science Corporation, Tokyo, Japan) with Cu-Kα radiation at 40 kV and 20 mA.

Scanning electron microscopy (SEM, SU8220, Hitachi hi-tech, Shanghai, China) was utilized in an air atmosphere to examine the morphological structure and grain size of the films coated on the glass. The vapor has its considerable impact on the dried sample before observation [29].

The UV-vis DRS measurements were recorded at room temperature for the dry-pressed disk samples using a UV-3600 UV-vis spectrophotometer (UV-3600 UV-vis, Shimadazu, Tokyo, Japan) equipped with an integrating sphere assembly within the range of 300–900 nm.

2.4. Measurement of Photocatalytic Activity

The experiments were carried out with the initial concentration of phenol equal to 10 mg/L at an ambient temperature (approximately 25 °C) and pressure. The photoreactor used an acrylic cuboid static opaque chamber (700 mm × 450 mm × 250 mm) equipped with a thermocouple to monitor the temperature during irradiation. The UV source was supplied by 6 ultraviolet tubes (20 W), which had a dominant wavelength of 254 nm. As for the visible light source, 6 ordinary fluorescent lamps (20 W) were employed to produce the longer wavelength light.

Each beaker contained 8 pieces of glass sheet supporting N-doped TiO_2 coatings of the same amount of doped nitrogen. The illuminant was about 15 cm from the bottom of the solution. The system was left in the dark for 30 min until reaching phenol adsorption equilibrium, and then a photocatalytic reaction was carried out under UV light or visible light. The samples were taken from the reactor for analysis every 30 min, where the samples were placed in a 2 cm quartz dish and the remaining concentrations were analyzed using 4-AAP extraction spectrophotometry and UV-vis spectrophotometry (Photo Lab 6600 UV–Vis, WTW, Munich, Germany) at 510 nm. The photocatalysis reaction lasted for 4 h.

3. Results and Discussion

3.1. XRD Analysis

Figure 1 shows the XRD patterns of the six powdery samples. For all samples, it can be observed that where the 2θ was 25.4° (101), the diffraction peak was especially distinct, and at 2θ = 30.7° (121), the relative intensity is quite small, which means that the anatase phase was dominant and the rutile phase was hardly existing. In addition, other characteristic peaks (2θ = 25.34°, 48.11° and 44.49°) were all accordant with JPDS-21-1272 [30] (anatase standard card). It can be confirmed that N-doped TiO_2 mainly existed in the form of anatase. With the increase in the N doping amount, the sample pattern almost did not change, indicating that N doping had little effect on the crystal structure. The difference being all samples on the 101 crystal plane may have been caused by the grinding of the samples with different N contents.

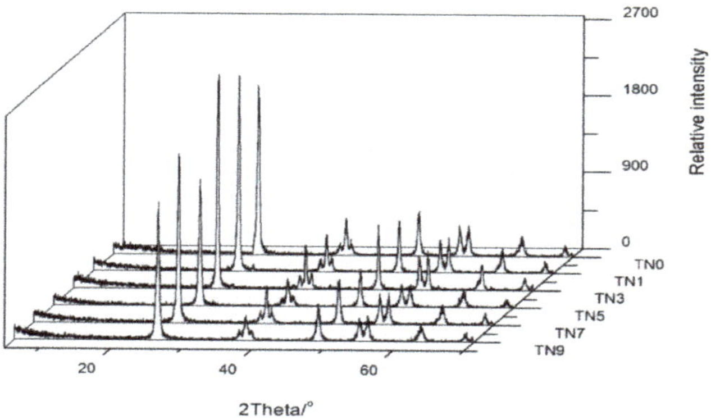

Figure 1. XRD patterns of TN0, TN1. TN3, TN5, TN7 and TN9.

3.2. SEM Analysis

Figure 2 shows the SEM photographs of a clean glass substrate and N-doped TiO_2 films. The glass substrate after being washed with acetone and eroded with hydrofluoric acid was quite clean and had an even roughness with a great transmission of light. A rough surface can increase the capacity of the load [27]. The microstructures of the six samples had no obvious differences, which indicated that the quantity of carbamide had little impact on the surface of the coatings. Figure 2d, e demonstrate that the films made using the sol–gel method were porous and the granules dispersed uniformly with particle sizes ranging from 30 nm to 80 nm.

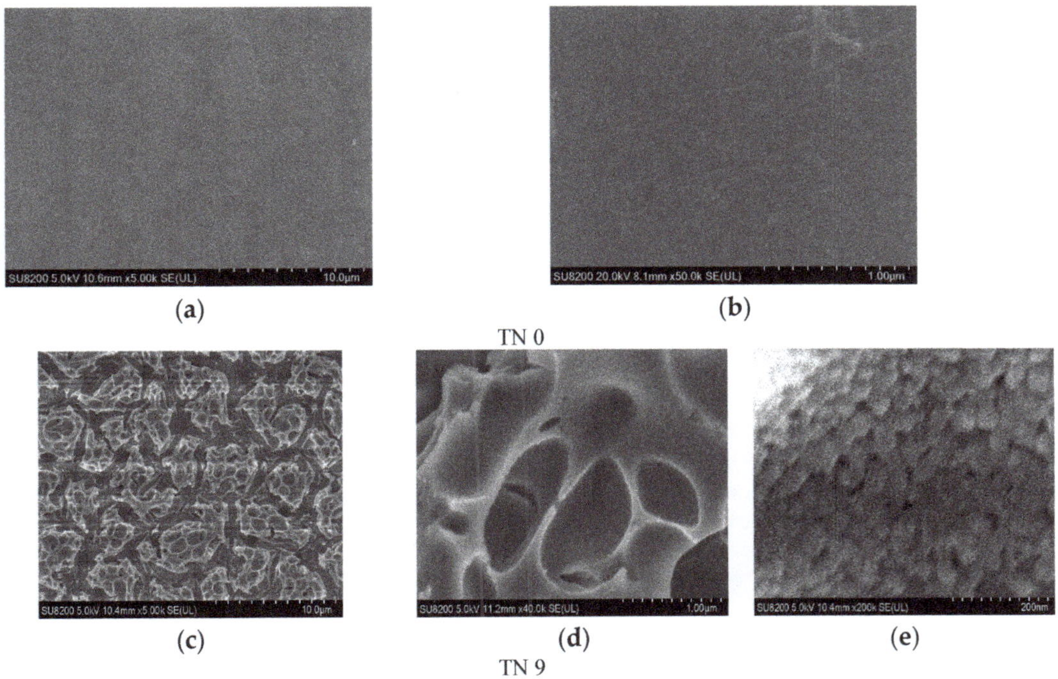

Figure 2. SEM photographs of a clean glass substrate (**a**,**b**) (TN0) and SEM photographs of sample TN9 (**c**–**e**).

3.3. *UV-Vis DRS Analysis*

UV-vis DRS results are shown in Figure 3. The maximum absorption wavelength of TN0 was about 380 nm, indicating the main components of samples were naked titanium dioxide with little impurities. Different amounts of N-doping replaced oxygen atoms in the TiO_2 crystal lattice with nitrogen atoms to different degrees, which was also the reason for different degrees of redshift when compared with the maximum absorption wavelength of TN0. It is worth noting that the nitrogen doping amount did not follow a "the more, the better" pattern within a certain range since the maximal redshift occurred for TN5, not TN9. Different amounts of N-doping can reduce the bandwidth of TiO_2, enhance the transfer of electrons from the valence band to the conduction band and improve the photocatalytic rate. However, when the amount of N-doping is too much, the nitrogen atom will become the center of electron recombination and accelerate the recombination rate of electrons and holes, thus affecting the photocatalytic rate [31,32].

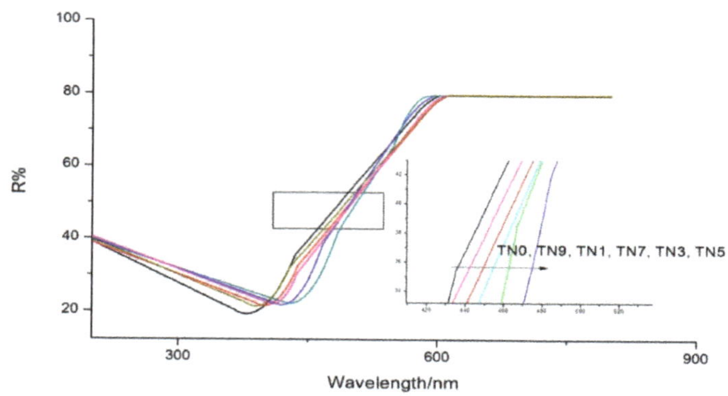

Figure 3. UV-vis DRS of TN0, TN1, TN3, TN5, TN7 and TN9.

3.4. *Photocatalytic Activity under UV and Visible Light Irradiation*

3.4.1. Photocatalytic Activity under UV Light Irradiation

With regard to absorption, the dark test showed that the variation of phenol concentration caused by absorption was lower than 3%, which meant that the absorption had little influence on the degradation ratio of the photocatalytic procedure.

The variation of each sample over time is shown in Figure 4. The phenol concentration at 0 min was measured just after absorption. It can be observed that the rank of the six samples regarding photocatalytic activity was TN5 > TN3 > TN7 > TN1 > TN9 > TN0, i.e., TN5 had the highest degradation over the others during the same period. The degradation ratios were 93.77%, 91.32%, 90.82%, 88.89%, 85.73% and 80.10%, respectively. The maximum absorption wavelength of the five N-doped samples were all redshifted to visible light to different degrees, while the photocatalytic activities of the five samples were also enhanced, which should have been caused by the doping nitrogen leading to the energy structure of titanium dioxide changed; that is to say, the optical energy band gap of TiO_2 diminished. A reduction in the optical energy band gap will enhance the transfer of electrons from the valence band to the conduction band under visible light, which may have been the reason for the better relative performance of the N-doped TiO_2 [33]. The kinetics of phenol removal followed the Langmuir–Hinshelwood kinetic equation [34]:

$$R = d_c/d_t = kK_c/(1 + K_c) \tag{1}$$

where R represents the reaction rate, c is the concentration of the substrate at the time, k is the reaction rate constant and K_C is the adsorption constant. When the concentration was very low, $K_c << 1$; therefore, the relevant equation turned out to be

$$\ln(c_0/c) = k_t + b \tag{2}$$

where c_0 is the initial concentration, t is the reaction time and k_t is the apparent reaction rate constant. Figure 5 shows the kinetics of the six samples. The apparent reaction rate constants ranged from 0.00614 to 0.01048.

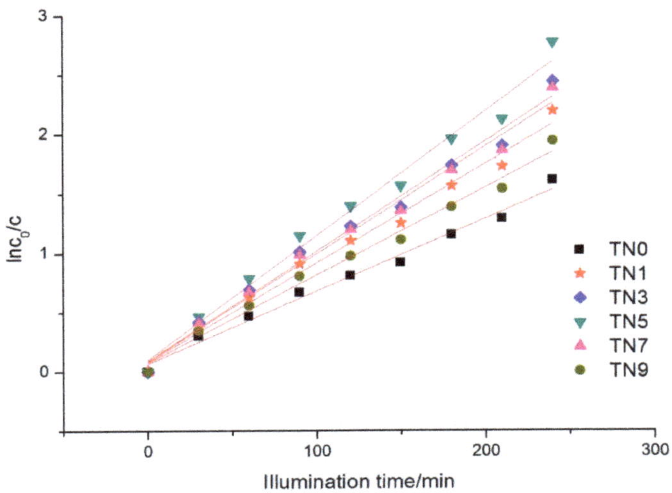

Figure 4. Phenol concentration trend under UV light.

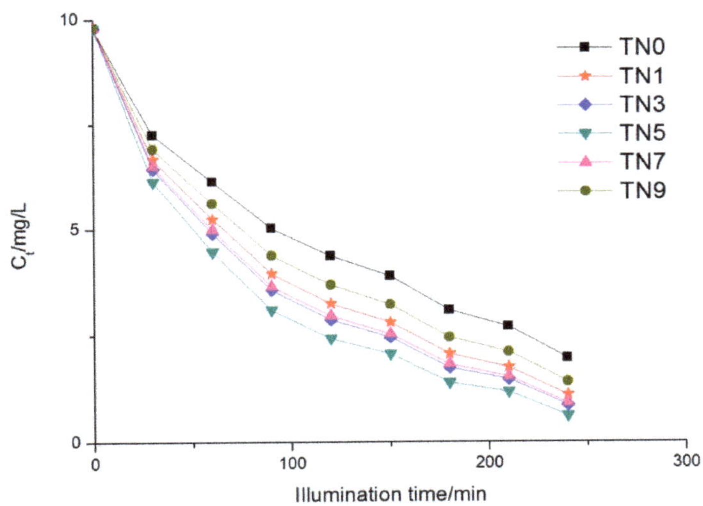

Figure 5. Reaction kinetics trend under UV light.

3.4.2. Photocatalytic Activity under Visible Light Irradiation

The absorption under visible light in the dark period made little difference compared with UV light, while the variation in phenol concentration during illumination time made a significant difference.

It can be observed from Figure 6 that the degradation rates of the N-doped samples were obviously better than that of the naked TiO_2 (TN0). TN5 still possessed the fastest degradation rate, followed by TN3, TN7, TN1 and TN9; that is to say, the best degradation efficiency appeared at the point where the molar ratio of N/Ti was 0.5, rather than the more nitrogen doping, the higher the degradation rate. The degradation results were also coincident with the UV-vis DRS results, where TN5 had the greatest redshift.

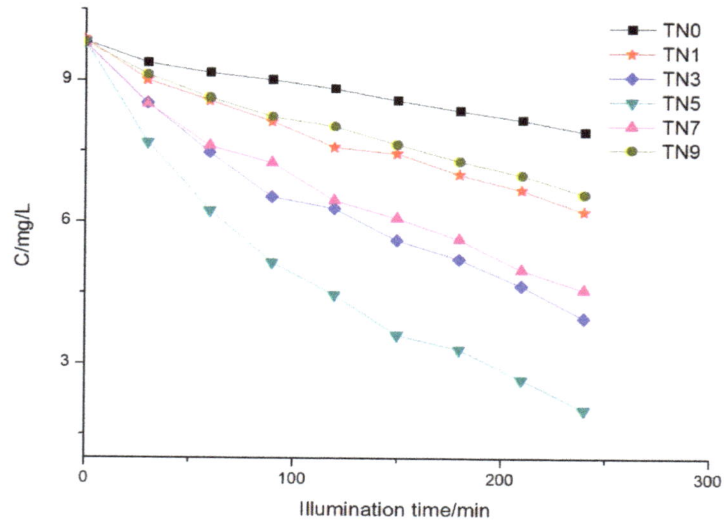

Figure 6. Phenol concentration trend under visible light.

The kinetics of phenol removal under visible light also followed the Langmuir–Hinshelwood kinetic equation (the kinetic Equation (1) in 3.4.1), as Figure 7 shows. The apparent reaction rate constants ranged from 0.00156 to 0.04893.

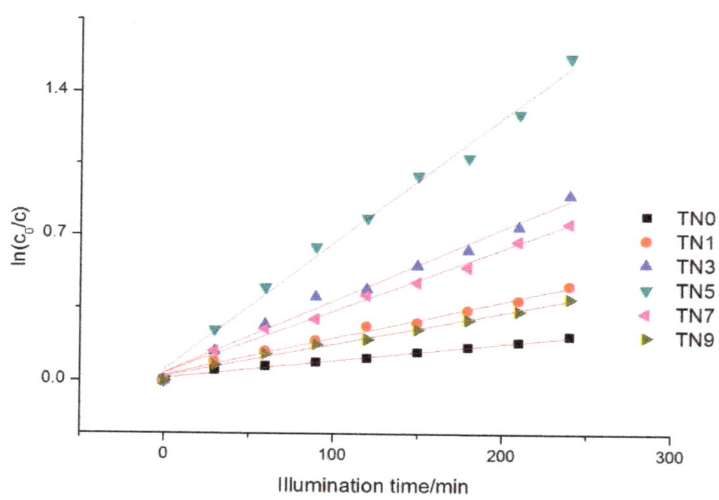

Figure 7. Reaction kinetics trend under visible light.

4. Conclusions

N-doped TiO_2 thin films were successfully immobilized on commercial glass substrates via the sol–gel method starting from tetrabutyltitanate dissolving in anhydrous ethanol as a precursor. The formulation of sol and an annealing temperature of 490 °C were optimal, as seen by the highly uniform lattice structure that was mainly constituted of anatase and the TiO_2 granules being evenly distributed with ultrafine nano-particle sizes ranging from 30 to 80 nm. The surface morphology of the coating was basically unaffected by different nitrogen contents. Under the same experimental conditions, the degradation efficiency of phenol in the experimental group under visible light irradiation reached about 90% of that under UV light, indicating that N-doping caused the optical energy band gap of TiO_2 to diminish; therefore, the maximum absorption wavelength had obvious redshifts, leading to the doped films having more efficient photocatalytic activity, both under UV light and visible light. The optimal photocatalytic efficiency was realized at an N-doping ratio of 0.5, rather than the more dopant, the better the photocatalytic efficiency since an excess N-doping ratio leads to an increased recombination ratio of electrons and holes, which reduces the photon utilization factor. The pollution absorption ability of the TiO_2 and glass sheet was quite feeble; thus, the kinetics of degradation followed the Langmuir–Hinshelwood kinetic equation, which describes first-order reaction kinetics.

Author Contributions: Y.D. and L.F. contributed substantially to the design of the study, the analyses, the writing of the manuscript and the original draft preparation. X.D. and J.L. contributed to the project administration, the funding acquisition and verifying the experimental data. Y.C. and B.Y. took responsibility for performing the experiments and validation and made the figures, replied to comments and corrected the draft. All authors have read and agreed to the published version of the manuscript.

Funding: This work was supported by the Project of Guizhou Education, Science and Technology Project of Guizhou Province ([2016]1163), the Project of Guizhou Education Department([2016]047), the Training Program for Young Backbone Teachers in Colleges and Universities of Henan Province, (no. 2019GGJS140), the Zunyi Science and Technology Planning Project (no. ZSKH(2021)199), the 2021 Rural Revitalization Project of Zunyi Normal University (Special Scientific Research Fund of Guizhou Provincial Department of Education) (QJHKY [2021]017-9), the Zunyi Normal University Serving Local Industrial Revolution project (N0.ZSCXY [2021]04) and the "2021 academic new seedling cultivation and innovation exploration project" of Zunyi Normal University (no. ZSXM [2021]1-10).

Institutional Review Board Statement: Not applicable.

Informed Consent Statement: Not applicable.

Data Availability Statement: Not applicable.

Conflicts of Interest: The authors declare no conflict of interest.

References

1. Kudo, A.; Miseki, Y. Heterogeneous photocatalyst materials for water splitting. *Chem. Soc. Rev.* **2009**, *38*, 253–278. [CrossRef]
2. Kuwahara, Y.; Yamashita, H. Efficient photocatalytic degradation of organics diluted in water and air using TiO_2 designed with zeolites and mesoporous silica materials. *J. Mater. Chem.* **2011**, *21*, 2407–2416. [CrossRef]
3. Liu, D.; Cui, W.; Lin, J.; Xue, Y.; Huang, Y.; Li, J.; Zhang, J.; Liu, Z.; Tang, C. A novel TiO_2—xNx/BN composite photocatalyst: Synthesis, characterization and enhanced photocatalytic activity for Rhodamine B degradation under visible light. *Catal. Commun.* **2014**, *57*, 9–13. [CrossRef]
4. Khan, H.; Berk, D. Characterization and mechanistic study of Mo^{+6} and V^{+5} codoped TiO_2 as a photocatalyst. *J. Photochem. Photobiol. A Chem.* **2014**, *294*, 96–109. [CrossRef]
5. Yang, Y.; Zhang, T.; Le, L.; Ruan, X.; Fang, P.; Pan, C.; Xiong, R.; Shi, J.; Wei, J. Quick and Facile Preparation of Visible light-Driven TiO_2 Photocatalyst with High Absorption and Photocatalytic Activity. *Sci. Rep.* **2014**, *4*, 7045. [CrossRef]
6. Ljubas, D.; Franzreb, M.; Hansen, H.C.B.; Weidler, P.G. Magnetizing of nano-materials on example of Degussa's P-25 TiO_2 photocatalyst: Synthesis of magnetic aggregates, characterization and possible use. *Sep. Purif. Technol.* **2014**, *136*, 274–285. [CrossRef]
7. Paul, T.; Machesky, M.L.; Strathmann, T.J. Correction to Surface Complexation of the Zwitterionic Fluoroquinolone Antibiotic Ofloxacin to Nano-Anatase TiO_2 Photocatalyst Surfaces. *Environ. Sci. Technol.* **2014**, *48*, 11736. [CrossRef]
8. Kamegawa, T.; Ishiguro, Y.; Kido, R.; Yamashita, H. Design of Composite Photocatalyst of TiO_2 and Y-Zeolite for Degradation of 2-Propanol in the Gas Phase under UV and Visible Light Irradiation. *Molecules* **2014**, *19*, 16477–16488. [CrossRef]
9. He, K.F.; Xu, E.N.; Liu, Y.; Chen, W.P. Hydrogenation of Nano-Structured TiO_2 Photocatalyst Through an Electrochemical Method. *J. Nanosci. Nanotechnol.* **2015**, *15*, 303–308. [CrossRef]
10. Naeimi, A.; Sharifi, A.; Montazerghaem, L.; Abhari, A.R.; Mahmoodi, Z.; Bakr, Z.H.; Soldatov, A.V.; Ali, G.A.M. Transition Metals Doped WO_3 Photocatalyst towards High Efficiency Decolourization of Azo Dye. *J. Mol. Struct.* **2022**, *1250*, 131800. [CrossRef]
11. Saad, A.M.; Abukhadra, M.R.; Ahmed, S.A.-K.; Elzanaty, A.M.; Mady, A.H.; Betiha, M.A.; Shim, J.J.; Rabie, A.M. Photocatalytic Degradation of Malachite Green Dye Using Chitosan Supported ZnO and Ce–ZnO Nano-Flowers under Visible Light. *J. Environ. Manag.* **2020**, *258*, 110043. [CrossRef]

12. Ethiraj, A.S.; Uttam, P.K.V.; Chong, K.F.; Ali, G.A.M. Photocatalytic Performance of a Novel Semiconductor Nanocatalyst: Copper Doped Nickel Oxide for Phenol Degradation. *Mater. Chem. Phys.* **2020**, *242*, 122520. [CrossRef]
13. Mandor, H.; Amin, N.K.; Abdelwahab, O.; El-Ashtoukhy, E.-S.Z. Preparation and Characterization of N-Doped ZnO and N-Doped TiO_2 Beads for Photocatalytic Degradation of Phenol and Ammonia. *Environ. Sci. Pollut. Res.* **2022**, *29*, 56845–56862. [CrossRef]
14. Fujishima, A.; Honda, K. Electrochemical Photolysis of Water at a Semiconductor Electrode. *Nature* **1972**, *238*, 37–38. [CrossRef] [PubMed]
15. Lui, G.; Liao, J.-Y.; Duan, A.; Zhang, Z.; Fowler, M.; Yu, A. Graphene-wrapped hierarchical TiO_2 nanoflower composites with enhanced photocatalytic performance. *J. Mater. Chem. A* **2013**, *1*, 12255–12262. [CrossRef]
16. Jaiswal, R.; Patel, N.; Dashora, A.; Fernandes, R.; Yadav, M.; Edla, R.; Varma, R.; Kothari, D.; Ahuja, B.; Miotello, A. Efficient Co-B-codoped TiO_2 photocatalyst for degradation of organic water pollutant under visible light. *Appl. Catal. B Environ.* **2016**, *183*, 242–253. [CrossRef]
17. Asahi, R.; Morikawa, T.; Ohwaki, T.; Aoki, K.; Taga, Y. Visible-Light Photocatalysis in Nitrogen-Doped Titanium Oxides. *Science* **2001**, *293*, 269–271. [CrossRef] [PubMed]
18. Yoshinaga, M.; Yamamoto, K.; Sato, N.; Aoki, K.; Morikawa, T.; Muramatsu, A. Remarkably enhanced photocatalytic activity by nickel nanoparticle deposition on sulfur-doped titanium dioxide thin film. *Appl. Catal. B Environ.* **2009**, *87*, 239–244. [CrossRef]
19. Ho, W.; Yu, J.C.; Lee, S. Synthesis of hierarchical nanoporous F-doped TiO_2 spheres with visible light photocatalytic activity. *Chem. Commun.* **2006**, *10*, 1115–1117. [CrossRef]
20. Gumy, D.; Rincon, A.G.; Hajdu, R.; Pulgarin, C. Solar photocatalysis for detoxification and disinfection of water: Different types of suspended and fixed TiO_2 catalysts study. *Sol. Energy* **2006**, *80*, 1376–1381. [CrossRef]
21. Robert, D.; Piscopo, A.; Heintz, O.; Weber, J.V. Photocatalytic detoxification with TiO_2 supported on glass-fibre by using artificial and natural light. *Catal. Today* **1999**, *54*, 291–296. [CrossRef]
22. Águia, C.; Ângelo, J.; Madeira, L.M.; Mendes, A. Photo-oxidation of NO using an exterior paint—Screening of various commercial titania in powder pressed and paint films. *J. Environ. Manag.* **2011**, *92*, 1724–1732. [CrossRef] [PubMed]
23. van Grieken, R.; Marugán, J.; Sordo, C.; Pablos, C. Comparison of the photocatalytic disinfection of E. coli suspensions in slurry, wall and fixed-bed reactors. *Catal. Today* **2009**, *144*, 48–54. [CrossRef]
24. Burda, C.; Lou, Y.; Chen, X.; Samia, A.C.S.; Stout, J.; Gole, J.L. Enhanced Nitrogen Doping in TiO_2 Nanoparticles. *Nano Lett.* **2003**, *3*, 1049–1051. [CrossRef]
25. Wang, J.; Tafen, D.N.; Lewis, J.P.; Hong, Z.; Manivannan, A.; Zhi, M.; Li, M.; Wu, N. Origin of Photocatalytic Activity of Nitrogen-Doped TiO_2 Nanobelts. *J. Am. Chem. Soc.* **2009**, *131*, 12290–12297. [CrossRef]
26. Park, J.H.; Kim, S.; Bard, A.J. Novel Carbon-Doped TiO2 Nanotube Arrays with High Aspect Ratios for Efficient Solar Water Splitting. *Nano Lett.* **2006**, *6*, 24–28. [CrossRef]
27. Pinho, L.X.; Azevedo, J.; Miranda, S.M.; Angelo, J.; Mendes, A.; Vilar, V.J.P.; Vasconcelos, V.; Boaventura, R.A.R. Oxidation of microcystin-LR and cylindrospermopsin by heterogeneous photocatalysis using a tubular photoreactor packed with different TiO_2 coated supports. *Chem. Eng. J.* **2015**, *266*, 100–111. [CrossRef]
28. Lv, H.; Li, N.; Zhang, H.; Tian, Y.; Zhang, H.; Zhang, X.; Qu, H.; Liu, C.; Jia, C.; Zhao, J.; et al. Transferable TiO_2 nanotubes membranes formed via anodization and their application in transparent electrochromism. *Sol. Energy Math. Sol. C* **2016**, *150*, 57–64. [CrossRef]
29. Su, Z.; Zhou, W. Formation mechanism of porous anodic aluminium and titanium oxides. *Adv. Mater.* **2008**, *20*, 3663–3667. [CrossRef]
30. Grilli, R.; di Camillo, D.; Lozzi, L.; Horovitz, I.; Mamane, H.; Avisar, D.; Baker, M. Surface characterisation and photocatalytic performance of N-doped TiO_2 thin films deposited onto 200 nm pore size alumina membranes by sol-gel methods. *Mater. Chem. Phys.* **2015**, *159*, 25–37. [CrossRef]
31. Kobayakawa, K.; Murakami, Y.; Sato, Y. Visible-light active N-doped TiO_2 prepared by heating of titanium hydroxide and urea. *J. Photochem. Photobiol. A Chem.* **2005**, *170*, 177–179. [CrossRef]
32. Matsushita, M.; Nosaka, A.Y.; Nishino, J.; Nosaka, Y. Preparation of Nitrogen Doped Titanium Dioxide by Using Guanidine and Its Characterization. In *Journal of the Ceramic Society of Japan, Supplement Journal of the Ceramic Society of Japan*; The Ceramic Society of Japan: Gifu, Japan, 2004; pp. S1411–S1413.
33. Senthilnathan, J.; Philip, L. Photocatalytic degradation of lindane under UV and visible light using N-doped TiO_2. *Chem. Eng. J.* **2010**, *161*, 83–92. [CrossRef]
34. Wang, X.; Meng, S.; Zhang, X.; Wang, H.; Zhong, W.; Du, Q. Multi-type carbon doping of TiO_2 photocatalyst. *Chem. Phys. Lett.* **2007**, *444*, 292–296. [CrossRef]

International Journal of
Environmental Research and Public Health

MDPI

Article

Preparation of High-Porosity B-TiO_2/C_3N_4 Composite Materials: Adsorption–Degradation Capacity and Photo-Regeneration Properties

Xiang Guo [1], Lei Rao [2,*] and Zhenyu Shi [1]

[1] College of Environment, Hohai University, Nanjing 210098, China; gx2015@hhu.edu.cn (X.G.); szy@hhu.edu.cn (Z.S.)
[2] College of Mechanics and Materials, Hohai University, Nanjing 211100, China
* Correspondence: raol@hhu.edu.cn

Abstract: Adsorption can quickly remove pollutants in water, while photocatalysis can effectively decompose organic matter. B-TiO_2/g-C_3N_4 ternary composite photocatalytic materials were prepared by molten method, and their adsorption–degradation capability under visible light conditions was discussed. The morphology of the B-TiO_2/g-C_3N_4 materials was inspected by SEM, TEM, BET, and EDS, and the results showed that close interfacial connections between TiO_2 and g-C_3N_4, which are favorable for charge transfer between these two semiconductors, formed heterojunctions with suitable band structure which was contributed by the molten B_2O_3. Meanwhile, the molten B_2O_3 effectively increased the specific surface area of TiO_2/C_3N_4 materials, thereby increasing the active sites and reducing the recombination of photogenerated electron–hole pairs and improving the photocatalytic degradation abilities of TiO_2 and g-C_3N_4. Elsewhere, the crystal structure analysis (XRD, XPS, FTIR) results indicated that the polar -B=O bond formed a new structure with TiO_2 and g-C_3N_4, which is not only beneficial for inhibiting the recombination of electron holes but also improving the photocatalytic activity. By removal experiment, the adsorption and degradation performances of B-TiO_2/g-C_3N_4 composite material were found to be 8.5 times and 3.4 times higher than that of g-C_3N_4. Above all, this study prepared a material for removing water pollutants with high efficiency and provides theoretical support and experimental basis for the research on the synergistic removal of pollutants by adsorption and photocatalysis.

Keywords: adsorption; degradation; B_2O_3; TiO_2/C_3N_4; molten

Citation: Guo, X.; Rao, L.; Shi, Z. Preparation of High-Porosity B-TiO_2/C_3N_4 Composite Materials: Adsorption–Degradation Capacity and Photo-Regeneration Properties. *Int. J. Environ. Res. Public Health* **2022**, *19*, 8683. https://doi.org/10.3390/ijerph19148683

Academic Editors: Paul B. Tchounwou, Xin Zhao, Xun Wang and Zhiyuan Wang

Received: 24 June 2022
Accepted: 15 July 2022
Published: 17 July 2022

1. Introduction

Photocatalysis is a technology that can use semiconductor materials to remove pollutants from the environment under different lighting conditions [1–3]. Most materials require ultraviolet conditions to produce redox effects. The visible part of sunlight irradiate that reaches the water surface only occupies 45%, while the ultraviolet part occupies less than 4% [4]. At the same time, water absorbs and reflects sunlight in different wavelengths, which seriously limits the removal efficiency of photocatalytic materials in a water environment [5–7].

g-C_3N_4 is a good semiconductor material with application potential that is metal free and has visible light responsiveness [8–10]. However, researchers found that g-C_3N_4 still has many disadvantages such as low utilization of visible light, poor photoelectric conversion efficiency, and low specific surface area [11]. Since Serpone et al. [12] first reported that a solid–solid heterojunction interface with good contact can be constructed from different coupled semiconductor materials to promote electron transfer between particles, more and more researchers have paid attention to different semiconductor materials. Among the many photocatalytic semiconductor materials, TiO_2 is widely used in sewage treatment [13], photocatalytic synthesis [14], and self-cleaning [15]. Due to its easy availability, low cost, stable chemical properties, corrosion resistance, non-toxicity, and strong oxidizing properties [16], its air purification [17] and antibacterial properties [18] have been extensively studied. Therefore, using the advantages of TiO_2 material properties to composite it with g-C_3N_4 to become a more competitive material has attracted the attention of more and more researchers [19–21].

In the treatment of water environment pollution, the adsorption performance and visible light response ability of photocatalytic materials are two important factors that determine whether the photocatalytic technology can be effectively promoted [22,23]. The high-efficiency adsorption and enrichment ability of the material reduce the concentration of pollutants in water and provide a high-concentration contact environment that is conducive to the photocatalytic reaction for the material [24,25]. The melting characteristics of B_2O_3-modified g-C_3N_4 are effective for improving its specific surface area and adsorption performance, but its response to visible light is limited, which affects the visible light photocatalytic activity [26].

Effectively improving the adsorption performance and visible light catalytic activity of materials at the same time is a key issue for researchers. Functional coupling through different materials is an idea to solve this problem. Here, TiO_2 and g-C_3N_4 were formed into a composite heterostructure to improve the photocatalytic oxidation ability of the material. Then, the melting characteristics of B_2O_3 during the heating process were used as a "reaction environment regulator", and TiO_2/C_3N_4 was co-calcined to prepare a composite material. The results of the experiment showed that B element doped the composite-modified materials, and, finally, the adsorption behavior and visible light catalytic degradation ability of B-TiO_2/C_3N_4 were analyzed by using MB and RhB organic pollutants. This method effectively increased the specific surface area of the material, increased the conjugated system, improved the adsorption capacity, and effectively improved the visible light catalytic effect of the material. This provides a technical idea for the promotion and application of a low-cost preparation of an efficient adsorption–degradation photocatalytic composite material.

2. Materials and Methods

2.1. Sample Preparation

(1) TiO_2: 17 mL of butyl titanate (Aladdin, 99% pure) was placed into 55 mL of ethanol solution (Aladdin, 99.7% pure), then 4.5 mL of glacial acetic acid (Aladdin, 99% pure) was added. After mixing using a magnetic stirrer, the solution was recorded as A. We measured 27.6 mL of ethanol, added 0.9 mL of distilled water, and adjusted the pH to 4 with nitric acid (Sinopharm, 65−68% pure), and this solution was recorded as B after thorough mixing. The solutions A and B were stirred for 30 min respectively, the B solution was added dropwise to the A solution at a rate of 10 drops per minute, and the final obtained mixed solution was TiO_2 sol. Then, the TiO_2 sol was stirred at room temperature and placed in an oven to dry after gelation. Then, the dried xerogel was ground into powder and placed in a crucible with a lid to heat at a rate of 5 °C/min until 550 °C and was maintained at this temperature for 2 h. After the temperature in the muffle furnace (XS2−10, Li Chen, China) dropped to room temperature, the calcined material was ground into powder, and the obtained material was denoted as TiO_2.

(2) g-C_3N_4: 10 g melamine (Aladdin, 99% pure) was placed into a 50 mL crucible and put into muffle furnace, and the heating rate was set to 5 °C/min until it reached the reaction temperature of 550 °C. Then, the reaction temperature was maintained for 2 h. After the material cooled to room temperature, we ground the calcined solid, and the obtained yellow powder was g-C_3N_4.

(3) TiO_2/C_3N_4 composite material: 1 g of melamine and 10 mL of TiO_2 sol were weighed in a beaker, stirred, and mixed thoroughly and then the mixture was placed into an oven for drying after gelation. The dried composite precursor was placed in a crucible with a lid and calcined at 550 °C for 2 h. The obtained material was ground into powder after cooling to room temperature to obtain TiO_2/C_3N_4 composite material.

(4) B-C_3N_4: The B_2O_3 (Aladdin, 99.9% pure) and g-C_3N_4 were mixed and ground evenly. Then, the composite powder was placed into a ceramic crucible. After that, the samples were heated to 550 °C at a heating rate of 5 °C/min and maintained for 2 h. When the furnace was cooled to room temperature, the powders were washed with water and ethanol several times. For convenience of description, the composite material was abbreviated as B-C_3N_4.

(5) B-TiO_2/C_3N_4: The TiO_2/C_3N_4 material and B_2O_3 were mixed and stirred in distilled water at a mass ratio of 1:1. After grinding evenly, we put it into an atmosphere furnace at 550 °C for 2 h. Then, the material was soaked and washed three times with ethanol and distilled water, respectively. The sample was denoted as B-TiO_2/C_3N_4. For comparison, another ternary composite material, B2-C_3N_4/TiO_2, was prepared: 1 g of B-C_3N_4 and 10 mL of TiO_2 sol were weighed with the same method.

2.2. Characterization

The structural analysis of the samples was carried out by X-ray diffraction (XRD, Rigaku Ultima III, Tokyo, Japan) and was recorded in the 2θ range of 5–80° with a scan rate of 0.02°/0.4 s using a Bruker AXSD8 system (Bruker, Billerica, MA, USA) equipped with a Cu Kα radiation source (λ = 0.15406 Å, in which the X-ray tube was operated at 40 kV and 40 mA). UV–Vis diffuse reflectance spectra (DRS) were obtained on a UV–visible (UV–Vis) spectrophotometer (PerkinElmer, Waltham, MA, USA) with $BaSO_4$ as the reference. The sample morphology of the different composite materials was examined using a scanning electron microscope (SEM, Hitachi, Tokyo, Japan) and a transmission electron microscope (TEM, JEM-2100, JEOL, Tokyo, Japan). X-ray photoelectron spectroscopy (XPS) measurements were performed on a Thermo Scientific ESCALAB 250 instrument (Thermo Fisher Scientific, Waltham, MA, USA) with an Al Kα source. Low-temperature N_2 adsorption/desorption measurements (Brunauer–Emmett–Teller (BET) method) were carried out using a Micromeritics ASAP 2020 system (Micromeritics, Norcross, GA, USA) at −196 °C following degassing of all samples at 120 °C for 2 h.

2.3. Adsorption and Photocatalytic Degradation of Organic Pollution

In order to evaluate whether composite materials have a better effect on the removal of pollutants, TiO_2, g-C_3N_4, B-TiO_2, B-C_3N_4, TiO_2/C_3N_4, B-TiO_2/C_3N_4, and B2-C_3N_4/TiO_2 were weighed in the removal experimental. A 30 mg amount of each of material was added to the MB solution with a volume of 100 mL and a concentration of 20 mg/L. Under the condition of magnetic stirring, the photocatalytic efficiency was measured after 30 min of dark reaction adsorption experiment. A 1.5 mL amount of solution was taken out for measurement each time, and the removal rate of pollutants was determined by the following formula:

$$R_t = \frac{C_0 - C_t}{C_0} \times 100\% \tag{1}$$

where R_t is the removal rate at time t after commencing the adsorption and photocatalytic degradation process, and C_0 and C_t are initial concentration and concentration at time t, respectively.

$$Qe = \frac{C_0 - C_e}{m} V \tag{2}$$

where Qe is the adsorbed quantity of samples at the equilibrium moment of adsorption and desorption. C_0, C_e, V, and m are the initial concentration, concentration at time t, initial volume of MB, and quantity of adsorbent, respectively.

2.4. Kinetics and Adsorption Isotherm Model

The adsorption behaviors were fitted to the materials by pseudo-first-order kinetic and pseudo-second-order kinetic models, respectively. Two kinetic linear models are shown below [27]:

Pseudo-first-order kinetic model:

$$ln(Q_e - Q_t) = lnQ_e - k_1 t \quad (3)$$

Pseudo-second-order kinetic model:

$$\frac{t}{Q_t} = \frac{1}{k_2 Q_e^2} + \frac{t}{Q_e} \quad (4)$$

where Q_t and Q_e in the formula are the adsorption amount of the sample at time t and equilibrium, respectively. k_1 and k_2 are pseudo-first-order kinetic constants and pseudo-second-order kinetic constants, respectively.

Langmuir and Freundlich are the two most common adsorption isotherm models. The Langmuir model assumes that the surface of the adsorbate is uniform and the adsorption capacity is the same everywhere, and it only occurs on the outer surface of the adsorbent; this is monolayer adsorption. The Freundlich adsorption equation can be applied to both monolayer adsorption and adsorption behavior on uneven surfaces. Freundlich, as an empirical adsorption isotherm for non-uniform surfaces, is more applicable to low-concentration adsorption and can interpret experimental results over a wider concentration range. Therefore, in this experiment, the experimental results are fitted by these two adsorption isotherm models, and the formulas are as follows [28]:

Langmuir:

$$\frac{C_e}{Q_e} = \frac{1}{Q_m K_L} + \frac{C_e}{Q_m} \quad (5)$$

Freundlich:

$$lnQ_e = \frac{1}{n} lnC_e + lnK_F \quad (6)$$

where Q_e and Q_m represent the adsorption capacity and the maximum adsorption capacity, respectively; C_e is the concentration of pollutants in the solution at adsorption equilibrium; K_L is the Langmuir adsorption equilibrium constant; and K_F is the Freundlich constants with the affinity coefficient.

3. Result and Discussion

3.1. Performance Testing

In order to study the physical and chemical properties of the materials, the materials were screened by the adsorption catalytic performance of removing pollutant MB (Figure S1). It can be clearly seen from the figure that the adsorption–catalytic efficiency of several different materials for removing MB under the dark reaction and visible light conditions was B-TiO_2/C_3N_4 > B-C_3N_4 > B2-C_3N_4/TiO_2 > TiO_2/C_3N_4 > B-TiO_2 > g-C_3N_4 > TiO_2. Among them, the adsorption and removal rate of MB on B-TiO_2/C_3N_4 in the dark reaction process reached 73.8%, while that of B2-C_3N_4/TiO_2 was only 17.8%. Then, under the visible light condition for 2 h, the removal rate of MB by B-TiO_2/C_3N_4 reached 97.3%, while that of B2-C_3N_4/TiO_2 only reached 66.5%. Therefore, according to the preliminary experimental results, the B-TiO_2/C_3N_4 material was selected as the composite photocatalyst with the highest removal efficiency for follow-up research.

The mixing ratio and calcination temperature of the composites are both important factors affecting the performance of the composite materials. Therefore, in order to further analyze the performance of the composite material, this experiment first selected B_2O_3 and TiO_2/C_3N_4 with different mass ratios to be mixed and calcined at 550 °C and then selected the material with the best pollutant removal effect for subsequent experiments.

According to the amount of doping of B_2O_3, the composite materials were marked as B-TiO_2/C_3N_4-x (x = 1, 2, 3, 4, 5). The removal efficiencies are shown in Figure S2. The removal rates of B-TiO_2/C_3N_4-1, B-TiO_2/C_3N_4-2, and B-TiO_2/C_3N_4-3 were 69.3%, 83.2%, and 73.4%, respectively, while the removal rates of B-TiO_2/C_3N_4-4 and B-TiO_2/C_3N_4-5 were only 50.9% and 23.9%.

Therefore, according to the above experimental results, B-TiO_2/C_3N_4-2 had the best effect and was selected as the follow-up research material. A 2 g amount of B_2O_3 and 1 g of TiO_2/C_3N_4 material were mixed uniformly, and the mixture temperature was heated at a rate of 5 °C/min for 2 h at different temperatures. The set target temperatures were 350 °C, 450 °C, 550 °C, 650 °C, and 750 °C. For convenience of description, B-TiO_2/C_3N_4 prepared at different calcination temperatures was named after its temperature (350 °C, 450 °C, 550 °C, 650 °C, and 750 °C).

3.2. Characterization and Analysis of B-TiO_2/C_3N_4 Composites

The morphology of B-TiO_2/C_3N_4 composites prepared under different calcination conditions at different temperatures are shown in Figure 1. It can be seen from Figure 1a that at 350 °C the surface of the structure was relatively smooth, which may have been due to molten B_2O_3 wrapping the base material. Figure 1b shows the surface layer of the calcined material at 450 °C was still partially covered, but a large number of loose structures also appeared. This may have been due to the increase in the calcination temperature, and the disorder degree increased due to the B_2O_3 entering the molten state, which contains polar -B=O groups. As seen in Figure 1c, many loose pores appeared on the surface of the composite under the 550 °C calcination condition. As the temperature continued to increase, the main structure of the g-C_3N_4 material began to decompose when the temperature was above 600 °C, and the pulverization of the g-C_3N_4 structure can be clearly seen in Figure 1d. At 750 °C, g-C_3N_4 was decomposed into nitrogen and nitrile-based fragments, and TiO_2 was gradually transformed from anatase to rutile; so, the layered structure of g-C_3N_4 cannot be observed in Figure 1e, while some agglomerates of particles can be seen.

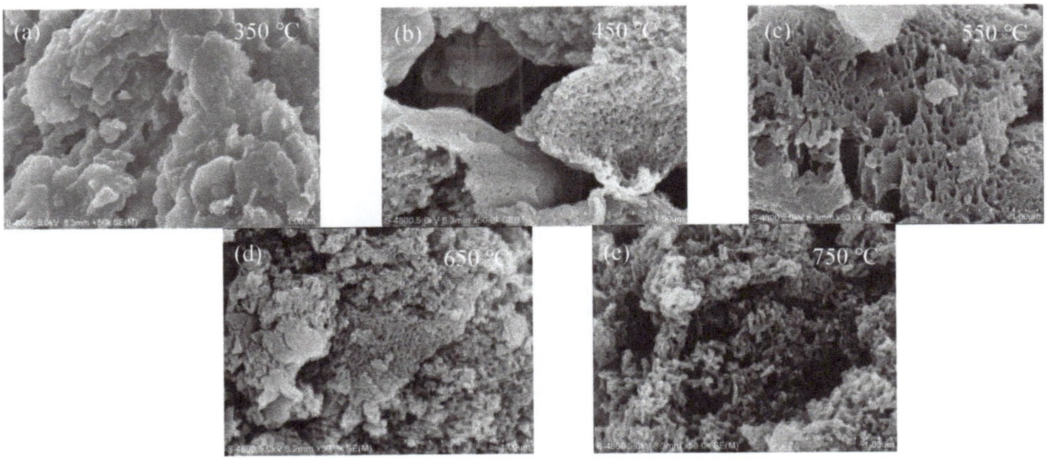

Figure 1. SEM of B-TiO_2/C_3N_4 under different calcination conditions: (**a**) 350 °C; (**b**) 450 °C; (**c**) 550 °C; (**d**) 650 °C; (**e**) 750 °C.

The specific surface area and porosity of materials are crucial factors affecting the adsorption and photocatalytic reactions. As shown in Figure S3, the representative N_2 adsorption–desorption isotherms of composite materials were of type IV [29]. According to the classification of IUPAC, the adsorption and desorption curve hysteresis loop types of the prepared B-TiO_2/C_3N_4 materials belonged to the "H3" hysteresis loop ($0.5 < P/P_0 < 0.98$:550 °C, 650 °C, 750 °C; $0.75 < P/P_0 < 0.97$:350 °C, 450 °C). The low-temperature calcination shown for loops located at high P/P_0 was caused by the aggregation of the 2D lamellar structure of g-C_3N_4. By contrast, the high temperature and B_2O_3 as a medium effectively resulted in more holes on the surface of g-C_3N_4. In Table 1, details regarding the specific surface areas, average pore diameters, and pore volumes of various materials are summarized. Among the temperature series, 550 °C showed the highest specific surface areas, average pore diameters, and pore volumes. This is because, at the calcination temperature of 550 °C, the molten B_2O_3 could fully contact the TiO_2/C_3N_4 composite as a reaction environment modifier. The lamellar structure of g-C_3N_4 was exfoliated in the molten environment while the pore structure was formed during the cooling process, so the specific surface area was significantly increased.

Table 1. BET surface area, average pore size, and pore volume of B-TiO_2/C_3N_4 material at different temperatures.

Sample	350 °C	450 °C	550 °C	650 °C	750 °C
BET (m^2/g)	13.381	13.472	58.654	34.622	25.795
Average pore size (nm)	6.311	7.809	6.960	12.733	14.263
Average pore volume (cm^3/g)	0.090	0.073	0.130	0.162	0.103

The surface morphologies and microstructure of B-TiO_2/C_3N_4 (550 °C) composite material are shown in Figure 2. From Figure 2a,b, it can be clearly observed that there were a lot of loose macropores and particle agglomeration on the surface of the g-C_3N_4-based material. In order to better analyze the morphological characteristics of the material, the high-resolution transmission electron microscope (HR-TEM) pattern was used for analysis. In Figure 2c, it can be seen that there were a lot of loose pores on the surface of the g-C_3N_4. Otherwise, a large number of granular material lattice fringes appeared on the g-C_3N_4 lamellar structure. The lattice spacing embedded particles were measured to be ~0.35 nm (Figure 2d), which is similar to the TiO_2 (101) crystal plane structure [30]. Through the analysis of specific surface area and the characterization of SEM and TEM, it was found that TiO_2 grew uniformly on the surface of the g-C_3N_4 layered structure, and the modification of the TiO_2/C_3N_4 composite structure by B_2O_3 produced a large number of loose pores on the g-C_3N_4 layered structure. Such a modification method not only improved the specific surface area of the material, but also preserved the composite structure of TiO_2/C_3N_4, which is more conducive to the charge transfer of photogenerated carriers between composite semiconductor materials and improves the photocatalytic activity of the material.

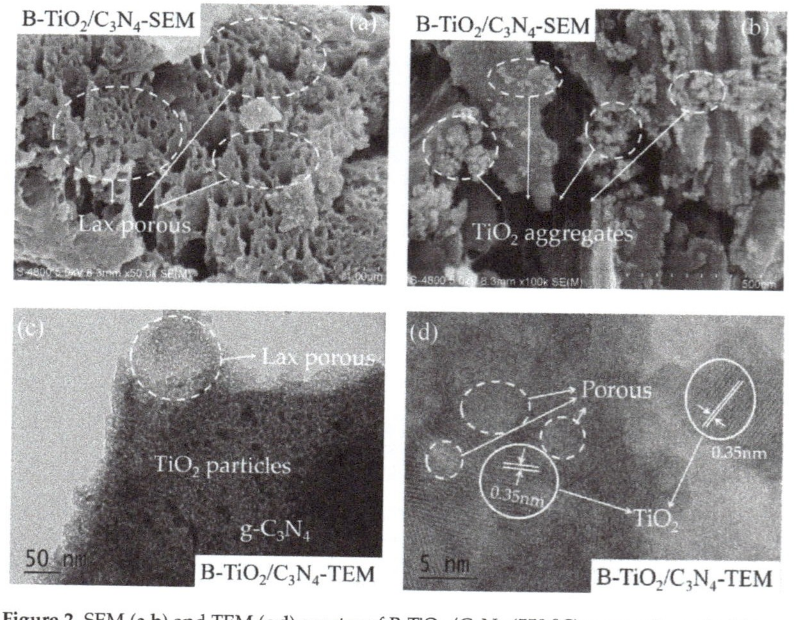

Figure 2. SEM (**a**,**b**) and TEM (**c**,**d**) spectra of B-TiO_2/C_3N_4 (550 °C) composite materials.

The XRD pattern of materials was recorded and is shown in Figure S4. It can be seen that B_2O_3 showed two obvious, characteristic peaks at 14.6°and 27.9° [31]. The composites prepared at 350 °C and 450 °C had only two obvious characteristic peaks of B_2O_3 but no characteristic peaks of TiO_2 and g-C_3N_4 [32]. This may have been because the surface of the TiO_2/C_3N_4 material was coated by B_2O_3 in the molten state at 350 °C and 450 °C, which is basically consistent with the SEM characterization shown in Figure 1. At the same time, due to the large amount of B_2O_3 added, there was no peak position of g-C_3N_4 and TiO_2 but only the characteristic peak of B_2O_3. At 550 °C, the two characteristic peaks of B_2O_3 disappeared in the XRD pattern, and the anatase (25.3°) and rutile (27.4°) diffraction peaks of TiO_2 appeared. It was caused by the entry of polar -B=O into the TiO_2/C_3N_4. When the temperature rose to 650 °C, several characteristic peaks (27.4°, 41.3°, etc.) of the rutile phase of TiO_2 became more and more obvious, which indicated that the crystal structure of TiO_2 gradually changed from anatase to rutile with the increase in temperature. It can be seen from XRD analysis that when the temperature gradually increased, the degree of disorder gradually increased, and a polar -B=O bond was formed to form a new structure with TiO_2 or g-C_3N_4 [33].

The Fourier-transform infrared spectroscopy of composite materials is shown in Figure S5. In this figure, it can be seen that the peak at 550 °C was the smallest, which was the Ti–O bond expansion joint and Ti stretching vibration of the -O–Ti bond. Compared with the B-TiO_2/C_3N_4 composites at different calcination temperatures (350 °C, 450 °C, 550 °C, 650 °C, and 750 °C), the obvious peaks appeared at 810 cm^{-1} and 1200–1600 cm^{-1}. The characteristic peak at 810 cm^{-1} was the stretching vibration of the triazine ring structure in g-C_3N_4, while the broad peak spectrum at 1200–1600 cm^{-1} was caused by the stretching vibration of C=N. Otherwise, the obvious differences of all composite materials were mainly due to the incomplete transformation of B_2O_3 from solid to molten state. At low temperatures, the peaks were not obvious, which was caused by molten B_2O_3 coating on the surface of composite material, resulting in limited exposure of the surface structure of the B-TiO_2/C_3N_4 composite. In addition, the absorption peaks of B-TiO_2/C_3N_4 composites at 3000–3300 cm^{-1} were mainly the stretching vibration of the amino group (-NH_2). Meanwhile, the two peaks at 2260 cm^{-1} and 2350 cm^{-1} were the C=O stretching vibration of CO_2 in the atmosphere [34].

Figure S6 shows the XPS spectra of several elements of B-TiO_2/C_3N_4: B1s, N1s, C1s, Ti2s and O1s. B1s shows three binding energies at 190.4 eV, 192.7 eV, and 193.5 eV, of which 192.7 eV belonged to the B–O bond, 190.4 eV was the binding energy of the B–N bond, and 193.5 eV may have been O–B formed by B entering the TiO_2 crystal structure—Ti or TiB_2 bond [35]. This indicates that B element entered into the structures of TiO_2 and g-C_3N_4. C1s showed three main binding energies, of which 284.8 eV was the typical sp2 hybridization of the graphite phase (C=C bond peak position), and 288.7 eV corresponded to the C–N–C bond in the structure. In addition, 286.5 eV corresponded to the exocyclic C–O of carbon–nitrogen polymer materials [36]. Figure S6c shows the N1s binding energy peaks at 397.3 eV, 398.2 eV, and 401.9 eV. Among them, 398.2 eV was attributed to the sp2 hybridization (C–N=C) of the N triazine ring structure, while the peak at 401.9 eV corresponded to the bridged N atom (N–$(C)^3$) [37], and the peak at 397.3 eV corresponded to N–B [38]. In addition, the characteristic peaks in the N spectrum were slightly shifted, possibly due to the increased delocalization between the large π bonds in the carbon nitride structure due to the doping of B element [29]. The two peaks of 458.7 eV and 464.8 eV in the binding energy of Ti 2s were the electron binding energies of Ti^{4+} $2p^{3/2}$ and $2p^{1/2}$ in TiO_2, respectively. The strong peak at 530.1 eV in the O1s spectrum was the O–Ti bond in TiO_2, while 531.0 eV and 532.7 eV were the surface-adsorbed water molecules, H_2O and surface hydroxyl groups (-OH), respectively [39].

Figure S7 shows the UV–Vis diffuse reflectance absorption spectra of the composite material (350 °C, 450 °C, 550 °C, 650 °C, 750 °C). It can be seen from the figure that under different temperature conditions, the UV–visible light absorption properties of the composites were different. The B-TiO_2/C_3N_4 prepared at the three temperatures of

350 °C, 450 °C, and 550 °C had better visible light absorption ability in both the ultraviolet and visible light regions, and the effect of 550 °C was the best. This result was caused by B element entering into the TiO_2 lattice and g-C_3N_4 structure. According to the low electronegativity of B element, it can be inferred that the 2p orbital of B may be different from the 2p orbital of O. Hybridization was generated, thus, enabling enhanced visible light absorption [40]. Otherwise, the 650 °C and 750 °C had high light absorption capacity in the ultraviolet regions of 200–350 nm and 200–400 nm, respectively. The light absorption ability of the visible light region greater than 400 nm was weak. This may have been due to the gradual decomposition of the g-C_3N_4 structure in the composite material with the increase in temperature, resulting in a decrease in the visible light absorption capacity of the material. At the same time, although the temperature increased, it could also have caused the transformation of the TiO_2 crystal phase, while the mixed-phase TiO_2 had visible light absorption ability, but the effect was not obvious. Therefore, according to the UV–Vis diffuse reflectance spectrum analysis, it is shown that the B-TiO_2/C_3N_4 composites prepared at 550 °C had good light absorption ability.

Based on the above characterization analysis description, we can roughly infer the preparation process of composite materials. The TiO_2/C_3N_4 material was exfoliated by using the molten reaction environment provided by B_2O_3 during the heating process. At the same time, the gas was decomposed by the trace amount of g-C_3N_4, thereby forming a large number of loose pore structures. With the increase in calcination temperature, the generated polar -B=O bond replaced the H atom of the amino group on carbon nitride melon and formed an O–B–Ti or TiB_2 structure, thereby finally forming a structurally stable B-TiO_2/C_3N_4 material. The specific reaction process inference diagram is shown in Figure S8.

3.3. Photocatalytic Efficiencies

Figure 3 shows the dark adsorption and light-driven photocatalytic degradation ability of MB (a) and RhB (b) of B-TiO_2/C_3N_4 composite materials (350 °C, 450 °C, 550 °C, 650 °C, and 750 °C). The specific experimental process was as follows: 30 mg of B-TiO_2/C_3N_4 composite was added into 100 mL of MB solution with a concentration of 20 mg/L and 10 mg/L of RhB solution, respectively. A half-hour dark reaction time was set in the experiment and then the photocatalytic degradation experiment was carried out. The removal ability of RhB and MB was analyzed by catalytically degrading RhB and MB under the irradiation of a xenon light source with a wavelength greater than 400 nm.

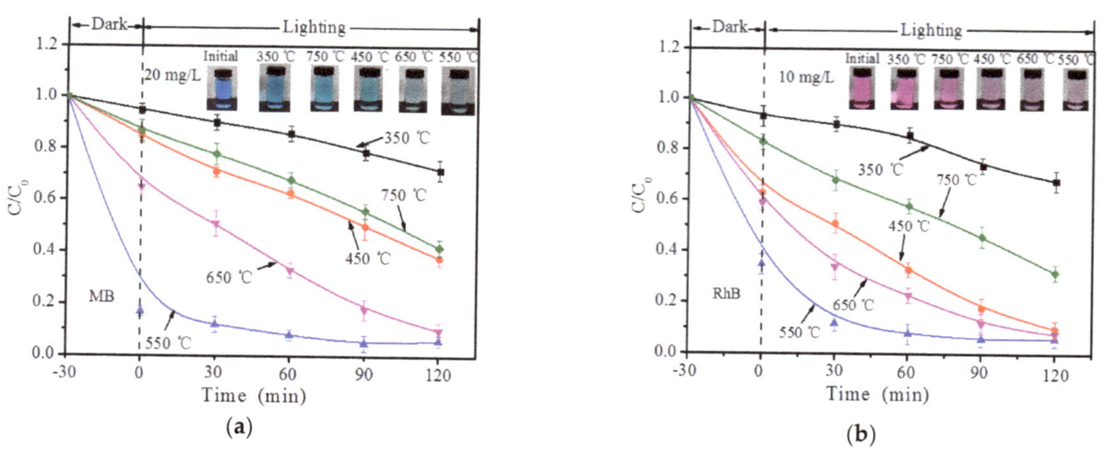

Figure 3. The adsorption–photocatalytic degradation curves of MB (**a**) and RhB (**b**) over B-TiO_2/C_3N_4.

The calculation of the removal efficiency of each pollutant by several catalyst materials was calculated by Formula (1). The experimental results of the dark reaction and the light reaction are shown in Figure 4a,b. The two figures show the adsorption–degradation curves of MB (20 mg/L) and RhB (10 mg/L) by the B-TiO_2/C_3N_4 composite, respectively. The relationship among the adsorption–degradation efficiencies can be clearly seen from the figure: 550 °C > 650 °C > 450 °C > 750 °C > 350 °C. The B-TiO_2/C_3N_4 prepared at the 550 °C calcination temperature had the best adsorption effect on the two dyes in the dark reaction adsorption process and the best removal efficiency (83.4% (MB), 64.1% (RhB)). At this temperature, B_2O_3 may undergo a complete molten state process, and become a good molten state reaction environment to fully contact with TiO_2/C_3N_4. Therefore, the prepared B-TiO_2/C_3N_4 had a larger specific surface area, exposing more adsorption sites and a reaction active site. Therefore, the doping of B element is beneficial to improve photocatalytic activity. When the temperature continued to rise above 600 °C, the g-C_3N_4 material began to decompose gradually, and the TiO_2 gradually transformed from anatase type to rutile phase with lower photocatalytic reaction activity, so the photocatalytic reaction efficiency decreased gradually.

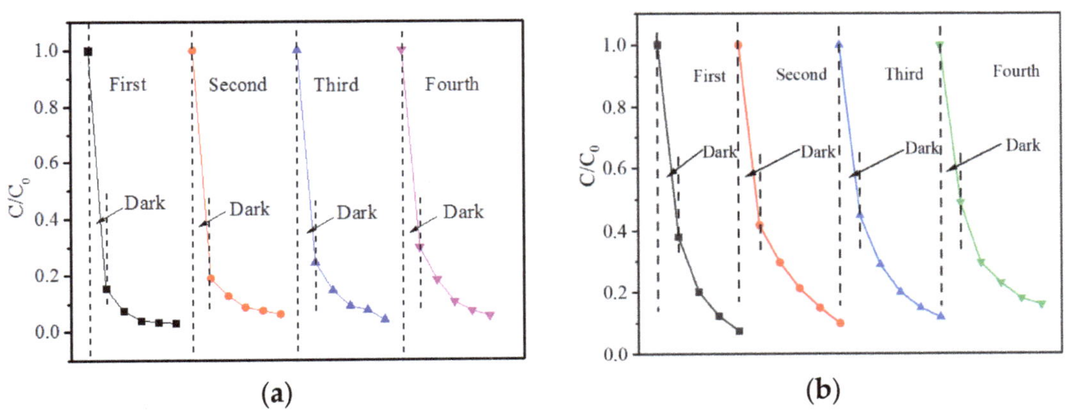

Figure 4. Cycling experiments of B-TiO_2/C_3N_4 for the adsorption–photocatalytic degradation of MB (**a**) and RhB (**b**).

Meanwhile, multiple cycle experiments were carried out with MB and RhB. It can be seen from Figure 4 that after four cycles of repeated tests, the adsorption and removal rate of B-TiO_2/C_3N_4 to 20 mg/L organic solution decreased from 83.4% to 70.1%(MB) and from 64.1% to 51.2% (RhB). The results showed that the adsorption effect of the material on MB and RhB was weakened gradually. This may have been because the pores on the surface of the material were damaged during the experiment. Meanwhile, the active sites were reduced due to the coverage of undegraded pollutants adsorbed on the surface of the material, resulting in a decrease in the adsorption and catalytic efficiency of the material after multiple experiments. Although the adsorption capacity of the material for the two pollutions was gradually weakened, the removal rate remained at around 95% (MB) and 80% (RhB) within two hours. This result shows that the B-TiO_2/C_3N_4 composite material had better stability and repeatability.

The TOC in the photocatalytic degradation system reflected the mineralization degree of organic pollutants. The results of the TOC are shown in Figure S9. The adsorption process effectively reduced the value. With the progress of visible light, the TOC value experienced small decreases with the prolongation of the lighting time. This is because, during the degradation process, the pollutant molecules were decomposed from macromolecules to small molecules containing organic carbon, and the pollutants adsorbed on the surface of the material were also desorbed. The resorption–degradation process resulted in insignificant changes in the organic carbon content in the solution [41,42].

3.4. Adsorption Property

To demonstrate the adsorption property of B-TiO_2/C_3N_4 (550 °C) composite materials, adsorption and removal rates of RhB and MB were chosen as the contaminant. The adsorption efficiency plays an important role in the rapid removal of pollutants.

The kinetic curve of adsorption of MB and RhB on B-TiO_2/C_3N_4 is shown in Figure 5. From this figure, it can be seen that B-TiO_2/C_3N_4 reached the adsorption equilibrium for the two pollutants within 30 min; the adsorption capacity of MB reached 56.25 mg/g; and the adsorption capacity of RhB reached 18.94 mg/g. Compared with other materials (as shown in Figure S1), the improvement of adsorption efficiency was due to the increase in specific surface area and the increase in adsorption sites, and the doping of B element in g-C_3N_4 expanded the large, π-bonded conjugated system, which was more conducive to the adsorption of MB [9].

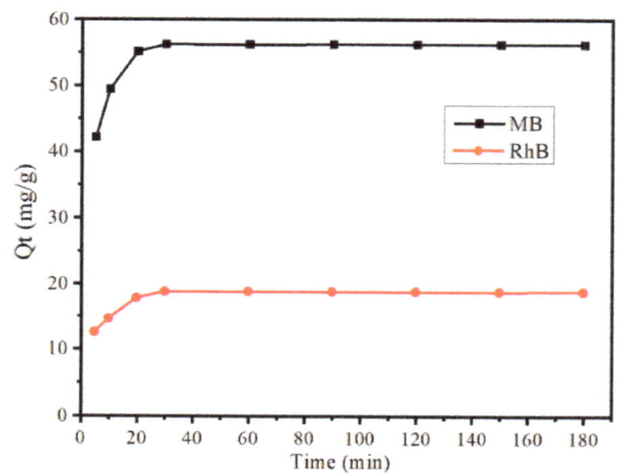

Figure 5. Adsorption kinetics curves of B-TiO_2/C_4 on MB and RhB.

The pseudo-first-order kinetic and pseudo-second-order kinetic models were fitted to the adsorption amounts of MB and RhB, respectively. The results are shown in Figure 6. The fitted correlation coefficients were: MB: ${R_1}^2$ = 0.5017 (a), ${R_2}^2$ = 0.9964 (b); RhB: ${R_1}^2$ = 0.6872 (c), ${R_2}^2$ = 0.9998 (d). According to the correlation coefficients fitted to the two pollutants, it can be seen that the pseudo-second-order kinetic model better fitted the adsorption behavior of B-TiO_2/C_3N_4.

Figure 7 shows the fitting of the Langmuir and Freundlich adsorption isotherm models for the adsorption of MB (a) (b) and RhB (c) (d) on B-TiO_2/C_3N_4, respectively. By comparing the correlation coefficients obtained by calculation, it was found that for the MB adsorption isotherm model, ${R_L}^2$ = 0.9946 was higher than that of the Freundlich model (${R_F}^2$ = 0.6883), while the simulated adsorption isotherm model for RhB had the same or similar results (${R_L}^2$: 0.9993 > ${R_F}^2$: 0.7395). This result showed that the adsorption of MB and RhB by B-TiO_2/C_3N_4 was monolayer adsorption, and the adsorption was more favorable with the increase in pollutant concentration. Therefore, the adsorption of these two pollutants by B-TiO_2/C_3N_4 was more inclined to the Langmuir adsorption isotherm model.

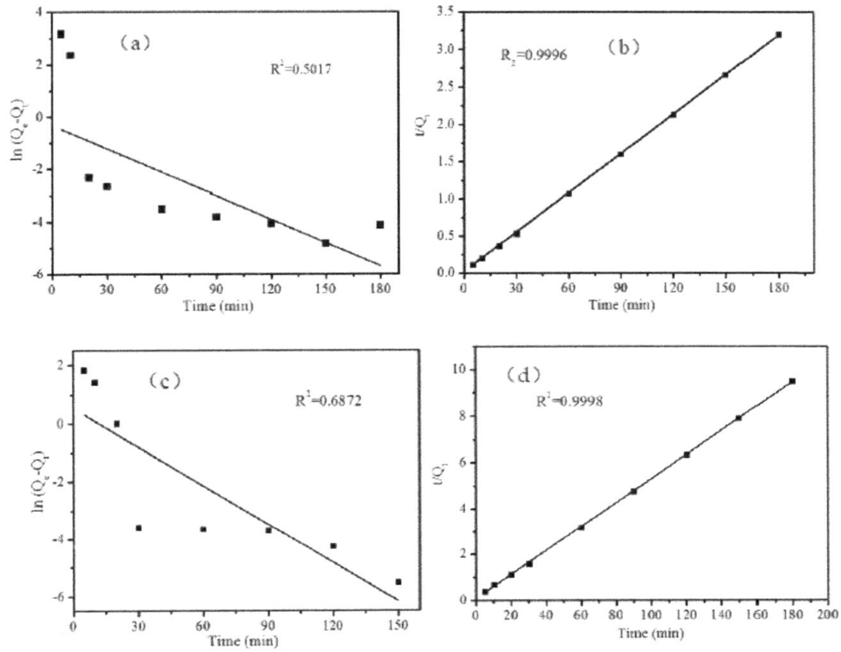

Figure 6. (**a**) Plots of the pseudo-first-order model (MB); (**b**) Plots of the pseudo-second-order model (MB); (**c**) Plots of the pseudo-first-order model (RhB); (**d**) Plots of the pseudo-second-order model (RhB).

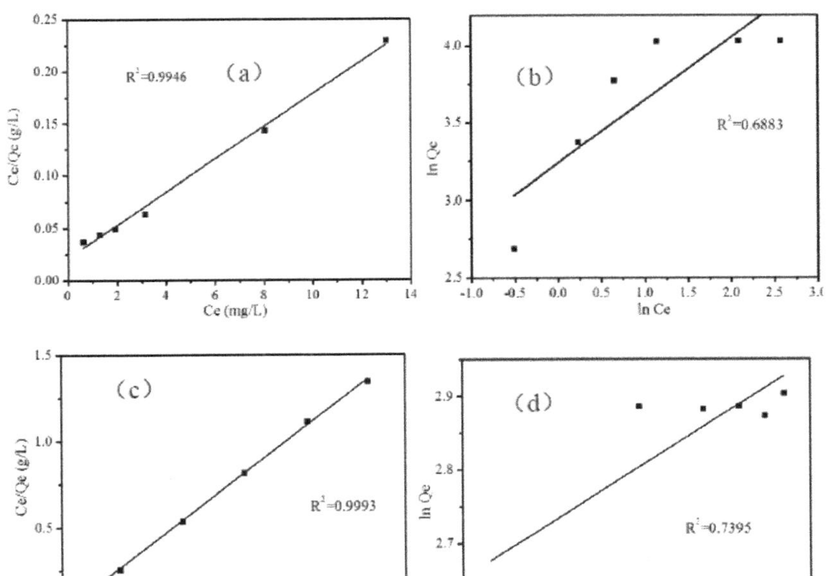

Figure 7. The linearized of Langmuir adsorption isotherm (**a**) MB (**c**) RhB and Freundlich adsorption isotherm (**b**) MB (**d**) RhB adsorption by B-TiO_2/C_3N_4.

3.5. Analysis of Photocatalytic Mechanism

Based on the above experimental results, MB was selected as the target pollutant with the best removal effect, and the main substances involved in the photocatalytic reaction were determined by the free radical capture experiment. The experimental results are shown in Figure 8. The addition of EDTA-2Na had less of an effect on the degradation effect of MB, which indicates that the role of holes is less important in the catalytic reaction. After the addition of p-benzoquinone, the degradation of MB was inhibited, indicating that superoxide radical ($\cdot O^{2-}$) is an important free radical involved in the photodegradation process. After the addition of tert-Butanol, the degradation of MB was also inhibited to a certain extent, so it can be concluded that hydroxyl radicals ($\cdot$OH) are also involved in the photocatalytic oxidation process. The above research results show that $\cdot O^{2-}$ and $\cdot$OH radicals were the active groups that mainly participated in the reaction during the degradation of MB by B-TiO_2/C_3N_4.

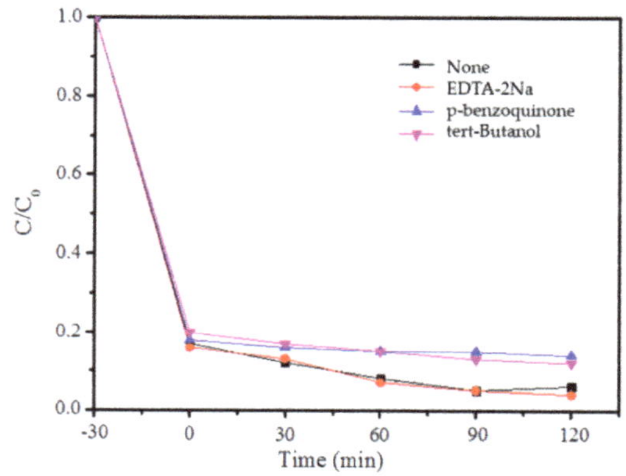

Figure 8. The capture experiments of MB by B-TiO_2/C_3N_4.

Through mechanism analysis, it was found that $\cdot O^{2-}$ and $\cdot$OH were the main active groups involved in the photocatalytic reaction. Therefore, DMPO ESR was used to detect the $\cdot O^{2-}$ and $\cdot$OH produced by B-TiO_2/C_3N_4 material under visible light above 400 nm. In addition, a comprehensive comparative analysis was conducted with other materials. It can be clearly seen from Figure 9a that after 10 min of light source irradiation, the order of $\cdot O^{2-}$ESR signal intensity was B-TiO_2/C_3N_4 > TiO_2/C_3N_4 > g-C_3N_4 > B-C_3N_4 > B-TiO_2 > TiO_2. The signal peak generated by B-TiO_2/C_3N_4 was slightly higher than that of TiO_2/C_3N_4 and significantly higher than that of the other three materials. In the ESR spectrum of $\cdot$Oh, as shown in Figure 9b, the signal peak generated by B-TiO_2/C_3N_4 was the strongest. According to the analysis results, the B-TiO_2/C_3N_4 material produced the most $\cdot O^{2-}$ and $\cdot$OH in the visible light photocatalytic system, so it had a better photocatalytic effect.

Otherwise, the XPS characterization showed that the B element in B_2O_3 entered the TiO_2 crystal structure. Therefore, in order to analyze the structural composition of B-TiO_2 and B-C_3N_4, the conduction band relationship of the forbidden band width of B-TiO_2 and B-C_3N_4 was analyzed. The specific results are shown in Figure 10a.

The photocatalytic reaction mechanism of B-TiO_2/C_3N_4 composite material was deduced according to the radical trapping experiment. As shown in Figure 10b, under visible light conditions, the conduction band position of B-C_3N_4 was -0.88 eV, which was more negative than that of B-TiO_2 (-0.31 eV) and E0 ($O_2/\cdot O^{2-}$ = -0.33 eV). The electrons could be captured by O_2 to form $\cdot O^{2-}$ and finally formed H_2O_2 on the surface of the material. The holes were trapped by H_2O and OH^- in the valence band of B-TiO_2 to form hydroxyl radicals ($\cdot$OH). These species are vital for photocatalytic degradation of organic pollution. Meanwhile, it can be determined that the B-TiO_2/C_3N_4 composite material formed a Z-type heterostructure [43,44].

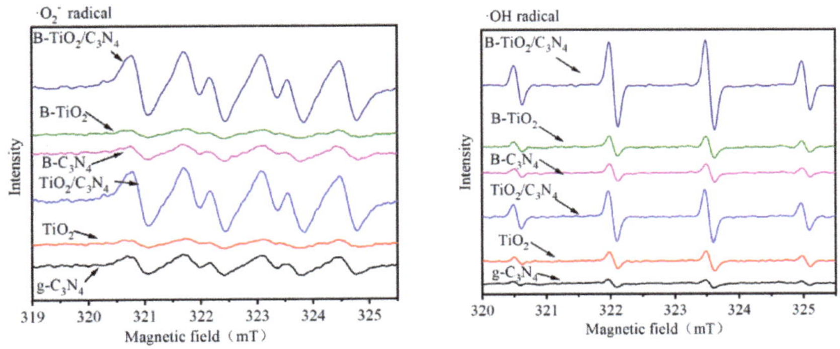

(**a**) (**b**)

Figure 9. Spectrum of DMPO spin trapping ESR of different materials (**a**) DMPO-O_2^- and (**b**) DMPO-·OH.

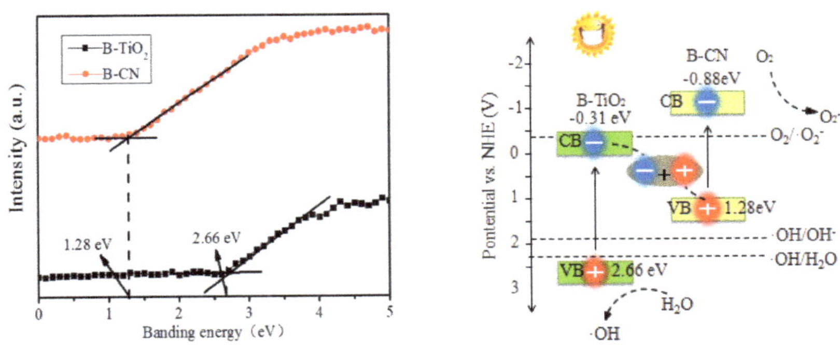

(**a**) (**b**)

Figure 10. The photocatalytic reaction mechanism of B-TiO_2/C_3N_4: (**a**) The VBXPS of B-TiO_2 and B-CN; (**b**) Scheme of proposed mechanism for degradation of B-TiO_2/C_3N_4.

4. Conclusions

Herein, B-TiO_2/C_3N_4 composites were prepared by modifying Z-type heterostructured TiO_2/C_3N_4 composites using the molten reaction environment created by B_2O_3. The physical-chemical properties of B-TiO_2/C_3N_4 composites prepared at different temperatures were analyzed by various characterization methods. Through characterization, it was found that the addition of B_2O_3 effectively increased the specific surface area of TiO_2/C_3N_4 materials, thereby increasing the active sites for adsorption and photocatalysis. Crystal analysis showed that the modification of TiO_2/C_3N_4 by B_2O_3 simultaneously doped B element into the TiO_2 lattice and g-C_3N_4, which helped to inhibit the recombination of electron holes and improved the photocatalytic activity. At the same time, the doping of B element expanded the π-conjugated system of the g-C_3N_4 material, which is beneficial for the improvement of MB adsorption performance. Above all, this paper started with consideration of the two aspects of improving the adsorption capacity and photocatalytic oxidation capacity of the material and integrated the preparation, characterization, and adsorption–degradation mechanism analysis of the material, providing theoretical support and experimental basis for research on the synergistic removal of pollutants by adsorption and photocatalysis.

Supplementary Materials: The following supporting information can be downloaded at: https://www.mdpi.com/article/10.3390/ijerph19148683/s1, Table S1: Comparison of MB removal rate of different materials, Figure S1: MB removal efficiency of several materials: TiO_2, g-C_3N_4, B-TiO_2, B-C_3N_4, TiO_2/C_3N_4, B2-C_3N_4/TiO_2 and B-TiO_2/C_3N_4, Figure S2: MB removal efficiency of B-TiO_2/C_3N_4 prepared with different doping amount, Figure S3: N_2 adsorption-desorption curves of B-TiO_2/C_3N_4 materials at different temperatures, Figure S4: XRD (X-ray diffraction) patterns of B_2O_3 and B-TiO_2/C_3N_4-2 Series materials, Figure S5: FT-IR spectra for different temperature B-TiO_2/C_3N_4, Figure S6: XPS spectra of B-TiO_2/C_3N_4 material (a)B1s, (b)C1s, (c) N1s, (d) Ti2p and (e) O1s, Figure S7. The Diffuse reflectance spectra for different temperature B-TiO_2/C_3N_4, Figure S8. Synthesis path of B-TiO_2/C_3N_4 material, Figure S9. Changes of TOC in the experiment of adsorption-catalytic degradation of pollutants (a) MB (b) RhB. References [45–47] are cited in the supplementary materials.

Author Contributions: X.G. contributed substantially to the design of the study, the analyses, writing of the manuscript, and the original draft preparation. L.R. contributed to the project administration, funding acquisition, and verified the experimental data, replied to comments, and corrected the proof. Z.S. took responsibility for performing experiments and validation and made the figures. All authors have read and agreed to the published version of the manuscript.

Funding: This work was funded by grants from the National Natural Science Foundation of China (no. 51775167) and the National Major Projects of Water Pollution Control and Management Technology (no. 2017ZX07204003).

Institutional Review Board Statement: Not applicable.

Informed Consent Statement: Not applicable.

Data Availability Statement: Not applicable.

Conflicts of Interest: The authors declare no conflict of interest.

References

1. Ratshiedana, R.; Kuvarega, A.T.; Mishra, A.K. Titanium Dioxide and Graphitic Carbon Nitride–Based Nanocomposites and Nanofibres for the Degradation of Organic Pollutants in Water: A Review. *Environ. Sci. Pollut. Res.* **2021**, *28*, 10357–10374. [CrossRef] [PubMed]
2. Xie, K.; Fang, J.; Li, L.; Deng, J.; Chen, F. Progress of Graphite Carbon Nitride with Different Dimensions in the Photocatalytic Degradation of Dyes: A Review. *J. Alloys Compd.* **2022**, *901*, 163589. [CrossRef]
3. Hu, C.; Tu, S.; Tian, N.; Ma, T.; Zhang, Y.; Huang, H. Photocatalysis Enhanced by External Fields. *Angew. Chem. Int. Ed.* **2021**, *60*, 16309–16328. [CrossRef] [PubMed]
4. Ahmaruzzaman, M.d.; Mishra, S.R. Photocatalytic Performance of g-C_3N_4 Based Nanocomposites for Effective Degradation/Removal of Dyes from Water and Wastewater. *Mater. Res. Bull.* **2021**, *143*, 111417. [CrossRef]
5. Oh, J.; Shim, Y.; Lee, S.; Park, S.; Jang, D.; Shin, Y.; Ohn, S.; Kim, J.; Park, S. Structural Insights into Photocatalytic Performance of Carbon Nitrides for Degradation of Organic Pollutants. *J. Solid State Chem.* **2018**, *258*, 559–565. [CrossRef]
6. Kuznetsov, V.N.; Serpone, N. Visible Light Absorption by Various Titanium Dioxide Specimens. *J. Phys. Chem. B* **2006**, *110*, 25203–25209. [CrossRef]
7. Xu, Q.; Zhang, L.; Cheng, B.; Fan, J.; Yu, J. S-Scheme Heterojunction Photocatalyst. *Chem* **2020**, *6*, 1543–1559. [CrossRef]
8. Ejeta, S.Y.; Imae, T. Photodegradation of Pollutant Pesticide by Oxidized Graphitic Carbon Nitride Catalysts. *J. Photochem. Photobiol. A Chem.* **2021**, *404*, 112955. [CrossRef]
9. Abe, M.; Kawashima, K.; Kozawa, K.; Sakai, H.; Kaneko, K. Amination of Activated Carbon and Adsorption Characteristics of Its Aminated Surface. *Langmuir* **2000**, *16*, 5059–5063. [CrossRef]
10. Zhou, Z.; Li, K.; Deng, W.; Li, J.; Yan, Y.; Li, Y.; Quan, X.; Wang, T. Nitrogen Vacancy Mediated Exciton Dissociation in Carbon Nitride Nanosheets: Enhanced Hydroxyl Radicals Generation for Efficient Photocatalytic Degradation of Organic Pollutants. *J. Hazard. Mater.* **2020**, *387*, 122023. [CrossRef]
11. Hossain, S.M.; Park, H.; Kang, H.-J.; Mun, J.S.; Tijing, L.; Rhee, I.; Kim, J.-H.; Jun, Y.-S.; Shon, H.K. Facile Synthesis and Characterization of Anatase TiO_2/g-CN Composites for Enhanced Photoactivity under UV–Visible Spectrum. *Chemosphere* **2021**, *262*, 128004. [CrossRef] [PubMed]
12. Serpone, N. Relative Photonic Efficiencies and Quantum Yields in Heterogeneous Photocatalysis. *J. Adv. Oxid. Technol.* **1997**, *2*, 203–216. [CrossRef]
13. Yu, L.; Wang, Z.; Shi, L.; Yuan, S.; Zhao, Y.; Fang, J.; Deng, W. Photoelectrocatalytic Performance of TiO_2 Nanoparticles Incorporated TiO_2 Nanotube Arrays. *Appl. Catal. B Environ.* **2012**, *113–114*, 318–325. [CrossRef]

14. Valero-Romero, M.J.; Santaclara, J.G.; Oar-Arteta, L.; van Koppen, L.; Osadchii, D.Y.; Gascon, J.; Kapteijn, F. Photocatalytic Properties of TiO_2 and Fe-Doped TiO_2 Prepared by Metal Organic Framework-Mediated Synthesis. *Chem. Eng. J.* **2019**, *360*, 75–88. [CrossRef]
15. Kamegawa, T.; Shimizu, Y.; Yamashita, H. Superhydrophobic Surfaces with Photocatalytic Self-Cleaning Properties by Nanocomposite Coating of TiO_2 and Polytetrafluoroethylene. *Adv. Mater.* **2012**, *24*, 3697–3700. [CrossRef]
16. Horikoshi, S.; Serpone, N. Can the Photocatalyst TiO_2 Be Incorporated into a Wastewater Treatment Method Background and Prospects. *Catal. Today* **2020**, *340*, 334–346. [CrossRef]
17. Zhang, F.; Hong, B.; Zhao, W.; Yang, Y.; Bao, J.; Gao, C.; Sun, S. Ozone Modification as an Efficient Strategy for Promoting the Photocatalytic Effect of TiO_2 for Air Purification. *Chem. Commun.* **2019**, *55*, 3757–3760. [CrossRef]
18. Yuzer, B.; Aydın, M.I.; Con, A.H.; Inan, H.; Can, S.; Selcuk, H.; Kadmi, Y. Photocatalytic, Self-Cleaning and Antibacterial Properties of Cu(II) Doped TiO_2. *J. Environ. Manag.* **2022**, *302*, 114023. [CrossRef]
19. Hossain, S.M.; Tijing, L.; Suzuki, N.; Fujishima, A.; Kim, J.-H.; Shon, H.K. Visible Light Activation of Photocatalysts Formed from the Heterojunction of Sludge-Generated TiO_2 and g-CN towards NO Removal. *J. Hazard. Mater.* **2022**, *422*, 126919. [CrossRef]
20. Ojha, D.P.; Karki, H.P.; Kim, H.J. Design of Ternary Hybrid ATO/g-C_3N_4/TiO_2 Nanocomposite for Visible-Light-Driven Photocatalysis. *J. Ind. Eng. Chem.* **2018**, *61*, 87–96. [CrossRef]
21. Wang, P.; Guo, X.; Rao, L.; Wang, C.; Guo, Y.; Zhang, L. A Weak-Light-Responsive TiO_2/g-C_3N_4 Composite Film: Photocatalytic Activity under Low-Intensity Light Irradiation. *Environ. Sci. Pollut. Res.* **2018**, *25*, 20206–20216. [CrossRef] [PubMed]
22. Qumar, U.; Hassan, J.Z.; Bhatti, R.A.; Raza, A.; Nazir, G.; Nabgan, W.; Ikram, M. Photocatalysis vs Adsorption by Metal Oxide Nanoparticles. *J. Mater. Sci. Technol.* **2022**. [CrossRef]
23. Bahrudin, N.N.; Nawi, M.A.; Zainal, Z. Insight into the Synergistic Photocatalytic-Adsorptive Removal of Methyl Orange Dye Using TiO_2/Chitosan Based Photocatalyst. *Int. J. Biol. Macromol.* **2020**, *165*, 2462–2474. [CrossRef] [PubMed]
24. Zhu, Y.; Ji, H.; He, K.; Blaney, L.; Xu, T.; Zhao, D. Photocatalytic Degradation of GenX in Water Using a New Adsorptive Photocatalyst. *Water Res.* **2022**, *220*, 118650. [CrossRef] [PubMed]
25. Liu, M.; Zhang, D.; Han, J.; Liu, C.; Ding, Y.; Wang, Z.; Wang, A. Adsorption Enhanced Photocatalytic Degradation Sulfadiazine Antibiotic Using Porous Carbon Nitride Nanosheets with Carbon Vacancies. *Chem. Eng. J.* **2020**, *382*, 123017. [CrossRef]
26. Guo, X.; Rao, L.; Wang, P.; Zhang, L.; Wang, Y. Synthesis of Porous Boron-Doped Carbon Nitride: Adsorption Capacity and Photo-Regeneration Properties. *Int. J. Environ. Res. Public Health* **2019**, *16*, 581. [CrossRef]
27. Rodríguez-Narciso, S.; Lozano-Álvarez, J.A.; Salinas-Gutiérrez, R.; Castañeda-Leyva, N. A Stochastic Model for Adsorption Kinetics. *Adsorpt. Sci. Technol.* **2021**, *2021*, 5522581. [CrossRef]
28. Zhu, Z.; Xie, J.; Zhang, M.; Zhou, Q.; Liu, F. Insight into the Adsorption of PPCPs by Porous Adsorbents: Effect of the Properties of Adsorbents and Adsorbates. *Environ. Pollut.* **2016**, *214*, 524–531. [CrossRef]
29. She, X.; Liu, L.; Ji, H.; Mo, Z.; Li, Y.; Huang, L.; Du, D.; Xu, H.; Li, H. Template-Free Synthesis of 2D Porous Ultrathin Nonmetal-Doped g-C_3N_4 Nanosheets with Highly Efficient Photocatalytic H2 Evolution from Water under Visible Light. *Appl. Catal. B Environ.* **2016**, *187*, 144–153. [CrossRef]
30. Guo, Q.; Ma, Z.; Zhou, C.; Ren, Z.; Yang, X. Single Molecule Photocatalysis on TiO_2 Surfaces. *Chem. Rev.* **2019**, *119*, 11020–11041. [CrossRef]
31. Li, L.; Hong, W.B.; Yan, X.J.; Chen, X.M. Preparation and Microwave Dielectric Properties of B_2O_3 Bulk. *Int. J. Appl. Ceram. Technol.* **2019**, *16*, 2047–2052. [CrossRef]
32. Photocatalytic Activity Enhancement of Core-Shell Structure g-C3N4@TiO2 via Controlled Ultrathin g-C3N4 Layer-ScienceDirect. Available online: https://www.sciencedirect.com/science/article/abs/pii/S0926337317307403 (accessed on 20 June 2022).
33. Mozzi, R.L.; Warren, B.E. The Structure of Vitreous Boron Oxide. *J. Appl. Cryst.* **1970**, *3*, 251–257. [CrossRef]
34. Danon, A.; Stair, P.C.; Weitz, E. FTIR Study of CO_2 Adsorption on Amine-Grafted SBA-15: Elucidation of Adsorbed Species. *J. Phys. Chem. C* **2011**, *115*, 11540–11549. [CrossRef]
35. Quiñones, D.H.; Rey, A.; Álvarez, P.M.; Beltrán, F.J.; Li Puma, G. Boron Doped TiO_2 Catalysts for Photocatalytic Ozonation of Aqueous Mixtures of Common Pesticides: Diuron, o-Phenylphenol, MCPA and Terbuthylazine. *Appl. Catal. B Environ.* **2015**, *178*, 74–81. [CrossRef]
36. Ahmed, T.; Ammar, M.; Saleem, A.; Zhang, H.; Xu, H. Z-Scheme 2D-m-$BiVO_4$ Networks Decorated by a g-CN Nanosheet Heterostructured Photocatalyst with an Excellent Response to Visible Light. *RSC Adv.* **2020**, *10*, 3192–3202. [CrossRef]
37. Wei, H.; McMaster, W.A.; Tan, J.Z.Y.; Cao, L.; Chen, D.; Caruso, R.A. Mesoporous TiO_2/g-C_3N_4 Microspheres with Enhanced Visible-Light Photocatalytic Activity. *J. Phys. Chem. C* **2017**, *121*, 22114–22122. [CrossRef]
38. Guo, Y.; Wang, R.; Wang, P.; Rao, L.; Wang, C. Developing a Novel Layered Boron Nitride–Carbon Nitride Composite with High Efficiency and Selectivity To Remove Protonated Dyes from Water. *ACS Sustain. Chem. Eng.* **2019**, *7*, 5727–5741. [CrossRef]
39. Zhu, J.; Chen, F.; Zhang, J.; Chen, H.; Anpo, M. Fe^{3+}-TiO_2 Photocatalysts Prepared by Combining Sol–Gel Method with Hydrothermal Treatment and Their Characterization. *J. Photochem. Photobiol. A Chem.* **2006**, *180*, 196–204. [CrossRef]
40. Barkul, R.P.; Sutar, R.S.; Patil, M.K.; Delekar, S.D. Photocatalytic Degradation of Organic Pollutants by Using Nanocrystalline Boron-Doped TiO_2 Catalysts. *ChemistrySelect* **2021**, *6*, 3360–3369. [CrossRef]
41. Beltrán, F.J.; Aguinaco, A.; García-Araya, J.F. Kinetic Modelling of TOC Removal in the Photocatalytic Ozonation of Diclofenac Aqueous Solutions. *Appl. Catal. B Environ.* **2010**, *100*, 289–298. [CrossRef]

42. Lee, J.C.; Park, S.; Park, H.-J.; Lee, J.-H.; Kim, H.-S.; Chung, Y.-J. Photocatalytic Degradation of TOC from Aqueous Phenol Solution Using Solution Combusted ZnO Nanopowders. *J. Electroceram.* **2009**, *22*, 110–113. [CrossRef]
43. Tatykayev, B.; Chouchene, B.; Balan, L.; Gries, T.; Medjahdi, G.; Girot, E.; Uralbekov, B.; Schneider, R. Heterostructured g-CN/TiO_2 Photocatalysts Prepared by Thermolysis of g-CN/MIL-125(Ti) Composites for Efficient Pollutant Degradation and Hydrogen Production. *Nanomaterials* **2020**, *10*, 1387. [CrossRef] [PubMed]
44. Karpuraranjith, M.; Chen, Y.; Rajaboopathi, S.; Ramadoss, M.; Srinivas, K.; Yang, D.; Wang, B. Three-Dimensional Porous MoS2 Nanobox Embedded g-C_3N_4@TiO_2 Architecture for Highly Efficient Photocatalytic Degradation of Organic Pollutant. *J. Colloid Interface Sci.* **2022**, *605*, 613–623. [CrossRef] [PubMed]
45. Feng, J.; Chen, T.; Liu, S.; Zhou, Q.; Ren, Y.; Lv, Y.; Fan, Z. Improvement of g-C_3N_4 Photocatalytic Properties Using the Hummers Method. *J. Colloid Interface Sci.* **2016**, *479*, 1–6. [CrossRef]
46. Kwon, D.; Kim, J. Silver-Doped ZnO for Photocatalytic Degradation of Methylene Blue. *Korean J. Chem. Eng.* **2020**, *37*, 1226–1232. [CrossRef]
47. Mo, Z.; She, X.; Li, Y.; Liu, L.; Huang, L.; Chen, Z.; Zhang, Q.; Xu, H.; Li, H. Synthesis of g-C_3N_4 at Different Temperatures for Superior Visible/UV Photocatalytic Performance and Photoelectrochemical Sensing of MB Solution. *RSC Adv.* **2015**, *5*, 101552–101562. [CrossRef]

International Journal of
Environmental Research and Public Health

MDPI

Article

Temporal and Spatial Distribution Analysis of Atmospheric Pollutants in Chengdu–Chongqing Twin-City Economic Circle

Ning Qi [1,*], Xuemei Tan [1], Tengfei Wu [2,*], Qing Tang [3], Fengshou Ning [1], Debin Jiang [1], Tengtun Xu [1], Hong Wu [1], Lingxiao Ren [4] and Wei Deng [5]

1 School of Environment and Resources, Chongqing Technology and Business University, Chongqing 400067, China; wonderland892@hotmail.com (X.T.); ningfs_00342@sohu.com (F.N.); jiangdebin@ctbu.edu.cn (D.J.); xutengtun@ctbu.edu.cn (T.X.); wuhong_1985@163.com (H.W.)
2 Institute of Agricultural Resources and Environment, Guangdong Academy of Agricultural Sciences, Guangzhou 510640, China
3 Chongqing Fushide Environmental Affairs Co., Ltd., Chongqing 401147, China; tangqingtq@163.com
4 Nanjing Institute of Technology, School of Environmental Engineering, Nanjing 211167, China; rlxjht@njit.edu.cn
5 Center of Yangtze River Ecological Protection and High Quality Development, Chongqing Academy of Environmental Science, Chongqing 401147, China; dengweidw0821@163.com
* Correspondence: 2019047@ctbu.edu.cn (N.Q.); wutengfei@gdaas.cn (T.W.); Tel.: +86-153-1099-6890 (N.Q.); +86-132-1020-1286 (T.W.)

Abstract: In order to study the temporal and spatial distribution characteristics of atmospheric pollutants in cities (districts and counties) in the Chengdu–Chongqing Twin-city Economic Circle (CCEC) and to provide a theoretical basis for atmospheric pollution prevention and control, this paper combined Ambient Air Quality Standards (AAQS) and WHO Global Air Quality Guidelines (GAQG) to evaluate atmospheric pollution and used spatial correlation to determine key pollution areas. The results showed that the distribution of atmospheric pollutants in CCEC presents a certain law, which was consistent with the air pollution transmission channels. Except for particulate matter with an aerodynamic diameter equal to or less than 2.5 μm ($PM_{2.5}$) and ozone (O_3), other pollutants reached Grade II of AAQS in 2020, among which particulate matter with an aerodynamic diameter equal to or less than 10 μm (PM_{10}), $PM_{2.5}$, sulfur dioxide (SO_2), nitrogen dioxide (NO_2) and carbon monoxide (CO) have improved. Compared with the air quality guidelines given in the GAQG, PM_{10}, $PM_{2.5}$, NO_2 and O_3 have certain effects on human health. The spatial aggregation of PM_{10} and $PM_{2.5}$ decreased year by year, while the spatial aggregation of O_3 increased with the change in time, and the distribution of NO_2 pollution had no obvious aggregation. Comprehensive analysis showed that the pollution problems of particulate matter, NO_2 and O_3 in CCEC need to be further controlled.

Keywords: spatial autocorrelation; Moran's I; air pollution transmission channels; ambient air quality standards; WHO Global Air Quality Guidelines

Citation: Qi, N.; Tan, X.; Wu, T.; Tang, Q.; Ning, F.; Jiang, D.; Xu, T.; Wu, H.; Ren, L.; Deng, W. Temporal and Spatial Distribution Analysis of Atmospheric Pollutants in Chengdu–Chongqing Twin-City Economic Circle. *Int. J. Environ. Res. Public Health* **2022**, *19*, 4333. https://doi.org/10.3390/ijerph19074333

Academic Editors: Xun Wang, Zhiyuan Wang, Xin Zhao and Paul B. Tchounwou

Received: 21 February 2022
Accepted: 31 March 2022
Published: 4 April 2022

1. Introduction

Currently, atmospheric pollution has become an important environmental problem, which places risk on human health. Many studies show that the release of atmospheric pollutants may cause many adverse health effects such as increased risks of cardiovascular and pulmonary diseases, decreased semen quality, and coronary heart disease [1–3]. The rapid developments of transportation and industry cause the discharge of atmospheric pollutants. The emissions of particulate matter with an aerodynamic diameter equal to or less than 2.5 μm ($PM_{2.5}$) and sulfur dioxide (SO_2) by coal-fired power plants accounted for 6% and 33% of national total emissions in 2010, respectively [4], and the usage of coal accounts for 69% of the total energy consumption [5]. Intense vehicular traffic causes the large emissions of nitrogen dioxide (NO_2) and carbon dioxide [6]. In order to assess the degree of air pollution, six pollutants such as particulate matter with an aerodynamic diameter equal to or less than 10 μm (PM_{10}), $PM_{2.5}$, SO_2, NO_2, ozone (O_3) and carbon monoxide (CO) have been selected to characterize the levels of air pollution. The World Health Organization (WHO) develops the WHO Global Air Quality Guidelines (GAQG) to reduce atmospheric pollutants in order to decrease the enormous health burden resulting from exposure to atmospheric pollution worldwide. In China, the Ambient Air Quality Standard (AAQS) is also set up to protect and improve the living environment and ecological health, and to ensure human health [7].

In January 2020, the construction of the "Chengdu–Chongqing twin-city Economic Circle" (CCEC) was first proposed by the sixth meeting of the Central Finance and Economics Commission, which aims to turn the Chengdu–Chongqing area into an economic circle with its own strengths and distinctive features, as well as a new driver and an important growth engine of the country's high-quality development. CCEC is made up of some cities in Sichuan province and some districts or counties in Chongqing municipality, which is the urbanization area with the highest development level and the greatest development potential in the western region of China. The high-quality development of CCEC can effectively enhance the economic development and the population-carrying capacity of the urbanization area, which is of great significance to the protection of the ecological environment in the upper reaches of the Yangtze River and of western China. It is also an important part of the implementation of the "Yangtze River Economic Belt" and "the Belt and Road Initiative". The whole area of the CCEC is 185,000 square kilometers, which includes 29 districts and counties in Chongqing and 15 cities in Sichuan province (Figure 1).

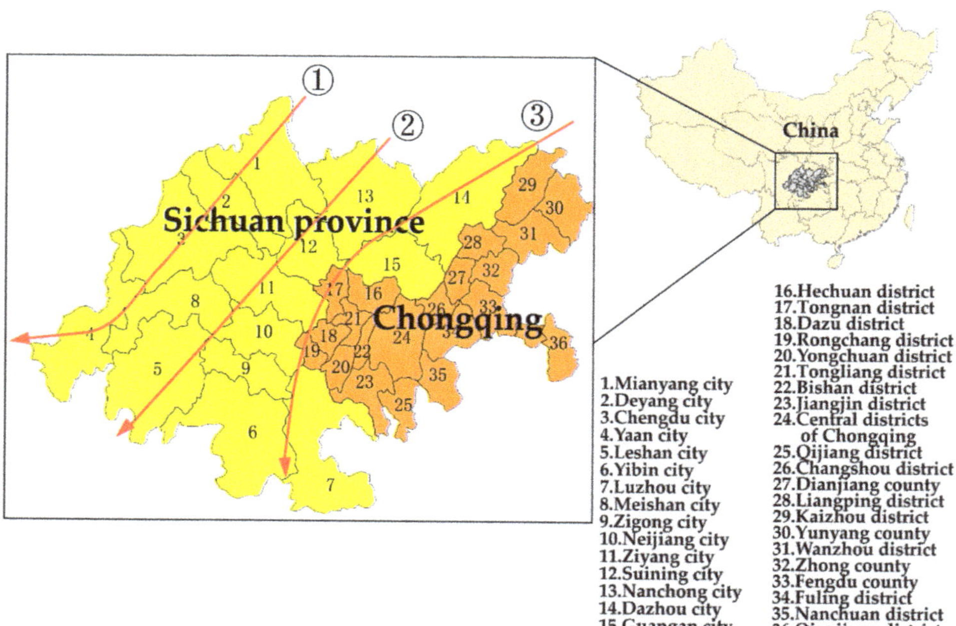

Figure 1. Map of air pollution transmission channels in the Chengdu–Chongqing twin-city Economic Circle (CCEC) [8–10].

For many years, the atmospheric pollution of CCEC has been particularly serious and complex. Due to the unique topography and climate in Sichuan Basin [11], the atmospheric pollutants accumulate in large quantities and cause the Sichuan Basin to become one of the most heavily polluted areas in China [12,13]. Furthermore, Chongqing is a mountainous city, and its environmental quality is significantly affected by the factors of pollution and dense built environment [14]. The research has shown that there are three air pollution transmission channels in Sichuan Basin due to the effect of the east Asian atmospheric circulation and the Qinghai Tibet Plateau flow field [8–10]: (1) Guangyuan → Mianyang → Deyang → Chengdu→ Ya'an; (2) Bazhong → Nanchong→ Suining →Ziyang → Meishan →Leshan; (3) Northern of Chongqing → Dazhou → Guang'an →Nanchong → Suining → Ziyang → Neijiang → Luzhou. Additionally, the pollution of O_3 has become more and more serious in Sichuan Province since 2015, while the particulate matter has shown characteristics of secondary pollutants [15,16]. The primary pollutants in the atmosphere of Chongqing were PM_{10}, nitrogen oxides and SO_2 in 2007–2014, which showed significant regional differences in air quality [17]. Environmental quality and public health determine the prospect of sustainable development [18]. The prevention and reduction of air pollution has become one of the key issues of current concern for high quality development of the economy in the CCEC.

The aim of this research is to explore the spatiotemporal distribution and pollution degrees of atmospheric pollutants from 2017 to 2020 in the CCEC, and to find the main pollution areas. Based on the results, the reasonable suggestions for pollution control in CCEC were propounded, and the theoretical basis for the coordinated governance of the atmosphere and water environment was provided.

2. Data Sources and Methods

2.1. Data Sources

In this study, the relevant data of ambient air quality in each area (city in Sichuan province or district and county in Chongqing) of CCEC from 2017 to 2020 were obtained from the ecological and environmental bulletin of Chongqing (http://sthjj.cq.gov.cn/hjzl_249/hjzkgb/, accessed on 31 January 2022) and from nthe ecological and environmental bulletin of each city in Sichuan Province (http://sthjt.sc.gov.cn/sthjt/c104157/hjglnew.shtml, accessed on 31 January 2022). The atmospheric pollutants that were chosen mainly include PM_{10}, $PM_{2.5}$, SO_2, NO_2, O_3 and CO.

2.2. Evaluation Standards

The evaluation standards for the concentration of atmospheric pollutants were based on the Grade II of AAQS [7,19] (http://www.mee.gov.cn/ywgz/fgbz/bz/bzwb/dqhjbh/dqhjzlbz/201203/t20120302_224165.shtml, accessed on 31 January 2022) and GAQG [20] (https://www.who.int/publications/i/item/9789240034433, accessed on 31 January 2022. The average annual concentrations of PM_{10}, $PM_{2.5}$, SO_2 and NO_2 were adopted, the 24 h average value of the 95th position for CO was adopted, and the 8 h average value of the maximum daily concentration of the 90th position for O_3 was adopted (Table 1). The areal interpolation was used to draw the figures of spatial distribution. The indexes in GAQG are stricter than that in AAQS, which is based on the latest evidence of human health caused by air pollution. The purpose of GAQG is to propose new air quality levels, and their interim targets play a role in guiding emission reduction and promoting air quality to reach the level of air quality guidelines. The purpose of the AAQS is based on air quality management. It aims at promoting harmonious and sustainable development between humans and nature [21]; thus, the indexes in AAQS are much closer to the interim target 1 of GAQG.

Table 1. Limit values of atmospheric pollutants in AAQS and GAQG [7,19,20].

Atmospheric Pollutants	Ambient Air Quality	WHO Global Air Quality Guidelines					
	The Secondary Limits	Interim Target 1	Interim Target 2	Interim Target 3	Interim Target 4	Air Quality Guideline	Unit
PM_{10}	70	70	50	30	20	15	μg/m^3
$PM_{2.5}$	35	35	25	15	10	5	μg/m^3
SO_2 *	60	-	-	-	-	-	μg/m^3
O_3	160	160	120	-	-	100	μg/m^3
NO_2	40	40	30	20	-	10	μg/m^3
CO	4	7	-	-	-	4	mg/m^3

* No average annual concentration of SO_2 was given in GAQG.

2.3. Methods

Spatial autocorrelation refers to the presence of systematic spatial variation in a mapped variable. The map shows positive spatial autocorrelation where adjacent observations have similar data values. The spatial autocorrelation is often used to detect whether the distribution of variables has spatial dependency, heterogeneity and constitutive properties. Moran's I is one of the important indexes used to analyze spatial correlation (Equation (1)) [22,23].

$$I = \frac{n\sum_{i=1}^{n}\sum_{j=1}^{n} w_{ij}(x_i-\overline{x})(x_j-\overline{x})}{\sum_{i=1}^{n}\sum_{j=1}^{n} w_{ij}\sum_{i=1}^{n}(x_i-\overline{x})^2} \tag{1}$$

where n represents the number of the cities and districts; w_{ij} represents the spatial relationship between region i and j; x_i and x_j, respectively, are the concentration values of certain atmospheric pollutant in each city; $\overline{x}$ is average concentration value of a certain atmospheric pollutant by study region, $\overline{x} = \frac{1}{n}\sum_{i=1}^{n} x_i$. The range of Moran's I lies between -1 and 1. If the Moran's I index >0, this implies a positive spatial correlation. Inversely, if the Moran's I index <0, this indicates a negative correlation [24]. The smaller the value, the stronger the spatial divergence [25].

An alternative approach to measure the relationship typology and intensity are provided by the local indicator of spatial association (LISA) (Equation (2)) [26]. It has four types of distributions, which includes high–high (HH) type, high–low (HL) type, low–high (LH) type and low–low (LL) type. High–high type or low–low type represents spatial clusters of similar high or low concentration values of atmospheric pollutants. Low–high type or high–low type indicates spatial outliers with low concentration values of atmospheric pollutants surrounding high concentration values of atmospheric pollutants or vice versa.

$$L_i = \frac{(x_i-\overline{x})}{S^2}\sum_{j=1}^{N} w_{ij}(x_i-\overline{x}) \tag{2}$$

where $S^2 = \frac{1}{n}\sum_{i=1}^{n}(x_i-\overline{x})^2$; S^2 is the concentration variance of a certain atmospheric pollutant. If $L_i > 0$, this implies the HH type or LL type. If $L_i < 0$, this indicates the HL type or LH type [27].

3. Result and Discussions

3.1. Temporal and Spatial Changes of the Concentrations of Atmospheric Pollutants

3.1.1. PM_{10}

Figure 2 shows the change in PM_{10} for each area in the CCEC during 2017–2020. The average annual concentration of PM_{10} for the whole CCEC was 72.0 μg/m^3, and the range and standard deviation were 42.0 μg/m^3 and 9.5 μg/m^3, respectively, in 2017. The areas with the highest annual average PM_{10} concentration were Zigong city in Sichuan province and Jiangjin district in Chongqing (89 μg/m^3), and the concentrations were 27.0% higher than Grade II of AAQS. The area with lowest annual average PM_{10} concentration was

Qianjiang district in Chongqing (47 μg/m^3). The number of areas exceeding Grade II of AAQS was about 42.0%. The areas with a higher concentration were mainly located in the cities of Chengdu, Deyang, Leshan, Zigong in Sichuan province and the districts of Jiangjin, Qijiang in Chongqing, and the distribution of PM_{10} was consistent with three air pollution transmission channels. In 2020, the average annual concentration of PM_{10} for the whole CCEC decreased to 50.1 μg/m^3, while the range and standard deviation were 32.0 μg/m^3 and 7.2 μg/m^3, respectively. All cities (districts and counties) met the Grade II of AAQS. The area with the highest concentration of PM_{10} was Chengdu city in Sichuan province (64 μg/m^3), while the area with the lowest concentration of PM_{10} was Ya'an city in Sichuan province (33 μg/m^3).

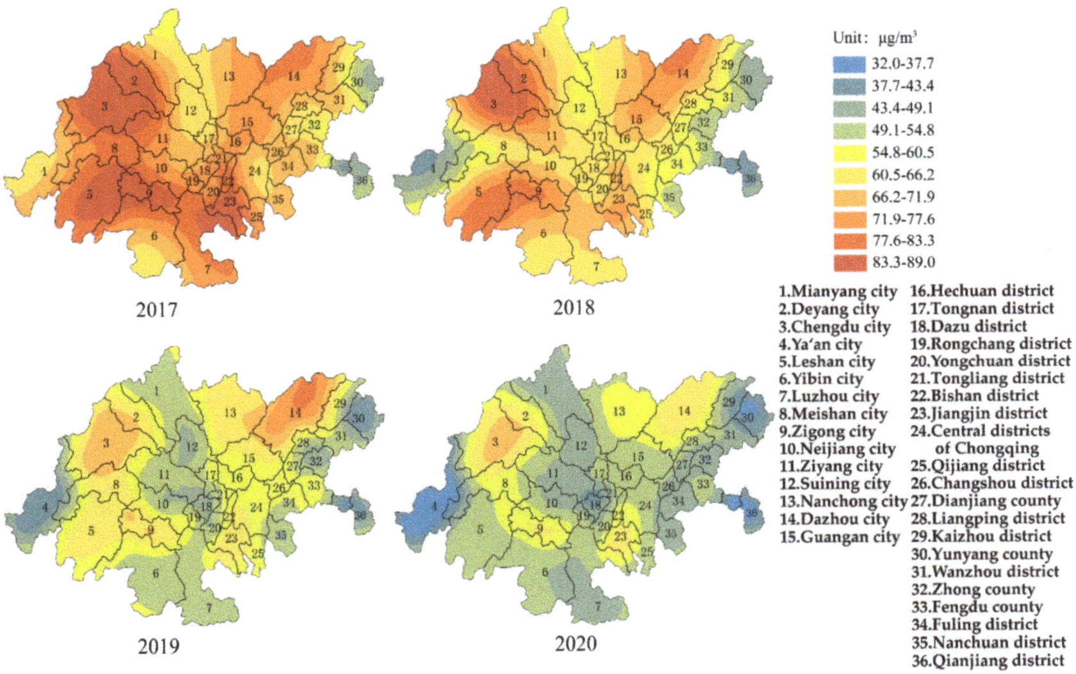

Figure 2. Spatial distribution of PM_{10} in CCEC from 2017 to 2020.

Studies have found that particulate matter in the atmosphere is harmful to human health [28,29]. Based on the interim targets and air quality guideline of PM_{10} in QAGQ, the average annual concentrations of PM_{10} for each area in the CCEC were higher than the air quality guideline (15 μg/m^3), which caused the potential risk to public health. The numbers of areas exceeding interim target 2 (50 μg/m^3) and interim target 3 (30 μg/m^3) were 55.6% and 44.4% respectively. If mortality in a population exposed to PM_{10} at the air quality guideline level was arbitrarily set at 100, then it would be 114 and 106 in populations exposed to PM_{10} at the interim target 2 and 3 levels.

3.1.2. $PM_{2.5}$

Figure 3 shows the change in $PM_{2.5}$ in the CCEC from 2017–2020. The average annual concentration of PM_{10} for the whole CCEC was 48.6 μg/m^3 (the range was 31.0 μg/m^3, and standard deviation was 7.2 μg/m^3) in 2017. In total, 97.0% of the cities failed to reach Grade II of AAQS. The area with the highest concentration was Zigong city in Sichuan province (66 μg/m^3, which was 89.0% higher than Grade II of AAQS), while the area with the lowest values was Yunyang county in Chongqing (35 μg/m^3). The heavily polluted areas were mainly located in Chengdu–Deyang and Leshan–Zigong, and the $PM_{2.5}$ distributions were consistent with air pollution transmission channels 1 and 2, which were consistent with the results of Luo et al. [9] and Liao et al. [30]. Some studies showed that air stagnation always happened in the Sichuan–Chongqing region in winter, which caused the high concentration of $PM_{2.5}$ [30]. Furthermore, as one of the most important agricultural regions in China, the combustion of biomass has also caused the severe pollution in Sichuan Basin [31]. The average annual concentration of $PM_{2.5}$ in CCEC decreased to 31.9 $\mu g/m^3$ (the range was 22.9 $\mu g/m^3$, and standard deviation was 5.0 $\mu g/m^3$) in 2020, and 75.0% of the cities met the Grade II of AAQS. The area with highest concentration of $PM_{2.5}$ was Rongchang district in Chongqing (44 $\mu g/m^3$, which was 25.7% higher than the Grade II of AAQS), while the area with lowest value was Ya'an city in Sichuan province (21 $\mu g/m^3$). Based on the mean values, ranges and standard deviations of PM_{10} and $PM_{2.5}$, the control of particulate matter had achieved remarkable improvement in 2020.

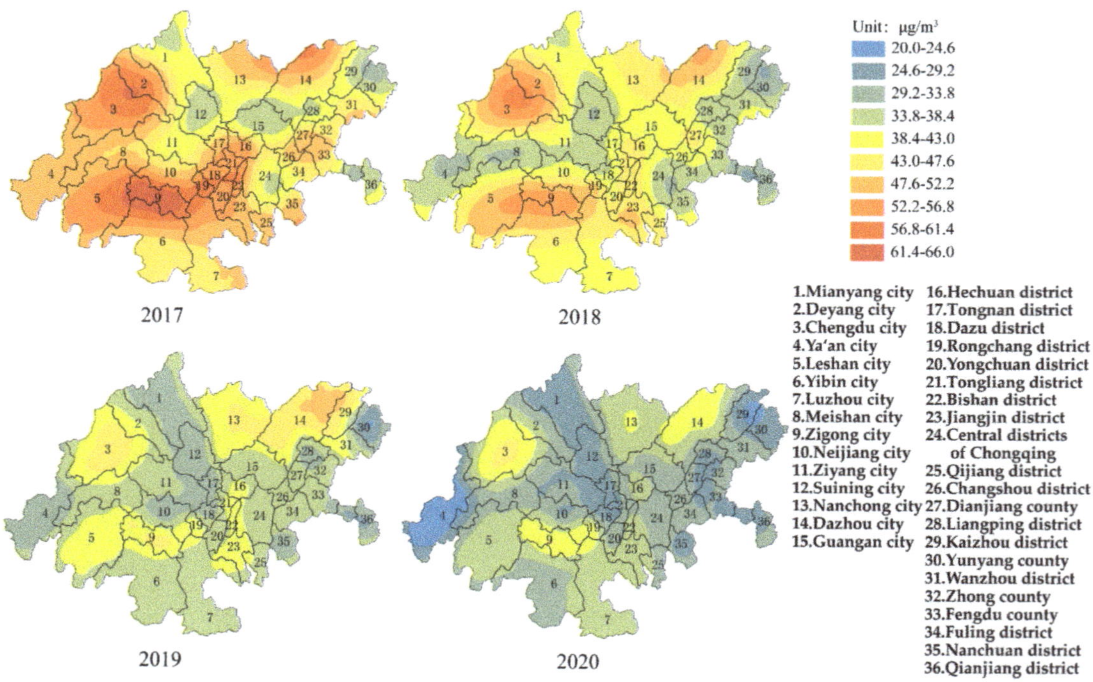

Figure 3. Spatial distribution of $PM_{2.5}$ in CCEC during 2017 to 2020.

It was said by QAGQ that priority should be given to the air quality guidelines of $PM_{2.5}$ when considering the impact of particulate matter. The average annual concentrations of $PM_{2.5}$ for each area in the CCEC were higher than the air quality guidelines (5 $\mu g/m^3$) in 2020. The number of areas exceeding the interim target 1 (35 $\mu g/m^3$), interim target 2 (25 $\mu g/m^3$) and interim target 3 (15 $\mu g/m^3$) were 55.6%, 69.4% and 2.8%, respectively. If mortality in a population exposed to $PM_{2.5}$ at the air quality guideline level was arbitrarily set to 100, then it would be 124, 116 and 108, respectively, in populations exposed to $PM_{2.5}$ at interim target 1, 2 and 3 levels. The results showed that the concentration level of $PM_{2.5}$ causes a high risk to public health and should be given priority to control.

It was found that the higher the ratio value of $PM_{2.5}/PM_{10}$, the heavier the influence on the atmospheric environment [32]. The mean value of the $PM_{2.5}/PM_{10}$ ratio in Beijing was 0.72, which ranged from 0.31~0.96 [32]. The $PM_{2.5}/PM_{10}$ ratio values of Kaohsiung and Hong Kong were around 0.62 and 0.63 [33,34], respectively. The ratio values of $PM_{2.5}/PM_{10}$

in the whole CCEC were 0.68, 0.66, 0.66 and 0.64 from 2017 to 2020, respectively. It was shown that the larger the proportion of $PM_{2.5}$ in particulate matter, the higher the risk to public health. The range of $PM_{2.5}/PM_{10}$ ratio in cities of Sichuan province in the CCEC were between 0.63–0.64. The ratio of $PM_{2.5}/PM_{10}$ showed a small decrease from 0.7 to 0.64 in the districts (counties) of Chongqing in the CCEC. The results showed that the pollution of particulate matter in Chongqing was much heavier than that in Sichuan province.

3.1.3. O_3

Figure 4 shows the change in O_3 in the CCEC from 2017–2020. In 2017, the concentration of O_3 for the whole CCEC was 139.6 $\mu g/m^3$ (the range was 60.0 $\mu g/m^3$, and standard deviation was 15.6 $\mu g/m^3$), and 11.0% of the areas failed to reach Grade II of AAQS. The area with the highest concentration of O_3 was the Jiangjing district in Chongqing (164 $\mu g/m^3$), which was 9.0% higher than Grade II of AAQS, and the area with the lowest concentration of O_3 was Mianyang city in Sichuan province (114 $\mu g/m^3$). The mean values of O_3 were slightly increased at 144.8 $\mu g/m^3$ in 2018 and 140.6 $\mu g/m^3$ in 2019. The distributions mainly located in Chengdu–Meishan, Chongqing, and Zigong were consistent with three air pollution transmission channels. Based on our results, the pollution of O_3 was severe in the area of Chengdu. Yang et al. found that the gasoline vehicle exhaust and the use of solvents was the main reason [35]. Due to the low atmospheric pressure, small pressure gradient, and the stable weather, the condition of horizontal diffusion was bad for the diffusion of pollutants [36]. The photochemical reactions were promoted in Chongqing due to the increase in oxidation after 2016, and the exogenous input of VOCs in nearby areas led to the increase in O_3 concentration [37]. The concentration of O_3 for the whole CCEC decreased to 133.8 $\mu g/m^3$ (the range was 65.0 $\mu g/m^3$, and standard deviation was 16.4 $\mu g/m^3$) in 2020. The area with the highest concentration of O_3 was Chengdu city in Sichuan province (169 $\mu g/m^3$), which was 6.0% higher than Grade II of AAQS, and the area with the lowest concentration of O_3 was Qianjiang district in Chongqing (104 $\mu g/m^3$). In total, 97.2% of the areas met the Grade II of AAQS.

Long-term exposure to high concentrations of O_3 could cause chronic damage to the human body. The concentrations of O_3 for each area in the CCEC were higher than the air quality guidelines (100 $\mu g/m^3$). The number of areas exceeding interim target 1 (160 $\mu g/m^3$) and interim target 2 (120 $\mu g/m^3$) were 2.8% and 72.2%, respectively. If mortality in a population exposed to ozone at the air quality guideline level was arbitrarily set at 100, then it would be 103 and 101 in the populations exposed to ozone at the interim target 1 and 2 levels. The results suggest that more strict measures are needed to be implemented for O_3 control in the future.

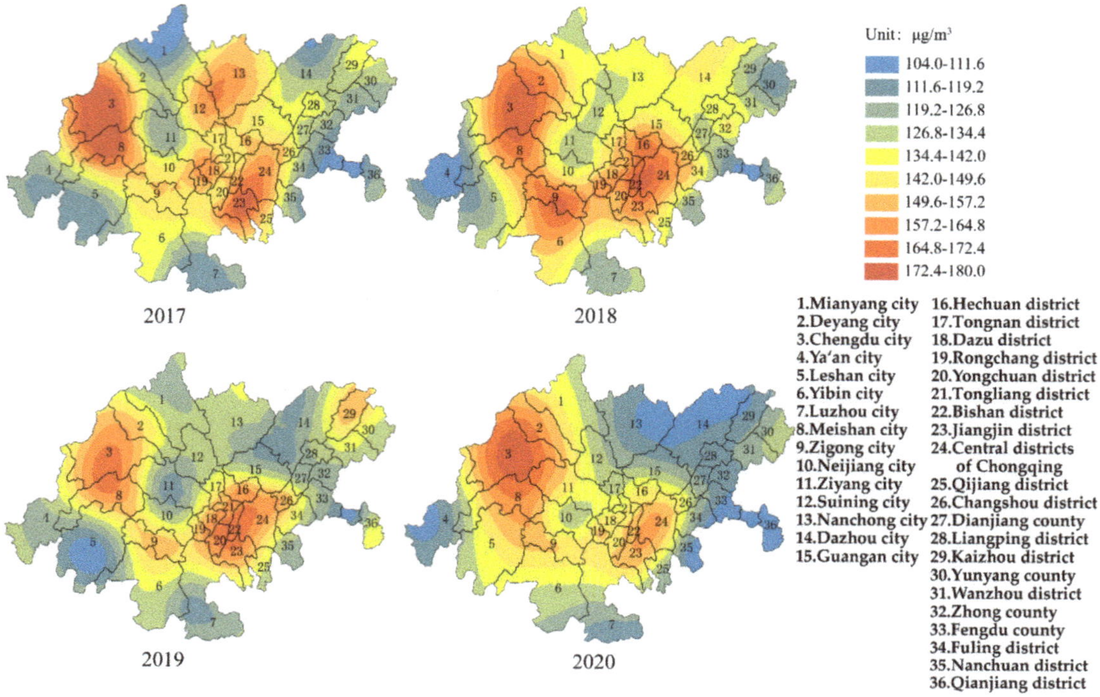

Figure 4. Spatial distribution of O_3 in CCEC during 2017 to 2020.

3.1.4. SO_2

The average annual concentration of SO_2 for the whole CCEC met Grade II of AAQS in 2017 (Figure 5). The mean value, range, and standard deviation were 15.4 μg/m^3, 26.0 μg/m^3, and 5.9 μg/m^3, respectively. The area with the highest annual average SO_2 concentration was Nanchuan district in Chongqing (34 μg/m^3), while the areas with the lowest annual average SO_2 concentration were the counties of Yunyang and Zhong in Chongqing (8 μg/m^3). The average annual concentration of SO_2 for the whole CCEC decreased to 10.3μg/m^3 in 2020; the range and standard deviation were 10.0 μg/m^3 and 2.8 μg/m^3, respectively. In recent years, the average annual concentration of SO_2 was far below Grade II of AAQS. The Sichuan Basin has suffered from pollution of SO_2 for the past decades, because 80% SO_2 concentrations in some areas of the Sichuan Basin was from industries such as chemical, textile, electronics, etc. [38]. Since the implementation of "the Division Plan of Acid rain and Sulfur dioxide Pollution Control Areas", the control of SO_2 achieved obvious improvement, basically removing the long-term problems of acid rain and sulfur dioxide pollution. The mean concentration of SO_2 in areas of Sichuan province in CCEC were reduced from 13.3 μg/m^3 in 2017 to 8.2 μg/m^3 in 2020, and the mean concentration of SO_2 in areas of Chongqing in the CCEC were reduced from 17.0 μg/m^3 in 2017 to 12.2 μg/m^3 in 2020. The concentration of SO_2 in the air depends on the consumption of coal [39]. The proportion of coal consumption in Sichuan province was 32.0% in 2019, while that in Chongqing was 53.0%. Therefore, the control of SO_2 in Sichuan province was better.

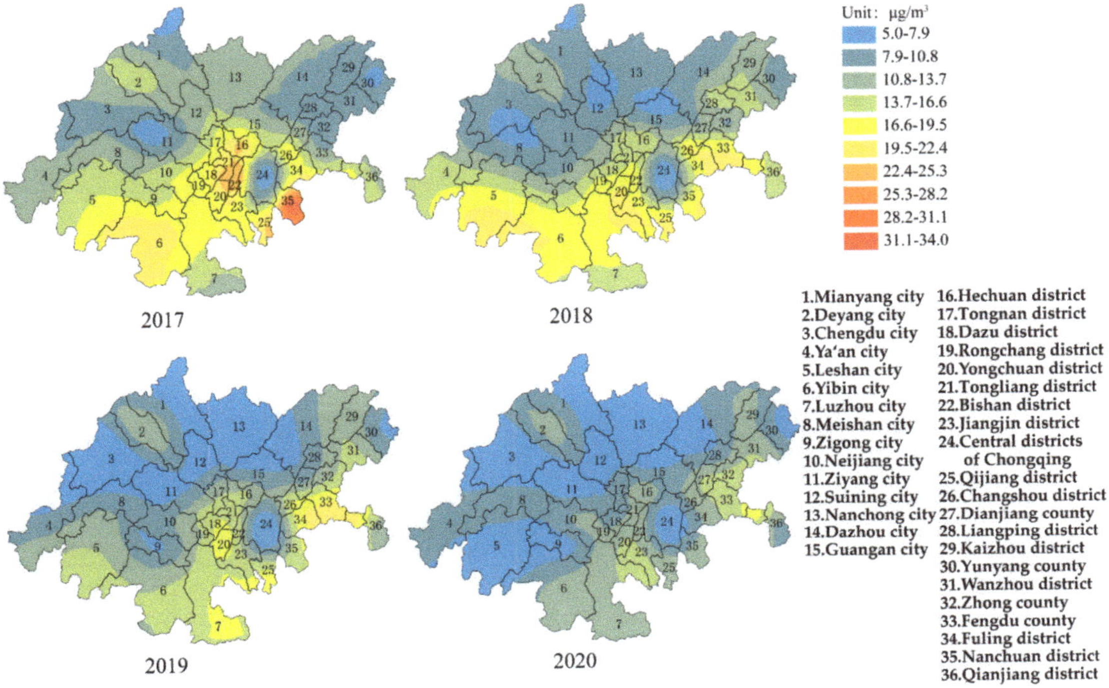

Figure 5. Spatial distribution of SO_2 in CCEC during 2017 to 2020.

3.1.5. NO_2

In 2017, the average annual concentration of NO_2 for the whole CCEC was 29.9 μg/m³, the range and standard deviation were 35.0 μg/m³ and 7.5 μg/m³, respectively (Figure 6). The number of areas exceeding Grade II of AAQS was about 11.0%. The area with the highest and lowest concentrations of NO_2 were Chengdu city in Sichuan province (53 μg/m³) and Dazu district in Chongqing (18 μg/m³). The distributions of NO_2 were consistent with air pollution transmission channels 1 and 3, which were mainly located in the cities of Chengdu, Meishan, Dazhou, the central districts and Jiangjin district of Chongqing. The rapid development of urbanization in these areas has led to a gradual increase in vehicles, and one of the main sources of NO_2 was traffic emissions [1]. The average annual concentration of NO_2 for the whole CCEC decreased to 24.1 μg/m³ in 2020, and the range and standard deviation were 24.0 μg/m³ and 5.9μg/m³, respectively. Each area in the CCEC reached Grade II of AAQS, which meant that the control measures of NO_2 in the CCEC also had certain effects in the past years.

The average annual concentrations of NO_2 for each area in the CCEC were lower than interim target 1 of QAGQ (40 μg/m³) in 2020. The number of areas exceeding interim target 2 (30 μg/m³), interim target 3 (20 μg/m³), and the air quality guidelines (10 μg/m³) were 11.1%, 58.3%, and 27.7%, respectively. If all-cause mortality in a population exposed to nitrogen dioxide at the AQG level was arbitrarily set at 100, then it would be 104 and 102 in populations exposed to nitrogen dioxide at the interim target 2 and 3 levels. The results showed that the potential risk to public health caused by NO_2 pollution might still exist.

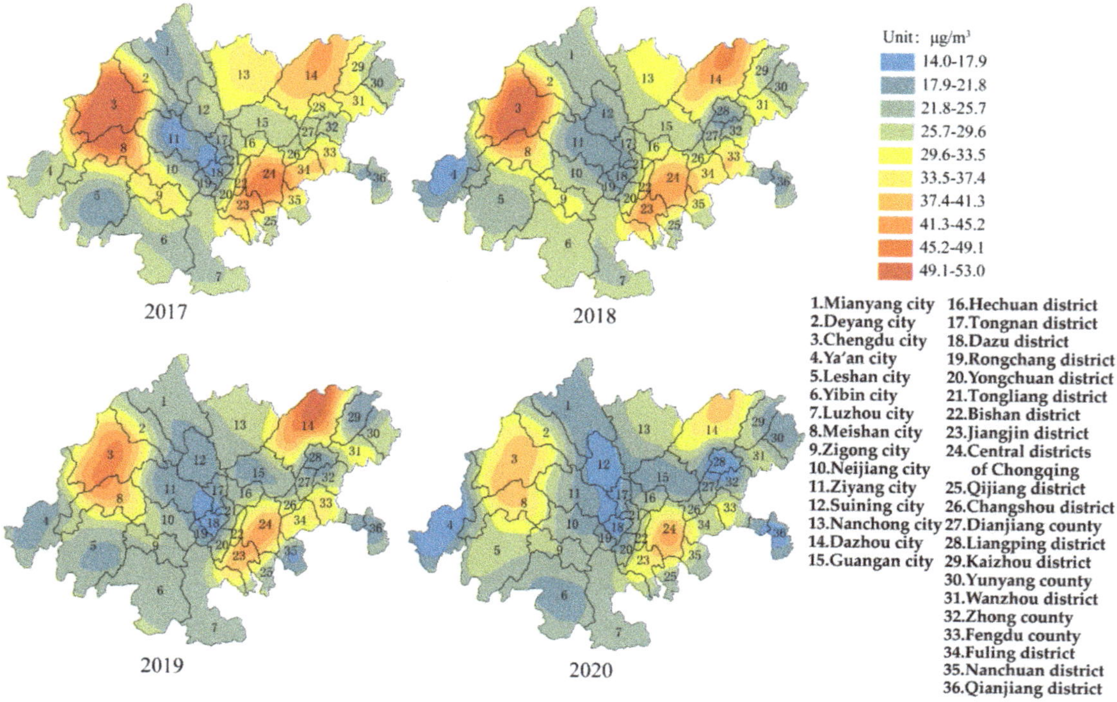

Figure 6. Spatial distribution of NO_2 in CCEC from 2017 to 2020.

3.1.6. CO

Figure 7 shows the change in CO in the CCEC from 2017–2020. The results showed that the control of CO in the CCEC had always been effective, since the concentration of CO for the whole CCEC was 1.4 mg/m^3 (the range and standard deviation were 0.9 mg/m^3 and 0.2 mg/m^3, respectively). The concentration of CO was far below Grade II of AAQS and the air quality guideline in QAGQ, which meant no potential risk to public health. The area with the highest and lowest concentrations of CO was Dazhou city (1.9 mg/m^3) and Mianyang city (1.0 mg/m^3) in Sichuan province, respectively. In 2020, the concentration of CO for the whole CCEC slightly decreased to 1.1 mg/m^3; the range and standard deviation were 0.5 mg/m^3 and 0.1 mg/m^3, respectively.

In summary, the controls of SO_2 and CO in CCEC were effective. The pollutions of PM_{10}, $PM_{2.5}$, and NO_2 had obvious improvement, while the control of O_3 was not obvious. The concentrations of PM_{10}, $PM_{2.5}$, O_3, and NO_2 in 2020 were still higher than the air quality guidelines in QAGQ, which meant that the potential risk to public health still exited. The terrain of CCEC was quite complex; the basin in Sichuan province and mountains in Chongqing caused the accumulation of atmospheric pollutants. The pollutions of PM_{10}, $PM_{2.5}$, and NO_2 were quite severe in 2017. Furthermore, the distributions of PM_{10}, $PM_{2.5}$, O_3 and NO_2 were consistent with three air pollution transmission channels, which verified the unique geographical and climatic factors influencing the distributions. Since the revised version of "The Environmental Protection Law of People's Republic of China" came into force in 2015, the concentrations of $PM_{2.5}$ and SO_2 have decreased over time [40]. There were 27 key tasks that had been completed to improve air quality, meet the capacity of atmospheric environment, and control the pollution of $PM_{2.5}$ and nitrogen oxide in Chongqing city since 2018. The air quality in Sichuan province was also improved by dividing the key areas of air pollution prevention and control, carrying out stricter environmental protection standards, and implementing environmental monitoring systems since 2019. Meanwhile, the interventions to control COVID-19 might improve air quality. Studies have found reductions in NO_2 and $PM_{2.5}$ concentrations during the pandemic [41,42]. The air quality of China was also significantly improved due to the anti-epidemic measures [43,44]. The reduction in human activities (traffic and industry) led to the decrease in atmospheric pollutants [45]. With the economic activities resumed, the effect of improvements on air quality will be offset in the short term. Some sustainable policies must be carried out to tackle air pollution in the post-pandemic era [44]. However, the heavily polluted areas still existed in 2020. Thus, the heavily polluted areas caused by PM_{10}, $PM_{2.5}$, O_3, and NO_2 were analyzed by spatial autocorrelation.

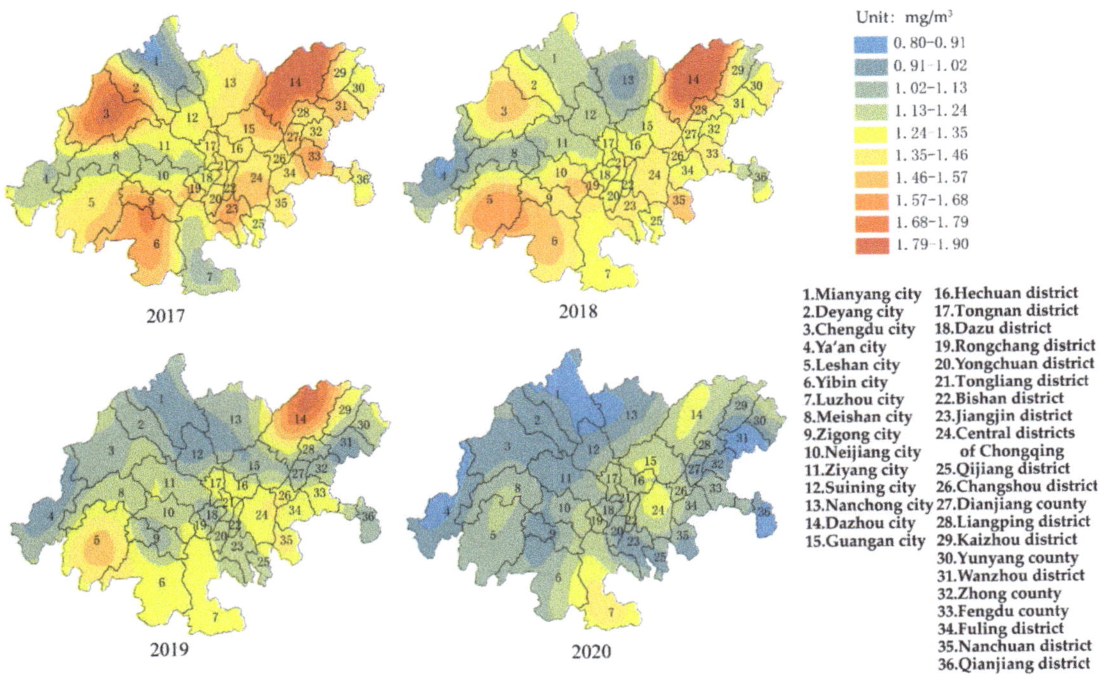

Figure 7. Spatial distribution of CO in CCEC during 2017 to 2020.

3.2. Spatial Autocorrelation of Air Pollution in CCEC

3.2.1. Global Spatial Autocorrelation

Table 2 shows the Global Moran's I of PM_{10}, $PM_{2.5}$, NO_2, and O_3 during 2017–2020. The Global Moran's I values of PM_{10} and $PM_{2.5}$ were 0.49 and 0.35 ($p < 0.05$, $Z > 1.65$) in 2017, respectively. The certain aggregation characteristics of PM_{10} and $PM_{2.5}$ with positive spatial correlation were shown in the CCEC, and there was an obvious tendency of aggregation in heavily polluted areas. The Global Moran's I values of PM_{10} and $PM_{2.5}$ decreased to 0.21 and 0.06 in 2020, which meant that the spatial aggregations of PM_{10} and $PM_{2.5}$ were changed from aggregation distribution to random distribution, due to the implementation of atmospheric control measures. The Global Moran's I value of O_3 increased from 0.21 in 2017 to 0.57 in 2020, and the spatial aggregation became significant. The Global Moran's I value of NO_2 did not show any significance, which meant that there was no significant spatial aggregation of NO_2 in CCEC.

Table 2. Global Moran's I of atmospheric pollutants.

Year	PM_{10}	$PM_{2.5}$	NO_2	O_3
2017	0.49	0.35	0.14	0.21
2018	0.43	0.31	0.08	0.33
2019	0.17	0.05	0.04	0.29
2020	0.21	0.06	0.10	0.57

3.2.2. Local Spatial Autocorrelation

The cities of Yibin, Neijiang, Luzhou, Meisan in Sichuan province and Yongchuan district in Chongqing showed the HH type of PM_{10} in 2017 (Figure 8), which meant that these areas suffered from heavy pollution of PM_{10}. The areas of Wanzhou district, Fengdu county, and Qianjiang district in Chongqing showed the LL type of PM_{10} in 2017, while the Ya'an city in Sichuan province showed the LH type. Meanwhile, the cities of Yibin, Neijiang, Luzhou in Sichuan province and the Yongchuan district in Chongqing showed the HH type of $PM_{2.5}$ in 2017. Nanchong city in Sichuan province and the districts of Wanzhou, Qianjiang in Chongqing showed the LL type of $PM_{2.5}$, while Dazhou city in Sichuan province showed the HL type. The results showed that the distribution of particulate matter had obvious regional aggregation; the heavily polluted areas were consistent with the distribution of air pollution transmission channels, which were at the end of the channels. The cities of Zigong, Yibin, Neijiang and Luzhou were traditional industrial bases, and the large-scale and intensification of heavy polluting industries also resulted in the aggregation of heavily polluted areas [27]. In 2020, with the implementation of atmospheric control measures, the number of cities with the HH type of PM_{10} decreased obviously, and the spatial aggregation became weak. Meanwhile, the number of cities with the HH type of $PM_{2.5}$ was barely changed. Based on the results, the cities with the HH type of PM_{10} and $PM_{2.5}$ located at the end of the three transmission channels and the control of PM_{10} were better than the control of $PM_{2.5}$. The southern part of the CCEC still deserved key attention in the future control of particulate matter.

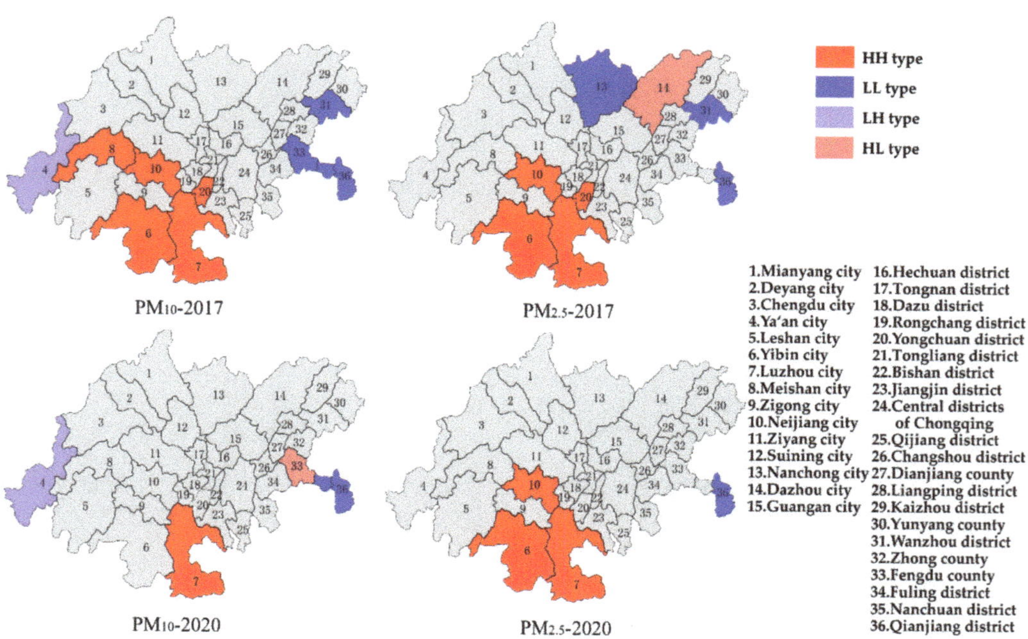

Figure 8. LISA cluster of PM_{10} and $PM_{2.5}$ in 2017 and 2020.

The areas of Wanzhou district, Liangpin district, Zhong county, and Fengdu county in Chongqing showed the LL type of O_3 in 2017 (Figure 9). These areas had relatively lighter O_3 pollution. Few cities showed the HH type of O_3. Meanwhile, the LH type of O_3 was shown in the cities of Ziyang, Luzhou in Sichuan province and the districts of Tongliang, Yongchuan in Chongqing. The distribution of O_3 in the CCEC was still random in 2017. The HH type of O_3 became more and more obvious year by year, which was mainly located in the cities of Chengdu, Deyang, Neijiang, Ziyang in Sichuan province and districts of Bisan, Tongliang, Yongchuan, and Dazu in Chongqing, which was consistent with the middle reach of the three air pollution transmission channels. Based on the distributions of particulate matter and O_3, the degree of pollution in Sichuan province was heavier, due to the low topography of the Sichuan Basin, which was not conducive to the discharge of pollutants. Furthermore, the relatively developed economy, large population density and high industrial density in Sichuan province caused the high emission of pollutants [13]. Therefore, the areas with a high concentration of O_3 should be controlled to prevent the expansion of heavy polluted areas.

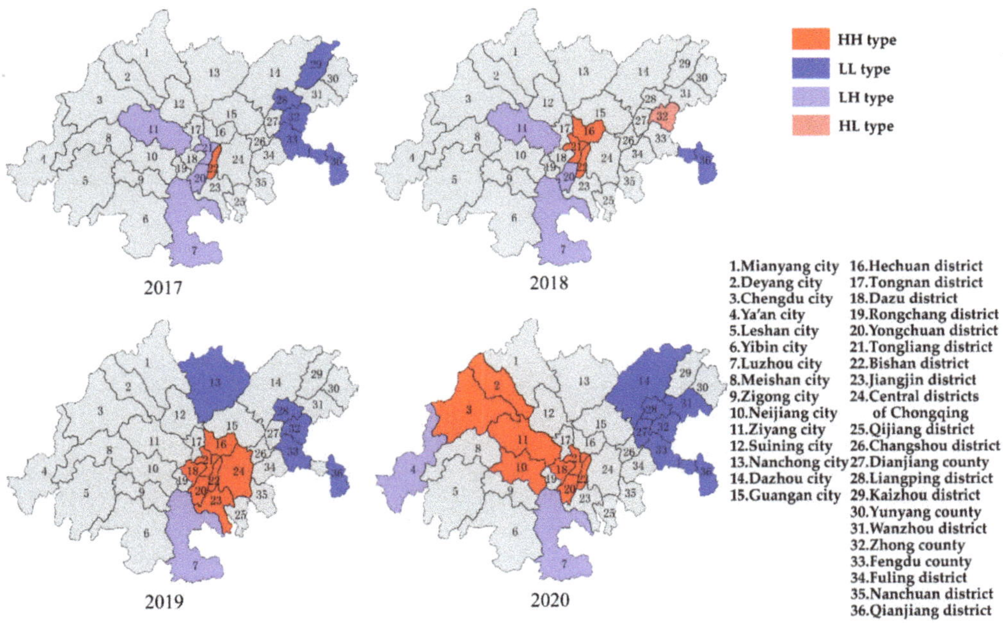

Figure 9. LISA cluster of O_3 from 2017 to 2020.

In 2017, the districts of Tongnan, Dazu, and Rongchang in Chongqing showed the LL type of NO_2, and the city of Ya'an in Sichuan province and district of Qijiang in Chongqing showed the LH type (Figure 10). In 2020, the distribution of NO_2 in CCEC was still random, yet the districts of Nanchuan and Qijiang in Chongqing showed the HH type. Based on the concentration of NO_2 in 2020, the average annual concentration of NO_2 for the whole CCEC was 24.1 $\mu g/m^3$, while the average annual concentration of NO_2 in the districts of Nanchuan and Qijiang in Chongqing was 25.5 $\mu g/m^3$. Since the obvious effect was achieved on NO_2 control in the CCEC, the average annual concentration of NO_2 decreased, and the high concentration of NO_2 in the districts of Nanchuan and Qijiang in Chongqing caused these areas to be the HH type. Therefore, the control of NO_2 in the districts of Nanchuan and Qijiang in Chongqing should be further strengthened in the future.

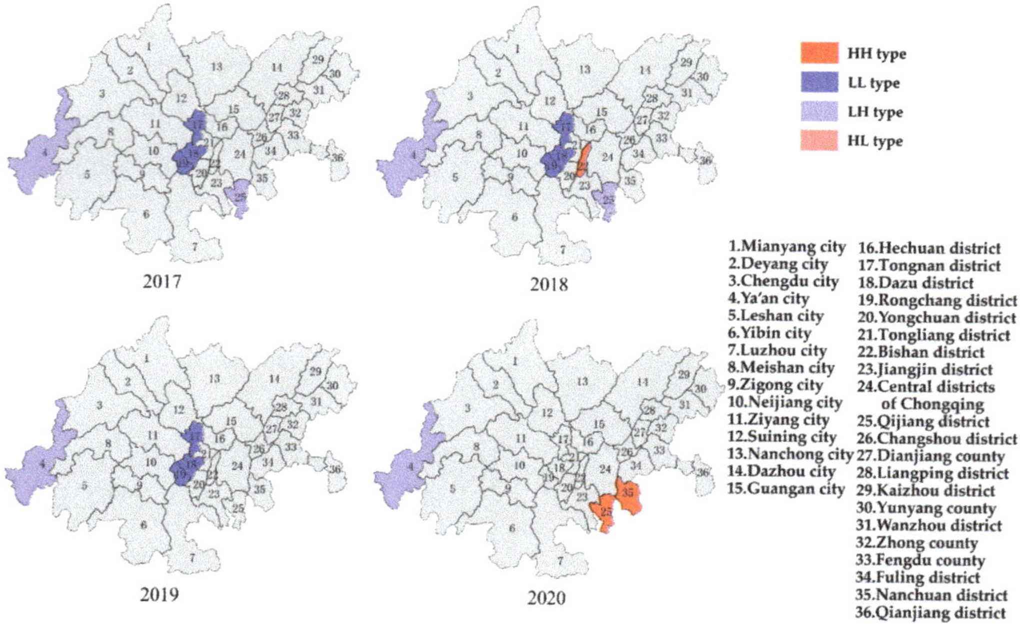

Figure 10. LISA cluster of NO_2 from 2017 to 2020.

4. Conclusions

We have investigated the temporal and spatial distribution of atmospheric pollutants in the CCEC from 2017 to 2020. The concentrations of PM_{10}, SO_2, NO_2, and CO met the Grade II of AAQS in 2020, due to the implementation of atmospheric control measures. The average annual concentration of SO_2 for the whole CCEC decreased from 15.4 $\mu g/m^3$ in 2017 to 10.3 $\mu g/m^3$ in 2020, and the long-term problems of acid rain and sulfur dioxide pollution were basically eliminated. The concentrations of PM_{10} and $PM_{2.5}$ also improved significantly; there were 30.4% and 34.4% reductions for the average annual concentration of PM_{10} and $PM_{2.5}$. The concentration change of O_3 was not obvious, yet 97.2% of the areas met the Grade II of AAQS in 2020. The distributions of PM_{10}, $PM_{2.5}$, O_3, and NO_2 were consistent with three air pollution transmission channels, which meant that the distribution of atmospheric pollutants was influenced by topographic and climatic conditions.

The concentrations of PM_{10}, $PM_{2.5}$, O_3, and NO_2 were far beyond the air quality guidelines in QAGQ in 2020. The purpose of the air quality guidelines was to illustrate the minimum impact of atmospheric pollutants on human health. The concentration levels of PM_{10}, $PM_{2.5}$, O_3, and NO_2 still had certain impacts on human health, and it is necessary to reduce the concentration of these atmospheric pollutants by using interim targets in QAGQ.

Based on the results of spatial autocorrelation of air pollution in the CCEC, the spatial aggregation of PM_{10} was significantly reduced, and the number of areas with the HH type of PM_{10} decreased in 2020. Meanwhile, the HH type of $PM_{2.5}$ was mainly located in the southern part of CCEC, and it barely changed in 2020. The spatial aggregation of O_3 became obvious in 2020, and the HH type of O_3 was shown in the central and northwest parts of the CCEC. The spatial aggregation of NO_2 was random during 2017–2020, yet the districts of Nanchuan and Qijiang in Chongqing showed the HH type of NO_2.

In summary, the key control areas of particulate matter should focus on the southern part of the CCEC, and the control of industrial pollution sources in the cities of Zigong, Yibin, Neijiang and Luzhou in Sichuan province should be strengthened. It was suggested that the growth rate of coal-fired power plants should be controlled strictly, the proportion of coal and gas in electricity needs to be optimized, and the transformation of the steel industry to achieve ultra-low emissions should be accelerated. The key control areas of O_3 should focus on the central and northwest parts of the CCEC. It was recommended to reduce the emission of NOx and VOCs in these regions, especially focusing on the sources of scattered polluted enterprises and key industry VOCs emissions. The distribution of NO_2 pollution was random to some extent, yet NO_2 pollution in the southern part of the CCEC is still worth paying attention to.

Author Contributions: Conceptualization, N.Q. and F.N.; methodology, H.W. and X.T.; investigation D.J. and T.W.; writing—original draft preparation, N.Q. and Q.T.; writing—review and editing, T.X.; project administration, N.Q.; funding acquisition, N.Q., L.R. and W.D. All authors have read and agreed to the published version of the manuscript.

Funding: This research was funded by the National Social Science Fund of China (20XSH015), Science and Technology Research Project of Chongqing Municipal Education Commission (KJQN202100831, KJQN202000834), Natural Science Foundation of Jiangsu Province (BK20210933), and the Start-up Foundation of High-level Talents (2056014) in Chongqing Technology and Business University.

Institutional Review Board Statement: Not applicable.

Informed Consent Statement: Not applicable.

Data Availability Statement: The data that support the findings of this study are available from the corresponding author upon reasonable request.

Conflicts of Interest: The authors declare no conflict of interest.

References

1. Zhou, N.; Cui, Z.; Yang, S.; Han, X.; Chen, G.; Zhou, Z.; Zhai, C.; Ma, M.; Li, L.; Cai, M.; et al. Air pollution and decreased semen quality: A comparative study of Chongqing urban and rural areas. *Environ. Pollut.* **2014**, *187*, 145–152. [CrossRef] [PubMed]
2. Anderson, J.O.; Thundiyil, J.G.; Stolbach, A. Clearing the Air: A Review of the Effects of Particulate Matter Air Pollution on Human Health. *J. Med. Toxicol. Off. J. Am. Coll. Med. Toxicol.* **2011**, *8*, 166–175. [CrossRef] [PubMed]
3. Nemery, B.; Hoet, P.H.; Nemmar, A. The Meuse Valley fog of 1930: An air pollution disaster. *Lancet* **2001**, *357*, 704–708. [CrossRef]
4. Wu, R.; Liu, F.; Tong, D.; Zheng, Y.; Lei, Y.; Hong, C.; Li, M.; Liu, J.; Zheng, B.; Bo, Y.; et al. Air quality and health benefits of China's emission control policies on coal-fired power plants during 2005–2020. *Environ. Res. Lett.* **2019**, *14*, 94016. [CrossRef]
5. Zhou, Q.; Yabar, H.; Mizunoya, T.; Higano, Y. Evaluation of Integrated Air Pollution and Climate Change Policies: Case Study in the Thermal Power Sector in Chongqing City, China. *Sustainability* **2017**, *9*, 1741. [CrossRef]
6. Silva, L.F.O.; Oliveira, M.L.S.; Neckel, A.; Maculan, L.e.S.; Milanes, C.B.; Bodah, B.W.; Cambrussi, L.P.; Dotto, G.L. Effects of atmospheric pollutants on human health and deterioration of medieval historical architecture (North Africa, Tunisia). *Urban Clim.* **2022**, *41*, 101046. [CrossRef]
7. Bai, Y.; Ni, Y.; Zeng, Q. Impact of Ambient Air Quality Standards revision on the exposure-response of air pollution in Tianjin, China. *Environ. Res.* **2021**, *198*, 111269–111275. [CrossRef] [PubMed]
8. Lin, N. The Research on Transport Law of Atmospheric Pollutant and Joint Prevention and Contral of Air Pollution Technology in Sichuan Province. Master's Thesis, Southwest Jiaotong University, Chengdu, China, 2015.
9. Luo, Q.; Liao, T.; Wang, B.; Jiang, H.; Wang, S.; Ouyang, Z.; Lin, D.; Sun, X. Statistical analysis of near-surface wind field and transport pathways of pollutants in Sichuan Basin. *Acta Sci. Circumstantiae* **2020**, *40*, 1374–1384.
10. Wang, W. Research on Establishment of Space-Air-Ground Integrated Monitoring Technology System for Air Pollution. Master's Thesis, Chongqing Technology and Business University, Chongqing, China, 2020.
11. Liao, T.; Wang, S.; Ai, J.; Gui, K.; Duan, B.; Zhao, Q.; Zhang, X.; Jiang, W.; Sun, Y. Heavy pollution episodes, transport pathways and potential sources of $PM_{2.5}$ during the winter of 2013 in Chengdu (China). *Sci. Total Environ.* **2017**, *584–585*, 1056–1065. [CrossRef]
12. Zhai, S.; Jacob, D.J.; Wang, X.; Shen, L.; Liao, H. Fine particulate matter ($PM_{2.5}$) trends in China, 2013–2018: Separating contributions from anthropogenic emissions and meteorology. *Atmos. Chem. Phys.* **2019**, *19*, 11031–11041. [CrossRef]
13. Feng, X.; Zhang, Y. Temporal and spatial distribution characteristice of air pollutant concentrations in Sichuan and Chongqing area. *China Sci.* **2018**, *13*, 1708–1715.
14. Liu, Y.; Yue, W.; Fan, P.; Zhang, Z.; Huang, J. Assessing the urban environmental quality of mountainous cities: A case study in Chongqing, China. *Ecol. Indic.* **2017**, *81*, 132–145. [CrossRef]
15. Shulan, W.; Fahe, C.; Yuanhang, Z.; Laidong, Z.; Qinling, W. Analysis on the sources and characters of particles in Chengdu. *Sci. Geogr. Sin.* **2004**, *24*, 488–492.
16. Cao, P.; Luo, B.; Zhang, W.; Li, B.; Li, R. Analysis on urban air quality of Sichuan province. *Sichuan Environ.* **2017**, *36*, 72–76.
17. Wang, P.; Wang, Y.; Liu, M.; Zhang, Y.; Li, G. Assessment of atmospheric environmental quality and countermeasures research in Chongqing. *Bull. Sci. Technol.* **2018**, *34*, 267–273.

18. Liu, Y.; Yang, C.; Tan, S.; Zhou, H.; Zeng, W. An approach to assess spatio-temporal heterogeneity of rural ecosystem health: A case study in Chongqing mountainous area, China. *Ecol. Indic.* **2022**, *136*, 108644. [CrossRef]
19. Zhang, F.; Shi, Y.; Fang, D.; Ma, G.; Nie, C.; Krafft, T.; He, L.; Wang, Y. Monitoring history and change trends of ambient air quality in China during the past four decades. *J. Environ. Manag.* **2020**, *260*, 110031. [CrossRef]
20. Zhao, B.; Shi, S.; Ji, J.S. The WHO Air Quality Guidelines 2021 promote great challenge for indoor air. *Sci. Total Environ.* **2022**, *827*, 154376. [CrossRef]
21. Wang, Z.; Wu, T.; Che, F.; Wang, S.; Zhou, Y.; Qian, X.; Wu, X. Comparison between domestic and international Ambient Air Quality Standards. *Res. Environ. Sci.* **2010**, *23*, 253–260.
22. Liu, H.; Du, G. Spatial pattern and distributional dynamics of urban air pollution in China—An empirical study based on aqi and six sub-pollutants of 161 cities. *Econ. Geogr.* **2016**, *36*, 33–38.
23. Chen, S.; Zhu, J.; You, T. Spatial autocorrelation analysis of air quality index—A case study of Fujian Province. *Wuyi Sci. J.* **2020**, *36*, 48–58.
24. Liu, Q.; Wang, S.; Zhang, W.; Li, J.; Kong, Y. Examining the effects of income inequality on CO_2 emissions: Evidence from non-spatial and spatial perspectives. *Appl. Energy* **2019**, *236*, 163–171. [CrossRef]
25. Miao, L.; Liu, C.; Yang, X.; Kwan, M.-P.; Zhang, K. Spatiotemporal heterogeneity analysis of air quality in the Yangtze River Delta, China. *Sustain. Cities Soc.* **2022**, *78*, 103603. [CrossRef]
26. Song, W.; Wang, C.; Chen, W.; Zhang, X.; Li, H.; Li, J. Unlocking the spatial heterogeneous relationship between Per Capita GDP and nearby air quality using bivariate local indicator of spatial association. *Resour. Conserv. Recycl.* **2020**, *160*, 104880–104887. [CrossRef]
27. Zheng, D.; Chen, C. Spatial-temporal characteristics and influence factors of $PM_{2.5}$ concentrations in Chengdu-Chongqing urban agglomeration. *Res. Environ. Sci.* **2019**, *32*, 1834–1843.
28. Ito, K.; Mathes, R.; Ross, Z.; Nádas, A.; Thurston, G.; Matte, T. Fine particulate matter constituents associated with cardiovascular hospitalizations and mortality in New York City. *Environ. Health Perspect.* **2011**, *119*, 467–473. [CrossRef] [PubMed]
29. Jin, Y.; Dai, T.; Yu, C.; Zheng, S.; Nie, Y.; Wang, M.; Bai, Y. Effects of ambient particulate matter (PM_{10}) on prevalence of diabetes and fasting plasma glucose. *Chin. J. Prev. Med.* **2021**, *55*, 7.
30. Liao, T.; Gui, K.; Jiang, W.; Wang, S.; Wang, B.; Zeng, Z.; Che, H.; Wang, Y.; Sun, Y. Air stagnation and its impact on air quality during winter in Sichuan and Chongqing, southwestern China. *Sci. Total Environ.* **2018**, *635*, 576–585. [CrossRef]
31. Luo, J.; Zhang, J.; Huang, X.; Liu, Q.; Luo, B.; Zhang, W.; Rao, Z.; Yu, Y. Characteristics, evolution, and regional differences of biomass burning particles in the Sichuan Basin, China. *J. Environ. Sci.* **2020**, *89*, 35–46. [CrossRef]
32. Zhao, C.; Wang, Y.; Wang, Y.; Zhang, H.; Zhao, B. Temporal and spatial distribution of $PM_{2.5}$ and PM_{10} pollution status and the correlation of particulate matters and meteorological factors during winter and spring in Beijing. *Environ. Sci.* **2014**, *35*, 418–427.
33. Lin, J.J. Characterization of water-soluble ion species in urban ambient particles. *Environ. Int.* **2002**, *28*, 55–61. [CrossRef]
34. Zhuang, H.; Chan, C.K.; Fang, M.; Wexler, A.S. Size distributions of particulate sulfate, nitrate, and ammonium at a coastal site in Hong Kong. *Atmos. Environ.* **1999**, *33*, 843–853. [CrossRef]
35. Yang, X.; Wu, K.; Wang, H.; Liu, Y.; Gu, S.; Lu, Y.; Zhang, X.; Hu, Y.; Ou, Y.; Wang, S.; et al. Summertime ozone pollution in Sichuan Basin, China: Meteorological conditions, sources and process analysis. *Atmos. Environ.* **2020**, *226*, 117392. [CrossRef]
36. Jiang, W.; Xie, W.; Wang, B.; Wang, S.; Long, Q.; Liao, T. Analysis on the characteristics of heavy pollution atmospheric circulation in the Sichuan Basin from 2014 to 2016. *Acta Sci. Circumstantiae* **2019**, *39*, 180–188.
37. Zhang, T.; Shen, N.; Zhao, X.; Wang, X.; Zhao, W. Spatiotemporal variation characteristics of ozone and its population exposure risk assessment in Chengdu-Chongqing urban agglomeration during 2015 to 2019. *Acta Sci. Circumstantiae* **2021**, *41*, 4188–4199.
38. Li, X.; Hussain, S.A.; Sobri, S.; Said, M.S.M. Overviewing the air quality models on air pollution in Sichuan Basin, China. *Chemosphere* **2021**, *271*, 129502. [CrossRef]
39. Wu, L.; Cheng, W.; Zhang, L. The relationship between energy consumption structure and the quality of atmospheric environment-an example in an area of Chengdu. *Environ. Sci. Manag.* **2009**, *34*, 166–169.
40. Yu, C.; Morotomi, T. The effect of the revision and implementation for environmental protection law on ambient air quality in China. *J. Environ. Manag.* **2022**, *306*, 114437. [CrossRef]
41. Berman, J.D.; Ebisu, K. Changes in U.S. air pollution during the COVID-19 pandemic. *Sci. Total Environ.* **2020**, *739*, 139864. [CrossRef]
42. Abdullah, S.; Mansor, A.A.; Napi, N.N.L.M.; Mansor, W.N.W.; Ahmed, A.N.; Ismail, M.; Ramly, Z.T.A. Air quality status during 2020 Malaysia Movement Control Order (MCO) due to 2019 novel coronavirus (2019-nCoV) pandemic. *Sci. Total Environ.* **2020**, *729*, 139022. [CrossRef]
43. Chen, K.; Wang, M.; Huang, C.; Kinney, P.L.; Anastas, P.T. Air pollution reduction and mortality benefit during the COVID-19 outbreak in China. *Lancet Planet. Health* **2020**, *4*, e210–e212. [CrossRef]
44. Qiu, W.; He, H.; Xu, T.; Jia, C.; Li, W. The air quality changes and related mortality benefits during the coronavirus disease 2019 pandemic in China: Results from a nationwide forecasting study. *J. Clean. Prod.* **2021**, *308*, 127327. [CrossRef] [PubMed]
45. Yuan, Q.; Qi, B.; Hu, D.; Wang, J.; Zhang, J.; Yang, H.; Zhang, S.; Liu, L.; Xu, L.; Li, W. Spatiotemporal variations and reduction of air pollutants during the COVID-19 pandemic in a megacity of Yangtze River Delta in China. *Sci. Total Environ.* **2020**, *751*, 141820. [CrossRef] [PubMed]

MDPI
St. Alban-Anlage 66
4052 Basel
Switzerland
www.mdpi.com

International Journal of Environmental Research and Public Health Editorial Office
E-mail: ijerph@mdpi.com
www.mdpi.com/journal/ijerph